Understanding Innovation in Canadian Industry

Understanding Innovation in Canadian Industry

Fred Gault, Editor

Published for the School of Policy Studies, Queen's University
by McGill-Queen's University Press
Montreal & Kingston • London • Ithaca

National Library of Canada Cataloguing in Publication

Understanding innovation in Canadian industry / edited by Fred Gault.

Includes bibliographical references and index.
ISBN 1-55339-030-X (pbk.).—ISBN 1-55339-031-8 (bound)

1. Technological innovations—Canada. 2. Industries—Canada.
I. Gault, Fred, 1942- II. Queen's University (Kingston, Ont.). School of Policy Studies.

HD45.U53 2003 338'.064'0971 C2003-905873-5

Table of Contents

Acknowledgements

This book is a collective work of many people and organizations: government departments and agencies, both federal and provincial; universities; research organizations; and, thousands of Canadian firms. The government departments and agencies included: Industry Canada; the National Research Council of Canada; Natural Resources Canada; Statistics Canada; and, the Institut de la Statistique du Québec. There were five universities represented: The University of Laval; the University of Maastricht in the Netherlands and specifically the Maastricht Economic Research Institute on Innovation and Technology (MERIT); the Université du Québec à Montréal (UQAM); the University of Sherbrooke; and, the University of Waterloo. The research institutes represented were: the Centre interuniversitaire de recherche en analyse des organisations (CIRANO); the Centre interuniversitaire de recherche sur la science et la technologie (CIRST); the Centre for the Study of Living Standards (CSLS); and, the Innovation Systems Research Network (ISRN).

Industry Canada hosted a workshop in November 2001 to discuss findings from the analysis of the results of the Survey of Innovation 1999 and commissioned a number of papers that eventually led to chapters in this book. The Honourable Perrin Beatty, from Canadian Manufacturers and Exporters, addressed the participants and stressed the importance of life long learning to the future of Canadian competitiveness, as well as the role of the small and medium sized firm.

Following the workshop, Industry Canada continued to be involved in the discussions and development of the text and contributed in many ways. Special thanks go to Marie Tobin, the Director General of the Innovation Policy Branch who helped to fund the survey and the subsequent analysis of the results, to Jeanne Inch who began this project as Director of Market Place Innovation, before her move to Canadian Heritage, and to Peter Boyd who saw it to completion as the new Director of Market Place Innovation. Within that division the hard work and creative contributions of Can Le and Pierre Therrien should be recognized.

Before the first and the last chapters were in their present form, all of the chapters of the book were edited by Jennifer Wilson. Her meticulous attention to detail, passion for consistency, and her determination, have combined to make this a much more readable text than it might otherwise have been.

The preparation of the final version of the text was overseen by Louise Earl with formatting and revisions done by Claire Racine-Lebel and Carol Poirier, of the Science, Innovation and Electronic Information Division at Statistics Canada. Frances Anderson of the same division provided useful comments on a number of chapters and, with Brian Nemes and Susan Schaan, was responsible for the Survey of Innovation 1999 on which most of the book is based. Yves Morin and his colleagues in the Business Survey Methods Division at Statistics Canada laid the strong foundation for the survey's sample design and provided methodological support throughout data production. Without Guy Sabourin, Adele St. Pierre and Brenda Hutchinson there would have been no tables of data.

The book would not have been possible without the funding of the federal Policy Research Initiative for the Information System for Science and Technology Project and the advocacy and support of Renée St-Jacques, the Chief Economist of Industry Canada

Tom Williams, the Acting Director of the School of Policy Studies at Queen's University, invited the book into the publication series of the School and Mark Howes, and his colleagues in the Publications Unit, produced the volume in a very short time.

The Editor
August 2003

Preface

This book is part of an on-going programme of statistical measurement and policy development that began in its present form in March 1996. This was just after the report of the federal Science and Technology Review that recognized the place for statistical measurement as part of the policy process. An outcome of the Review was the funding, by the Minister of Industry, of indicator development in the area of technological change until March 1999. This made it possible to expand the scope of the Statistics Canada surveys of research and development, to establish a survey of commercialization of intellectual property from universities and government laboratories, to conduct a survey of innovation in selected service industries, and to develop indicators of technology use for advanced manufacturing technologies and for biotechnologies.

The close collaboration between the Industry Canada and Statistics Canada continued beyond March 1999 thanks to the support of the federal Policy Research Initiative which recognized the need to support policy development in the areas of technological and organizational change with timely and relevant data. The indicator programme expanded to include a pilot survey of the use and planned use of knowledge management practices and tentative work on institutional learning. The surveys of innovation in manufacturing and selected natural resource industries and of construction and related industries were also launched in this period and many of their results have found their way into this book.

The book captures some of an on-going discussion among policy makers, statisticians, students of technological and organizational change, and survey respondents. As it goes to press, there is a survey of innovation in selected service industries going out to respondents to continue the discussion. That survey builds on what has been learned from previous surveys and their analysis, and also from the respondents who play a key role in understanding technological and organizational change.

Learning from respondents occurs when they respond to surveys, but there are also opportunities for the respondents to learn and to change. All new surveys go through a stage of cognitive testing to see if the relevant people in firms are able to answer the questions. In the course of testing a pilot survey on the use of knowledge management practices, it became evident that the people selected for the cognitive testing could not answer the questions, but more than one wanted copies of the questionnaire as a useful guide to things the management should be thinking about. The questionnaire was improved, so the statisticians learned and conducted a successful survey, but so also did the firms learn from the interaction and they may have changed their behaviour.

Policy is a collective process which may lead to a change in behaviour of the players in the system, or of the way in which they interact. The Innovation Strategy is an example of an initiative in which some goals are set for individual players (doubling the federal government expenditure on research and development (R&D) by 2010) while others can only be achieved by the collective (becoming one of the top five OECD countries, based on R&D spending, or developing the most skilled and talented labour force in the world.). Statistical measurement of activities, such as innovation or technology use, of linkages, such as those resulting from the commercialization of intellectual property, and of outcomes, such as changes in employment levels, or skill levels, informs the actors in the system about how it is changing and allows them to react.

Of course, indicators of activities, linkages and outcomes do not come just from statistical offices, there are patent and bibliographic statistics that are in the public domain, as well as annual reports of firms, and private sector data bases. To monitor Canada's Innovation performance, the Government of Canada is committed to work with stakeholders to develop a set of indicators to monitor Canada's innovation performance in absolute terms and in relation to major competitors. To fulfill this commitment, the Government of Canada will contract with an independent third party for the preparation of a report to the Minister of Industry that uses indicators to monitor Canada's innovation performance. The material in this book, and the further work it will stimulate, are contributions to the development of some of those indicators.

At the same time, the Organisation for Economic Co-operation and Development is revising the *Oslo Manual* which provides guidelines for the collection and interpretation of data on innovation and the material in the book also contributes to that process which will, in turn, influence future surveys and the interpretation of the results.

Systems of innovation, whether they are international, national, or local, are complex. Understanding systems of innovation takes many players and a co-ordinated effort and it is difficult. But, better understanding may lead to better performance and that is the goal.

Some of the chapters in the book have been written by colleagues from Industry Canada and from Statistics Canada. In each case they are presenting their own views and not those of their respective agencies.

Marie Tobin
Director General, Innovation Policy Branch, Industry Canada

Fred Gault
Director, Science, Innovation and Electronic Information Division, Statistics Canada

August 2003

Contributors

NABIL AMARA, University of Laval

FRANCES ANDERSON, Statistics Canada

ANTHONY ARUNDEL, MERIT (Maastricht Economic Research Institute on Innovation and Technology), University of Maastricht

PETER BOYD, Industry Canada

VANESSA CHANG, Industry Canada

BRIAN P. COZZARIN, University of Waterloo

FRED GAULT, Statistics Canada

DAOOD HAMDANI, Statistics Canada

PETR HANEL, University of Sherbrooke and CIRST (Centre interuniversitaire de recherche sur la science et la technologie)

JEANNE INCH, Canadian Heritage, formerly of Industry Canada

RÉJEAN LANDRY, University of Laval and the ISRN (Innovation Systems Research Network)

CAN D. LE, Industry Canada

PIERRE MOHNEN, MERIT, University of Maastricht, CIRANO (Centre interuniversitaire de recherche en analyse des organisations) and UQAM (Université du Québec à Montréal)

BRIAN NEMES, Statistics Canada

SUSAN SCHAAN, Statistics Canada

ANDREW SHARPE, Centre for the Study of Living Standards (CSLS)

JIANMIN TANG, Industry Canada

PIERRE THERRIEN, Industry Canada

Canadian Heritage, Industry Canada, and Statistics Canada are departments or agencies of the federal government of Canada.

Tables

Figures

Appendices

Introduction

1

Innovation Surveys and Their Contribution to Policy Development

Peter Boyd and Jeanne Inch

1. INTRODUCTION

"Innovation is the process through which new economic and social benefits are extracted from knowledge" (Government of Canada 2002a). Engaging the actors who make innovation happen is part of improving the process as is understanding how the process works. This book examines what can be found from the statistical measurement of innovation in Canada, but that examination has to be conducted in light of the policy context in order to show how intertwined are policy and measurement.

1.1 The Innovation Strategy of the Government of Canada

In February 2002 the Government of Canada released two papers that comprise Canada's Innovation Strategy, *Achieving Excellence: Investing in People, Knowledge and Opportunity* (Government of Canada 2002a) and *Knowledge Matters: Skills and Learning for Canadians* (Government of Canada 2002b). An extensive consultation with Canadians followed and was reported in *Canadians Speak on Innovation and Learning* (Government of Canada 2002c). This document summarized the different perspectives and shared priorities of the participants in the consultation and helped to articulate the emerging consensus.

In November 2002, people from all sectors of the economy met at the National Summit on Innovation and Learning in Toronto to discuss, and recommend priorities and next steps for, the Innovation Strategy. *Canadians Speak on Innovation and Learning* was an input to these discussions and the findings were reported in the document *National Summit on Innovation and Learning: Summary* (Government of Canada 2002d). In 2003, many recommendations from the National Summit appeared in the Government of Canada's Budget 2003 and stakeholders from all areas of the economy continue to work together to identify and implement ideas to make Canada more innovative.

1.2 Developing a Survey

Surveys are not done in a vacuum. The Survey of Innovation 1999 started in 1998 with the first discussions between statisticians and policy analysts on what the policy issues were that could be better informed by statistical measurement and analysis. Much of what had to be done flowed directly from *Science and Technology for a New Century: A Federal Strategy* (Government of Canada 1996: 5) where it was stated that: "We must take a more deliberate approach to building the Canadian innovation system, by understanding how it functions, playing on its strengths and reducing its weaknesses, engaging all participants, and getting the federal government's role right."

Discussions in 1998 and 1999 led to a series of questions on the environment for innovation, the activity of innovation, linkages of the firms with other actors, knowledge generation and protection in the firm, and the impacts of innovation. Then came the testing of the questions and the preparation of the final questionnaire which was used to collect data in the autumn of 1999 for the years 1997-1999. The final questionnaire can be found in Schaan and Anderson (2001).

The advantage of creating a questionnaire collaboratively was that both policy analysts and statisticians understood the limitations of the other and the time constraints. As well, the policy analysts were in a better position to oversee and contribute to the analytical work which started when the data set was released in 2001. The benefit of this was evident when analysis from the survey appeared in the Industry Canada contribution to the Innovation Strategy, *Achieving Excellence*.

Now that the results have been incorporated in Canada's Innovation Strategy and inform the on-going discussion about its implementation, there is a need to make the findings accessible to a wider community. This book is an opportunity to show how statistical measurement and policy interact with one another and it makes a valuable contribution to informing the ongoing policy debate and research agenda.

2. CONTRIBUTING TO THE POLICY DEBATE

Understanding Innovation in Canadian Industry focuses on the activity of innovation in the private sector, the links of that activity to other sectors of the economy, and the outcomes of the activity. To do this it takes a narrow view of innovation as the introduction of new or significantly improved products to the market by private sector firms, or the introduction by those firms of new or significantly improved processes for the production or delivery of products. A product here is either a good or a service. This narrower definition is consistent with what is used in other countries to measure the activity, linkages and outcomes of innovation in industry, and especially in the European Union (EU) where three rounds of the Community Innovation Survey (CIS) have contributed to our knowledge of how innovation works there.

The public policy debate goes beyond Canada and Europe as many of the non-European member countries of the Organization for Economic Co-operation and Development (OECD) have been measuring innovation for years, using the OECD/Eurostat *Oslo Manual* (OECD/Eurostat 1997) as a guideline for the collection and interpretation of the data. Australia and Hungary are examples, and New Zealand is conducting such a survey. As Eurostat, the statistical office of the EU, prepares for the fourth round of the CIS, Eurostat and the OECD are working on the second revision of the *Oslo Manual* which is expected in 2005 or 2006. This revision illustrates that measuring and interpreting the activity of innovation changes with deeper understanding, both of the activity and of policy needs.

Industry Canada has worked with Statistics Canada to develop innovation surveys and to use their findings in policy development. The Survey of Innovation and Advanced Technology 1993 looked at firms in manufacturing, the Survey of Innovation 1996 covered the communications, financial, and technical business services industries, and the Survey of Innovation 1999 covered manufacturing and selected natural resource industries. In addition to innovation surveys, the collaboration includes the surveys of research and development, commercialization of intellectual property from universities and government laboratories, and the use and planned use of technologies (advanced manufacturing technologies, biotechnologies, and information and communication technologies).

Findings of these surveys have been used in the policy papers and briefings related to technological change over the last decade. However, the policy issues around innovation have been expanding as social and economic policy have become more closely coupled.

3. SOME ISSUES ARISING

The evolution of the measurement of technological innovation in industry has been driven in part by the need to have a social as well as an economic understanding of innovation. In 1997 the federal government of Canada made the point that: "Innovation is critical not only to economic success, but also to preserving the environment and renewing social fabric" (Industry Canada 1997b: 36) which made clear that innovation, and innovation policy, were broad. More recently, Maxwell (2003) has stressed the social aspects of innovation.

The first edition of the *Oslo Manual* (OECD 1992) was focused on technological innovation in manufacturing industries. However, about seven out of ten jobs in OECD countries are in public or private service industries and this led to a broadening of the scope of the manual to all sectors of the economy in its first revision, and especially services. Innovation was no longer seen as something that occurred in factories, it was pervasive in the economy, set in an economic and social context, and it had social as well as economic outcomes. Innovation in a firm could change the skill levels required in the labour force, for example, with implications for staffing, training and retraining, and it could alter the level of employment. For the firm to be innovative it needs skilled people, a risk accepting culture, access to venture capital, and a regulatory environment that supports entrepreneurial activity.

Non-technological innovation includes the organization of the firm and the use of management practices as ways of changing the production or delivery process, or even the product. While there is an extensive literature on firm organization, it is not often linked to the activity of innovation. Looking at management practices, there has been little study of the adoption of, for example, knowledge management practices and their role in innovation. This contrasts with the preoccupation in the 1980s with the use and planned use of technologies, especially advanced manufacturing technologies and office automation technologies.

If the adoption of technologies or practices is part of the activity of innovation it raises a question about the degree of novelty of innovation that is being measured and, perhaps, encouraged by policy. In the Survey of Innovation 1999 on which many of the chapters in this book are based, the top of the novelty scale is "world-first" followed by "Canada-first" and then by "firm-first." The last category is capturing the adoption and diffusion of technologies, while the first two are indicative of more radical change. The characteristics of firms that are world-first or Canada-first innovators are also quite different from those that are either firm-first innovators or non-innovators.

While novelty is an important factor in analysing the activity of innovation, so also is the period in which it is observed. The period of three years is recommended in the *Oslo Manual* and is used in the CIS and in Canadian surveys. However, if the change is made through capital investment and the equipment, or the software system, lasts longer than three years, a leading edge firm could be classified as a non-innovator if it made the investment more than three years ago.

The issues just raised arise in various chapters and, again, in the last section on Next Steps.

4. THE OUTLINE OF THE BOOK

Each section with more than one chapter begins with a note on key points. What follows is a short introduction to each of the seven sections which follow this.

In the section, *Measuring and Analysing*, Nemes and Schaan introduce the Survey of Innovation 1999 and the methods used in designing and conducting the survey. This is followed by a chapter by Arundel and Mohnen on the econometric models used in the book. The two chapters combine to show what has been measured, how it is analysed, and the extent to which inferences can be drawn from the analysis. The objective is to equip the reader with some tools for an informed assessment of what is to follow.

The section *First to the Market* has three chapters on firms that are world-first or Canada-first innovators. In the first, Landry and Amara look at the relationship between the use of different sources of information and the novelty of innovation and they build a taxonomy where sources of information are aggregated into four different networks: internal; market; research; and, information networks. They find that firms which use a larger variety of sources of information from within the internal network, as well as from the research networks, are more likely than other firms to introduce world-first innovations.

In the second, Therrien and Chang look at the relationship between collaboration, and more specifically local collaboration, and novelty of innovation to show that collaboration is an important factor in world-first innovation. The significance of local collaboration has implications for policy that were reflected in *Achieving Excellence* in terms of seeking to improve the innovation capacity of communities.

In the third, Cozzarin finds a correlation between firms that are first to market (world-first or Canada-first) and higher performance in terms of market share, labour productivity and price cost margin.

The section *Objectives, Outcomes and Firm Characteristics* has two chapters by Tang and by Le and Tang. Tang links business objectives, innovation inputs, and innovation outcomes to find that the objectives of CEOs have a significant impact on innovation outcomes and that different business objectives will drive the firm towards what he terms "technological invention" or "technology adoption."

Le and Tang show that innovation input activities that firms have to undertake to innovate vary according to whether the firm wants a product or a process innovation. They find that research and development activities are more important for product innovation than for process innovation and that acquisition of technologies is only important for process innovation.

The policy implications are that there is no one policy to promote innovation, rather there has to be a mix of policies in such areas as R&D programmes, technology promotion and a regulatory environment that supports commercialization of intellectual property.

The section *Managing Knowledge* has two papers. The first by Hanel that deals with intellectual property protection, and implicitly with codified knowledge, and the second by Hamdani which deals with people and tacit knowledge. Hanel notes that large firms engage in R&D, use government support programmes, and intellectual property protection, and have a higher propensity to be world-first innovators. By contrast, small and medium-sized firms make less use of intellectual property protection and this may be related to the cost of obtaining, maintaining, and enforcing this protection. This may suggest a need to consider this issue in policy development related to intellectual property protection.

Hamdani, using the Survey of Innovation 1996 looks at engineering services and the interaction between knowledge and innovation. Firms that are more knowledge intensive are more likely to innovate. Where firms with a lower knowledge intensity do innovate, it is because they contract for the necessary expertise outside of the firm, or enter into alliances with other firms. This link between alliances and the propensity to innovate is also seen by Therrien and Chang in the *First to Market* section and has implications for policies related to improving the innovation performance at the community level.

The section *Canada and Europe* consists of a single chapter by Mohnen and Therrien which compares the results of the Canadian Survey of Innovation 1999 with those of the second EU Community Innovation Survey (CIS.2). They limit their observations to product innovations and combine the Canadian novelty categories of world-first and Canada-first into one for comparison with the CIS.2 category of first to the market. They find that while Canada leads in the percentage of firms with first to the market innovations, it does not when it comes to comparing the percentage of sales due to first to the market innovation. This raises a policy question around the commercialization of intellectual property and its effectiveness and whether it is linked to the point raised by Hanel that small and medium-sized firms may not have the resources to protect the intellectual property that they need to commercialize if they are to be first to the market innovators. This also relates to the issue, prominent in *Achieving Excellence*, of encouraging the commercializing — and capturing the benefits of — intellectual property from universities and from government laboratories.

The section *Beyond Manufacturing* provides information on innovation in industries outside of manufacturing and raises some measurement questions, some of which have already been noted in this chapter. There are two chapters and the first, by Schaan examines innovation in the Canadian forest sector. Using a systems approach, she analyses the characteristics of, and the relationships between, firms in this sector. She concludes that the forest sector is part of a larger system of innovation including resource management industries, primary manufacturing, service industries, and manufacturers of machinery and equipment. The policy implication of such an approach is that it is the entire system that must be addressed if the activity of innovation is to be improved. The firms, or actors, engaged in the activity are important, but so also are the linkages with other actors in the system and regulatory environment in which the system behaves.

Anderson looks at the construction sector and the flow of building products from the manufacturing to the construction industries. As in the Schaan chapter, the finding is that a systems approach is required if innovation is to be understood in the construction sector. Both chapters deal with the adoption of technologies as part of the process of innovation and this raises the question, already noted, that a three year period of observation may be insufficient if the capital equipment lasts longer than that and the investment was made outside the period of observation.

In the final section, *Next Steps*, Sharpe reviews the findings in the previous chapters, identifies gaps in the present work and opportunities for new work and then raises some policy issues and their implications.

5. POLICY, AWARENESS, MEASUREMENT AND LEARNING

This book is a contribution to the ongoing discussion of the implementation and evolution of the Innovation Strategy as it unfolds in 2003 and 2004 and beyond. It is also a means of developing awareness of the activity of innovation, its place in an innovation system, and the social and economic consequences of innovation.

It provides input to the work on the revision of the OECD/Eurostat *Oslo Manual* which provides guidelines for the collection and interpretation of data on innovation. It also is an input to the Statistics Canada

Survey of Innovation that will be conducted in the autumn of 2003 for selected service industries and released in 2004.

The measurement of innovation is still, and will continue to be for some years, a learning experience for statisticians, for students of economic and social change, and for policy analysts and decision makers. This book makes a significant contribution and we would like to thank Fred Gault for the considerable time and effort he has put into making this book possible.

Measuring and Analysing

Section Overview

Fred Gault

The two chapters in this section set the stage for the analysis of the Survey of Innovation 1999 which is a focus for much of the rest of the book.

The chapter by Schaan and Nemes presents the survey, its industrial coverage, sample size and how it was drawn, the design of the questionnaire, data collection, editing and imputation, and quality measures. Finally, there is a section on lessons learned, especially when innovation is measured outside of the manufacturing sector. The questionnaire can be found in the paper by Schaan and Anderson (2001) which is available at www.statcan.ca.

Measuring the activity of innovation, and its linkage to other activities in the innovation system, is not a well established practice, even though people have been doing it for over thirty years. In Canada, Christian DeBresson (1980) and Dennis De Melto and his colleagues (De Melto, McMullen and Wills 1980) pioneered the work on innovation surveys.

Those who do surveys learn from each attempt and this is evident from Schaan and Nemes who reveal difficulties as well as the strengths in the measurement process. Both can be traced back to the design of the questionnaire which took many months of intense discussion between the policy analysts, who were to be the ultimate users of the information, and the statisticians who were to produce it. Questions that survived this process were then subjected to the cognitive testing required by the statistical office before a survey is launched, and not all of those questions survived.

Testing survey questions on willing respondents is part of keeping the burden on respondents to a minimum. If the question is clear, and takes little time to answer, it contributes not just to a good response rate, but also to the quality of the analytical output from the survey. Time, a precious commodity in business, is not wasted. A recommendation of Schaan and Nemes is that a sample of respondents should be visited after the survey to confirm that their responses did reflect a clear understanding of the question and described what was actually happening in the firm.

Arundel and Mohnen review the econometric methods used in other chapters to analyse the data on innovation and to draw inferences. The paper is both a tutorial on techniques and a warning about the rules of evidence and their application. The reader is cautioned about leaping to conclusions about causal links, and about drawing policy inferences which may not be rooted in the data. The reader is also challenged to go beyond the analysis presented in the other chapters to establish a deeper understanding of the process of innovation.

While this book presents the best picture possible about innovation in Canada, it deliberately raises questions about all aspects of the processes of measurement and analysis to enable the reader to make informed judgements on the chapters that follow and the inferences drawn by the authors. The objective is to encourage an understanding of the activity of innovation among people engaged in the activity, policy analysts, statisticians, and students of technological change. The rewards of better understanding are great as they can contribute to more effective business practices, and policies at all levels of government.

2

Survey of Innovation 1999 – Methodological Framework: Decisions Taken and Lessons Learned

Susan Schaan and Brian Nemes

1. INTRODUCTION

1.1 Why Survey Innovation?

Innovation may be thought of as transforming knowledge into economic activity. From this perspective, innovation therefore performs a vital role in contributing to economic growth and development. It is through innovation that new products are introduced into the marketplace, new production processes are developed and organizational changes are made.

The information compiled from innovation surveys can be analyzed to determine the characteristics of innovative and non-innovative businesses. It can also be used to provide indicators on how policies are regarded by the business sector. Studies of innovation activity can be undertaken to examine impacts, collaboration, problems and obstacles. Finally, governments can utilize this information to develop national and regional policies.

The Statistics Canada Survey of Innovation 1999 was designed to address these and other issues. In this chapter, the methods used to develop, conduct and analyze the various elements of the survey process are described. A more detailed treatment, and a copy of the questionnaire, can be found in Schaan and Anderson (2001).

1.2 The Oslo Definition of Innovation

The *Oslo Manual* (OECD/Eurostat 1997) outlines proposed guidelines for collecting and interpreting innovation data at the level of the firm. The manual identifies two types of technological innovation — product and process. In the case of product innovation, the product must have been introduced to the market. The term "product" includes both goods and services as innovation outputs. A process innovation must have been used within the production process. An innovative firm is one that has offered a new or significantly improved product or introduced a new or significantly improved production/manufacturing process during the previous three years. Changes to a firm's existing products that are purely aesthetic or that involve only minor modifications are not considered to be innovations.

New production/manufacturing processes are those that are new to a firm. They involve the introduction of new production/manufacturing methods, procedures, systems, machinery or equipment into the firm. These must differ significantly from the firm's previous processes. Significantly improved production/manufacturing processes involve significant changes to existing processes which may be intended to produce new or significantly improved products or processes. Minor or routine changes to production/manufacturing processes are not considered to be innovations.

The first edition of the *Oslo Manual* (OECD 1992) was based on experiences with manufacturing. The second edition (OECD/Eurostat 1997) was updated to incorporate survey experience, an improved understanding of the innovation process, and to include a wider range of industries, more particularly in manufacturing, construction, utilities and marketed services. The purpose of the manual is to "provide a framework within which existing surveys can evolve towards comparability; and to assist newcomers to this important field" (OECD/Eurostat 1997). It allows the production of internationally comparable, meaningful indicators of innovation.

1.3 Innovation Surveys in Canada

Statistics Canada has conducted several surveys of innovation since 1993 to gain a better understanding of innovation in Canada.

- The Survey of Innovation and Advanced Technology 1993 surveyed manufacturing firms.
- The Survey of Innovation 1996 surveyed the communications, financial services and technical business services industries.
- The Survey of Innovation, Advanced Technologies and Practices in the Construction and Related Industries 1999 was the first survey of advanced technologies and practices in the construction sector.
- The Survey of Innovation 1999 surveyed manufacturing firms and was the first innovation survey of selected natural resource industries.

1.4 Structure of the Questionnaire for the Survey of Innovation 1999

The questionnaire designed for the Survey of Innovation 1999 consisted of 13 sections. The following topics were covered:

- competitive environment;
- factors for firm success;
- new and significantly improved products and processes;
- sources of information;
- objectives;
- problems and obstacles;
- impact;
- co-operative and collaborative arrangements;
- most important new or significantly improved product or process;
- building and construction products;
- natural resource products;
- R&D, intellectual property and human resources; and
- government support programs.

2. SAMPLE UNIT AND STRATIFICATION

2.1 Target Population

The target population for the Survey of Innovation 1999 was all firms in the manufacturing sector (NAICS or North American Industry Classification System codes 31-33) (Statistics Canada 1998a) or in selected natural resource industries (NAICS codes 1133, 212, 2211).

Statistics Canada has many annual surveys that collect production data. To reduce the response burden for the Survey of Innovation 1999, the questions were restricted to those exploring the nature of innovation. These types of questions are best suited to response by a person at the corporate level. For this reason, the chief executive officer (CEO) of a firm or a person designated by the CEO was targeted as the respondent.

2.2 Production Data

To provide production data to accompany the innovation data, a subsample of existing responses to surveys conducted by Statistics Canada's Manufacturing, Construction and Energy Division (MCED) and Natural Resources Canada was chosen. The surveys included MCED's Annual Survey of Manufacturers, Coal Mines Annual Survey and Annual Survey of Electric Power Generation, and Natural Resources Canada's Annual Census of Mines, Quarries and Sandpits. The use of the subsample of respondents to these existing surveys meant that new businesses (founded after 1997) were not part of the target population. At the time of sample selection, the most current data for a common year were chosen. Only businesses that responded to the production surveys were considered for the population for the Survey of Innovation 1999.

The data used from these responses constituted the principal statistics. The principal statistics included shipments, materials purchased, inventories, labour data (including the number of production and related workers, and employees' salaries and wages), value added[1] and commodity information.

2.3 Provincial Enterprise

The requirement to produce provincial statistics was a major criterion in defining the sampling unit. Following the recommendations of the *Oslo Manual*, the enterprise was chosen. Whenever possible, provincial data are to be an output of surveys at Statistics Canada. Accordingly, all establishments in the same province and industry at the four-digit NAICS level were grouped to form one sample unit or "provincial enterprise." To reduce the response burden on small businesses, only provincial enterprises with a gross business income of at least $250,000 and at least 20 employees were considered in sample selection.[2]

2.4 Stratification

Thirty-six industry strata were identified based on the groupings by four-digit NAICS codes. The sample was randomly drawn from the population of provincial enterprises stratified by province and the 36 industry strata (31 industry strata for manufacturing, four industry strata for the selected natural resource industries and one for utilities). Details of the industry codes used are found in Appendix 1.

As part of a federal-provincial agreement, the sample in Quebec was augmented. A sample of 5,944 provincial enterprises in manufacturing and 674 provincial enterprises in selected natural resource industries was drawn, for a total sample size of 6,618 provincial enterprises.

3. QUESTIONNAIRE DESIGN

The questionnaire was designed by the Science, Innovation and Electronic Information Division of Statistics Canada in collaboration with Industry Canada, the National Research Council of Canada's Institute for Research in Construction and Natural Resources Canada. A small sample of individual firms was interviewed to test the questionnaire and ensure that the questions were well understood. Feedback from these firms was incorporated into the questionnaire.

4. DATA COLLECTION

4.1 Data Capture

Computer-assisted telephone interviewing (CATI) was used to collect the survey data. All sample units were contacted ahead of time to determine the name and correct mailing address for the respondent. Questionnaires were then mailed out. Mail, telephone and fax follow-ups were carried out to elicit a response from non-respondents.

4.2 Response Rate

The response rate for the Survey of Innovation 1999 was calculated as the total number of completed questionnaires as a percentage of the total active, in-scope survey sample. For manufacturing industries, the total number of completed questionnaires was 5,455, and the overall response rate was 95%. For the selected natural resource industries, the total number of completed questionnaires was 582, and the overall response rate was 94%.

Given the low rate of non-response to the survey, it seemed reasonable to assume that the characteristics of the non-respondent population were the same as those of the respondent population. Accordingly, the contribution of non-responses to the estimates was accounted for by adjusting the sample weights of the respondent population.

Estimates based on the responses to survey questions are population estimates — that is, they represent the percentage of businesses in the population that exhibit a particular characteristic. Population estimates are generated through the accumulation of the product of the response variable and the sample weight for the defined tabulation cells.

4.3 Sampling Error

As the sample drawn for this survey was only one of many possible samples that could have been drawn, a sampling error was attributed to it. Standard errors were used to provide a guide as to the reliability of the results. The reliability of the data was reported using the following convention:

Symbol	*Meaning*	*Standard Error*
A	Very reliable	≤2.5%
B	Reliable	>2.5% and ≤7.5%
C	Use with caution	>7.5 and ≤15%
D	Too unreliable to be published	>15%

Estimates with very poor reliability (D) were suppressed.

Not all industries operate in all provinces. Consequently, it was impossible to produce estimates for all industries in all provinces. For example, small populations restricted the production of any data for Audio and Video Equipment Manufacturing. However, this small population did contribute to estimates at the aggregate level for manufacturing in Canada. Similarly, where the number of records contributing to an estimate brought the quality of representation of the data into question, the estimate was suppressed.

5. EDITS AND IMPUTATION

5.1 Edits

Validity and flow edits were built into the data capture system and were applied during data collection and data entry. Validity edits ensured that responses to particular questions fell within a limited range of possible values. Post-collection consistency edits were applied to complete questionnaires.

5.2 Imputation Strategy

There were several cases where the relevance of a set of questions relied on a response to a preceding question. The ability to proceed along a path of questioning was reliant on the nature of the response, and the subsequent responses were influenced by the firm behaviour indicated by the response to the preceding question. Block imputation (one donor) was used for these correlated questions as a means to avoid edit failures.

6. LINKAGES

The Survey of Innovation 1999 covered the manufacturing, logging, mining (coal mining, metal ore mining and non-metallic mineral mining) and electric power generation, transmission and distribution industries, requiring a number of separate linkage operations, one set at a time.

Linkage of each survey data set followed the same logic as that utilized to create the provincial enterprise. Establishment data for respondents in the same province and industry were accumulated to form the provincial enterprise created for the collection phase of the survey. Production data were accumulated to create the value of each principal statistic for each provincial enterprise. This value then became a characteristic of the firm in the Survey of Innovation database for analytical purposes.

7. KEY LESSONS LEARNED

7.1 Choice of Provincial Enterprise

It was assumed that different business characteristics existed between provinces and industries, thus the CEO was mailed one questionnaire for each provincial enterprise. The nature of the sample unit, the provincial enterprise, meant that one enterprise could potentially receive more than one questionnaire. The characteristics of the provincial enterprise were assumed to be different for provincial enterprises of the same firm. In the case of a multi-location and/or multi-industry business, did the responses received from the CEO in one location apply equally to all locations? Could it be assumed that the opinions of a CEO located in Toronto were equally representative of the business in northern Ontario or southwestern Manitoba?

CEOs who received more than one questionnaire could potentially report only once, indicating this during data collection. Statistics Canada's requirement for provincial estimates was that CEOs report for each industry in each province for their business. The appropriateness of adding these responses at the national level raised the issue of multiplicity due to the number of questionnaires sent to one respondent.

Therrien and Mohnen (2003) have analyzed the issue of multiplicity resulting from the use of a provincial enterprise. They compared innovation rates calculated for single-location firms with innovation rates calculated for the entire sample. The innovation rate for the whole population was estimated at 80.3%, whereas the innovation rate for single-location firms was 79.4%. The conclusion was that the presence of multi-location firms did not result in an upward bias in the innovation rate. Tests were carried out on other variables of interest, and only minor and non-significant differences were observed.

The absence of an upward bias in innovation rates supports the assumption that responses provided by CEOs for one location applied equally to all locations in cases where the CEOs completed only one questionnaire.

Post-survey data validation is a necessary step to understanding how respondents viewed the questions. An examination of how well respondents understood concepts and how they may have reacted to completing more than one questionnaire for their enterprise can aid in data interpretation. This would allow for hands-on evaluation of potential multiplicity and insight into the characteristics of each provincial enterprise, to evaluate whether characteristics are indeed the same or different.

7.2 Non-Innovators with Innovation Activities

Two questions were included to allow an assessment of those firms that were not innovative but had undertaken innovation activities and thus had the potential to be innovative. If the results of the survey had shown a lower innovation rate, it would have been possible to use this to assess the "innovative potential" of firms; however, with such high innovation rates this is no longer an issue.

7.3 Novelty of Innovation

The finding of an innovation rate of 80% may suggest that the definition of innovation used in the survey was too broad. For the Survey of Innovation 1999, there were three categories of "novelty" of innovation: world-first, Canada-first and firm-first. The 80% result corresponds to the weakest category (firm-first), and it may be more useful to focus on the more restrictive categories.

In this book, Brian Cozzarin discusses the relationship between novelty of innovation and firm performance, and Réjean Landry and Nabil Amara examine the effects of sources of information on novelty of innovation. To draw comparisons with the European Community Innovation Survey (CIS), Pierre Mohnen and Pierre Therrien combine the world-first and Canada-first categories and compare the results with those from the CIS question about being first to the market.

7.4 Definition of R&D

There were two questions in the Survey of Innovation 1999 that inquired about the performance of R&D. The wording of each and the responses varied. In the first question concerning R&D performance, respondents were asked to indicate if their firm had engaged in R&D linked to new or significantly improved products or production/manufacturing processes. Sixty-five per cent of manufacturing firms indicated that they had engaged in such R&D. The other question asked if a firm had undertaken R&D activities. Fifty-nine per cent of manufacturing firms indicated that they had undertaken R&D activities. The meaning of R&D was open to interpretation in both these cases, since R&D was not defined in the questionnaire.

8. CONCLUSION

Innovation surveys are relatively new and evolutionary. The Survey of Innovation 1999, designed to gain insights into the innovation process, has been judged to be an important contribution to the understanding of innovation systems in manufacturing and selected natural resource industries. It tested many of the concepts outlined in the *Oslo Manual* and found them to be workable. It also determined that some of the methods chosen to measure innovation did not work as well as expected. On balance, however, the successful features of this survey outweighed the unsuccessful. The lessons learned will find their way into the next round of innovation surveys.

Appendix 1: Industry Stratification

Table A1.1: Population, Survey of Innovation 1999

Stratum Number	NAICS	Description	NF	PE	NS	NB	QC	ON	MB	SK	AB	BC	YT	NT	Total
1	1133	Logging	8	1	5	16	80	37	3	6	25	160	0	0	341
	212	Mining													
2	2121	Coal Mining	0	0	4	1	0	0	0	1	4	4	0	0	14
3	2122	Metal Ore Mining	3	0	0	3	21	21	5	4	1	14	5	6	83
4	2123	Non-Metallic Mineral Mining and Quarrying	20	4	31	40	242	296	35	46	115	91	2	0	922
	221	Utilities													
5	2211	Electric Power Generation, Transmission and Distribution	4	2	1	2	6	298	2	2	8	5	2	1	333
	Total Selected Natural Resource Industries		**35**	**7**	**41**	**62**	**349**	**652**	**45**	**59**	**153**	**274**	**9**	**7**	**1,693**
6	311	Food Manufacturing	29	18	50	46	294	203	34	22	70	90	0	0	856
7	312	Beverage and Tobacco Product Manufacturing	3	2	5	6	32	22	3	3	10	13	0	0	99
8	313	Textile Mills	0	0	4	4	115	44	5	0	0	2	0	0	174
9	314	Textile Product Mills	0	1	5	2	55	46	7	2	9	12	0	0	139
10	315	Clothing Manufacturing	0	0	4	8	380	104	21	5	13	36	0	0	571
11	316	Leather and Allied Product Manufacturing	1	0	1	2	45	17	4	1	4	5	0	0	80
	321	Wood Product Manufacturing													
12	3211	Sawmills and Wood Preservation	6	2	16	15	133	25	5	6	14	67	0	1	290
13	3212	Veneer, Plywood and Engineered Wood Product Manufacturing	0	0	3	7	37	24	3	3	14	26	0	0	117
14	3219	Other Wood Product Manufacturing	0	1	5	8	125	81	7	3	22	53	0	0	305
15	322	Paper Manufacturing	4	1	8	14	106	100	11	4	16	35	0	0	299
16	323	Printing and Related Support Activities	1	1	12	7	189	233	27	9	46	73	0	0	598
17	324	Petroleum and Coal Products Manufacturing	1	0	3	3	13	16	3	3	8	7	0	0	57
	325	Chemical Manufacturing													
18	3251 & 3252 & 3253 & 3255 & 3256 & 3259	Basic Chemical Manufacturing & Resin, Synthetic Rubber, and Artificial and Synthetic Fibres and Filaments Manufacturing & Pesticide, Fertilizer and Other Agricultural Chemical Manufacturing & Paint, Coating and Adhesive Manufacturing & Soap, Cleaning Compound and Toilet Preparation Manufacturing & Other Chemical Product and Preparation Manufacturing	1	3	3	8	153	162	13	7	39	29	0	0	418
19	3254	Pharmaceutical and Medicine Manufacturing	0	1	1	0	33	20	2	1	1	3	0	0	62

... continued

Table A1.1 (continued)

Stratum Number	*NAICS*	*Description*	*NF*	*PE*	*NS*	*NB*	*QC*	*ON*	*MB*	*SK*	*AB*	*BC*	*YT*	*NT*	*Total*
20	326	Plastics and Rubber Products Manufacturing	3	0	9	10	218	213	20	5	33	55	0	0	566
21	327	Non-metallic Mineral Product Manufacturing	7	2	15	15	116	114	13	9	30	57	0	0	378
22	331	Primary Metal Manufacturing	0	1	2	3	73	92	15	4	16	21	0	0	227
23	332	Fabricated Metal Product Manufacturing	5	3	13	17	389	521	41	23	127	122	0	0	1,261
	333	Machinery Manufacturing													
24	3331 & 3332	Agricultural, Construction and Mining Machinery Manufacturing & Industrial Machinery Manufacturing	0	1	2	4	87	62	9	17	40	22	0	0	244
25	3333 & 3334 & 3335 & 3336 & 3339	Commercial and Service Industry Machinery Manufacturing & Ventilation, Heating, Air-Conditioning and Commercial Refrigeration Equipment Manufacturing & Metalworking Machinery Manufacturing & Engine, Turbine, and Power Transmission Equipment Manufacturing & Other General Purpose Machinery Manufacturing	0	1	6	3	190	322	17	10	39	43	0	0	631
	334	Computer and Electronic Product Manufacturing													
26	3341	Computer and Peripheral Equipment Manufacturing	0	0	0	0	19	16	3	1	0	7	0	0	46
27	3342	Communications Equipment Manufacturing	1	0	0	1	27	21	2	0	7	9	0	0	68
28	3343	Audio and Video Equipment Manufacturing	0	0	0	0	0	6	0	0	0	0	0	0	6
29	3344	Semiconductor and Other Electronic Component Manufacturing	0	0	1	1	24	19	0	0	3	4	0	0	52
30	3345 & 3346	Navigational, Measuring, Medical and Control Instruments Manufacturing & Manufacturing and Reproducing Magnetic and Optical Media	1	0	4	1	49	51	4	2	10	13	0	0	135
31	335	Electrical Equipment, Appliance and Component Manufacturing	0	0	4	1	84	95	4	3	12	19	0	0	222
	336	Transportation Equipment Manufacturing													
32	3361 & 3362 & 3363	Motor Vehicle Manufacturing & Motor Vehicle Body and Trailer Manufacturing & Motor Vehicle Parts Manufacturing	0	0	6	3	68	168	14	6	25	34	0	0	324
33	3364	Aerospace Product and Parts Manufacturing	0	3	3	0	30	22	4	0	3	7	0	0	72
34	3365 & 3366 & 3369	Railroad Rolling Stock Manufacturing & Ship and Boat Building & Other Transportation Equipment Manufacturing	2	1	9	1	22	19	2	1	0	23	0	0	80
35	337	Furniture and Related Product Manufacturing	1	1	3	8	219	153	18	6	44	47	0	0	500
36	339	Miscellaneous Manufacturing	3	0	9	8	140	160	14	4	35	52	0	1	426
	Total Manufacturing Industries		**69**	**43**	**206**	**206**	**3,465**	**3,151**	**325**	**160**	**690**	**986**	**0**	**2**	**9,303**
	Total Manufacturing and Selected Natural Resource Industries		**104**	**50**	**247**	**268**	**3,814**	**3,803**	**370**	**219**	**843**	**1,260**	**9**	**9**	**10,996**

Source: Statistics Canada, Survey of Innovation 1999.

Table A1.2: Sample, Survey of Innovation 1999

Stratum Number	NAICS	Description	NF	PE	NS	NB	QC	ON	MB	SK	AB	BC	YT	NT	Total
1	1133	Logging	8	1	5	16	78	16	3	6	13	22	0	0	168
	212	Mining													
2	2121	Coal Mining	0	0	3	1	0	0	0	1	4	4	0	0	13
3	2122	Metal Ore Mining	3	0	0	3	21	15	4	4	1	10	3	5	69
4	2123	Non-Metallic Mineral Mining and Quarrying	4	2	5	6	234	40	3	7	21	21	2	0	345
	221	Utilities													
5	2211	Electric Power Generation, Transmission and Distribution	4	2	1	2	6	44	2	2	8	5	2	1	79
	Total Selected Natural Resource Industries		**19**	**5**	**14**	**28**	**339**	**115**	**12**	**20**	**47**	**62**	**7**	**6**	**674**
6	311	Food Manufacturing	17	18	17	17	246	94	17	22	52	62	0	0	562
7	312	Beverage and Tobacco Product Manufacturing	3	2	5	6	28	20	3	3	9	13	0	0	92
8	313	Textile Mills	0	0	3	4	94	33	4	0	0	1	0	0	139
9	314	Textile Product Mills	0	1	2	2	38	28	5	2	2	10	0	0	90
10	315	Clothing Manufacturing	0	0	3	6	232	66	14	5	7	31	0	0	364
11	316	Leather and Allied Product Manufacturing	1	0	1	1	35	12	2	1	2	3	0	0	58
321		Wood Product Manufacturing													
12	3211	Sawmills and Wood Preservation	4	2	15	15	103	24	5	6	14	51	0	1	240
13	3212	Veneer, Plywood and Engineered Wood Product Manufacturing	0	0	2	6	27	19	3	3	14	26	0	0	100
14	3219	Other Wood Product Manufacturing	0	1	4	5	84	43	6	3	17	41	0	0	204
15	322	Paper Manufacturing	4	1	8	14	96	65	10	4	16	35	0	0	253
16	323	Printing and Related Support Activities	1	1	7	6	82	75	24	8	24	47	0	0	275
17	324	Petroleum and Coal Products Manufacturing	1	0	3	3	13	16	2	3	8	7	0	0	56
	325	Chemical Manufacturing													
18	3251 & 3252 & 3253 & 3255 & 3256 & 3259	Basic Chemical Manufacturing & Resin, Synthetic Rubber, and Artificial and Synthetic Fibres and Filaments Manufacturing & Pesticide, Fertilizer and Other Agricultural Chemical Manufacturing & Paint, Coating and Adhesive Manufacturing & Soap, Cleaning Compound and Toilet Preparation Manufacturing & Other Chemical Product and Preparation Manufacturing	1	3	3	8	133	90	11	7	39	29	0	0	324
19	3254	Pharmaceutical and Medicine Manufacturing	0	1	1	0	28	20	2	1	1	3	0	0	57
20	326	Plastics and Rubber Products Manufacturing	3	0	9	10	148	100	20	5	33	43	0	0	371
21	327	Non-metallic Mineral Product Manufacturing	7	2	13	15	90	60	11	8	26	44	0	0	276

... continued

Table A1.2 (continued)

Stratum Number	NAICS	Description	NF	PE	NS	NB	QC	ON	MB	SK	AB	BC	YT	NT	Total
22	331	Primary Metal Manufacturing	0	1	2	3	69	60	14	4	16	21	0	0	190
23	332	Fabricated Metal Product Manufacturing	4	3	8	15	224	120	15	15	67	71	0	0	542
	333	Machinery Manufacturing													
24	3331 & 3332	Agricultural, Construction and Mining Machinery Manufacturing & Industrial Machinery Manufacturing	0	1	1	4	59	50	8	17	35	21	0	0	196
25	3333 & 3334 & 3335 & 3336 & 3339	Commercial and Service Industry Machinery Manufacturing & Ventilation, Heating, Air-Conditioning and Commercial Refrigeration Equipment Manufacturing & Metalworking Machinery Manufacturing & Engine, Turbine, and Power Transmission Equipment Manufacturing & Other General Purpose Machinery Manufacturing	0	1	4	3	121	100	12	9	35	34	0	0	319
	334	Computer and Electronic Product Manufacturing													
26	3341	Computer and Peripheral Equipment Manufacturing	0	0	0	0	12	16	3	1	0	7	0	0	39
27	3342	Communications Equipment Manufacturing	1	0	0	1	21	21	2	0	5	9	0	0	60
28	3343	Audio and Video Equipment Manufacturing	0	0	0	0	0	6	0	0	0	0	0	0	6
29	3344	Semiconductor and Other Electronic Component Manufacturing	0	0	1	1	18	19	0	0	3	4	0	0	46
30	3345 & 3346	Navigational, Measuring, Medical and Control Instruments Manufacturing & Manufacturing and Reproducing Magnetic and Optical Media	1	0	4	1	26	48	3	2	9	13	0	0	107
31	335	Electrical Equipment, Appliance and Component Manufacturing	0	0	4	1	68	57	4	3	11	19	0	0	167
	336	Transportation Equipment Manufacturing													
32	3361 & 3362 & 3363	Motor Vehicle Manufacturing & Motor Vehicle Body and Trailer Manufacturing & Motor Vehicle Parts Manufacturing	0	0	4	3	47	87	12	6	17	33	0	0	209
33	3364	Aerospace Product and Parts Manufacturing	0	3	3	0	21	16	4	0	3	7	0	0	57
34	3365 & 3366 & 3369	Railroad Rolling Stock Manufacturing & Ship and Boat Building & Other Transportation Equipment Manufacturing	2	1	4	1	19	10	1	1	0	19	0	0	58
35	337	Furniture and Related Product Manufacturing	1	1	3	7	116	70	9	4	31	33	0	0	275
36	339	Miscellaneous Manufacturing	1	0	6	5	84	60	5	3	13	34	0	1	212
	Total Manufacturing Industries		**52**	**43**	**140**	**163**	**2,382**	**1,505**	**231**	**146**	**509**	**771**	**0**	**2**	**5,944**
	Total Manufacturing and Selected Natural Resource Industries		**71**	**48**	**154**	**191**	**2,721**	**1,620**	**243**	**166**	**556**	**833**	**7**	**8**	**6,618**

Source: Statistics Canada, Survey of Innovation 1999.

Notes

[1]Value added is calculated without the cost of purchased services being taken into account (Statistics Canada 1985). When the survey was a census of manufacturers, this was referred to as "census value added" (Statistics Canada 1979).

[2]The number of employees was derived from Statistics Canada's Business Register.

3

Analytical Methods and Interpretation of Innovation Surveys

Anthony Arundel and Pierre Mohnen

1. INTRODUCTION

Complex cross-sectional innovation surveys, such as Statistics Canada's Survey of Innovation 1999, create unique challenges for economic analysis and interpretation that must be met by a careful evaluation of the research question, its policy significance, the appropriate econometric techniques, and the types of interpretation that are supported by the data. This chapter uses seven chapters that analyzed the Survey of Innovation 1999 to illustrate the application and interpretation of econometric techniques for analyzing data from cross-sectional surveys. The major challenge for econometric analysis is to evaluate the factors that are correlated with innovation quality. Several minor changes to future innovation surveys would help in this task.

The Survey of Innovation 1999 of Statistics Canada follows in its scope and design the guidelines set out in the *Oslo Manual* (OECD/Eurostat 1997). It contains valuable information about the inputs, the outputs and the organization of innovative activities by Canadian enterprises producing manufactured goods, building and construction products, and natural resource products. It obtains some data that are not collected by the comparable European Community Innovation Surveys (CIS), including information on intellectual property rights, government support programs, perceived characteristics of the competitive environment, perceived firm success factors, and the novelty of the firm's most important innovation. The Statistics Canada survey also differs from the CIS by collecting as much information as possible from non-innovative firms. However, unlike the European CIS, the Statistics Canada survey does not collect quantitative information on the amount that innovative firms spend on a range of innovation activities.

The survey allows one to explore a number of questions regarding the determinants, the outcomes and the organization of innovation. This volume contains several studies that use descriptive and econometric methods to analyze the survey results. This chapter introduces the reader to the appropriate econometric techniques, given the nature of the data, and guides the reader through the econometric chapters by highlighting some of the main results and by comparing them in a critical way.

The chapter is organized as follows. Section 2 evaluates aspects of the survey data that must be taken into consideration in econometric analysis, while Section 3 provides the alternative view, which is the relevance of econometric techniques to analyses of survey data and how econometric and descriptive analyses can complement each other. Section 4 summarizes five classes of econometric techniques that are used to analyze discrete dependent variables, which are the most common types of variables in innovation surveys. Section 5 provides an overview of the econometric analyses in the seven relevant chapters in this volume. The overview is organized around analyses of three types of dependent variables, all of which are a measure of the innovative capacity of the firm: whether or not the firm innovates at the broadest definitional level, the novelty of the firm's most important innovation, and the firm's sales share from innovative products. A fourth approach uses a least squares technique to explore a latent variable for innovation as either an adoptive or creative process. Section 6 looks at some of the difficulties in drawing policy-relevant conclusions from the econometric models, while the final section provides a few general conclusions.

2. NATURE OF THE DATA

A common feature of all innovation surveys is that they are complex and full of potential pitfalls for the unwary analyst. The Survey of

Innovation 1999 of Statistics Canada is no exception. It contains a number of characteristics of which the applied researcher should be aware. There is also one enormous advantage of the Canadian Survey of Innovation 1999 compared to comparable surveys, such as the CIS, in other countries. This is the very high (95%) response rate to the Canadian survey. Consequently, simple descriptive results, once appropriately weighted for the sampling fractions, provide an accurate measure of the prevalence of different innovative activities among Canadian manufacturing firms. Although not immediately apparent, the accuracy of descriptive statistics is also of great relevance to an evaluation of econometric techniques and their use, as explained in Section 3.

2.1 Cross-sectional Survey Design

The problems that face econometric analysis begin with the cross-sectional survey design of the Survey of Innovation 1999. Most of the survey questions provide data about the same period of time, which is between 1997 and 1999. This makes it very difficult to detect causal relationships. For example, one question asks about the number of innovations introduced between 1997 and 1999. Other questions ask about the use of different information sources during the same time period. It is possible to correlate the use of specific information sources with the number of innovations, but we do not know if the use of specific internal sources was causally related to the number of innovations, or if these internal sources simply tend to be found more frequently among firms that develop more innovations. Many of the variables in the innovation survey are likely to be determined simultaneously and not in a chronological order from which we could infer causality.

There are several exceptions. The survey was matched to the Annual Survey of Manufacturing (ASM) for 1997, which provides 1997 data for several firm characteristics, such as employment and labour productivity, while two survey questions only refer to conditions in 1999: the sales share from innovative products and the number of employees. It is possible to assume that firm conditions in 1997 could have had some influence on later innovative activities, although the time lag is very short and there is some overlap between 1997 performance and innovative activities over the 1997-1999 time period. Similarly, there could be a causal relationship between 1999 sales and innovative activities over this period, but our ability to infer causality will be limited by a lack of information on the timing of each innovative activity over the three-year period covered by the survey.

If we assume no causality, the best that we can do with econometric analysis is to assume that our models are exploring correlations after controlling for the effect of other variables. However, it is still important to make sure that the dependent variable could potentially be influenced by all independent variables and that there is a distinction between the two. A related point can be made regarding exogeneity. As an example, to what extent are innovation activities and the use of intellectual property rights (IPRs) jointly determined? If they are, then we must be careful about simultaneity bias, in which the variables are correlated because of a common correlation with another variable.

2.2 Question Formats

Many of the survey questions obtain subjective responses based on a five-point importance scale, such as the respondent's perceptions of competition or factors of success. The measure is not objective, which means that the ordered responses from different firms on what is "important" are not fully comparable. Even within the same enterprise different respondents could give different evaluations. One should therefore be careful not to interpret these responses as entirely accurate and comparable between firms, but rather to check the sensitivity of the results to another scaling such as a simple binary distinction between "more" and "less" important.

Most of the remaining survey questions obtain nominal "yes" or "no" data, coded as a "1" or a "0." An example is the key questions for defining an innovative firm, which ask "during the last three years, did your firm introduce new or significantly improved products (or processes)?" For these variables, a link has to be made between the discrete responses and areas under a certain cumulative distribution function (above and below certain thresholds), which can be handled as continuous variables.

Several questions are asked sequentially. Respondents are first asked if they have introduced a new (or technologically improved) product or process, or attempted to do so. Only firms that answer "yes" to any of these questions are asked to complete subsequent questions on the use of different sources of information, objectives of innovation and so on. Because the questionnaire is set up in this way, we do not have information on, for instance, objectives of innovation for non-innovating firms, although a few of these firms could have innovation objectives. We are thus in the presence of a censoring problem. The applied researcher needs to account for and correct possible selection biases when variables are only available for a subset of the population of interest.

Censoring is not a problem if we limit both the target population and the research question to factors that only concern the target population. For instance, censoring is less of a problem if the analysis is confined to innovative firms. However, this is not feasible if our research questions require the inclusion of non-innovative firms. For example, our interest in the use of information sources might extend to non-innovative firms. These firms could have attended trade fairs or read professional journals without finding anything of interest to encourage them to innovate.[1]

2.3 Innovation Quality

The survey questionnaire uses the broad *Oslo Manual* definition of an innovation, which includes both radical innovations that cost hundreds of millions of dollars and years of R&D to develop and new production technology that a firm buys "off-the-shelf." Both the CIS and the Canadian Survey of Innovation 1999 deal with the disparate quality of an innovation by including additional questions that can be used to assess the "innovativeness" of the firm. The Canadian innovation survey includes questions on the number of product innovations introduced in the past three years, the sales share from new or improved products, questions on product and process innovation, and a question on the novelty of the firm's most important innovation. In addition, questions on R&D performance and patenting provide alternative measures of innovative intensity. One of the main difficulties is to separate innovation as a creative activity in which the firm expends effort on developing its own new or improved products and processes and innovation as a process of adoption, in which new products and processes are acquired from external sources. The chapter by Tang explores this division of firms into technology "inventors" and technology "adopters."

Several of the econometric analyses (see Table 2 below) focus on the definition of the firm's most important innovation. Three options are provided: world-first, Canada-first, and firm-first. A major difficulty for analysis is that the outcome only applies to the firm's most important innovation. However, all of the analyses in this volume assume that other variables, which apply to *all* of a firm's innovative activities, can be linked to this specific innovation. The potential for a mismatch will create some classification error and reduce the significance of correlations, compared to a theoretical case in which we have an accurate measure of the global innovativeness of each firm. This is an important issue, which is discussed at greater length in Section 5.2.

2.4 Question Interpretation

Good questionnaire design requires that each question be interpreted literally, with no room for ambiguity. The equivalent for survey analysis is that we must first ensure that the respondents have correctly understood each question. When in doubt, the analyst may need to use alternative questions. Furthermore, we must literally interpret questions that appear to have been correctly understood. This can create problems for the interpretation of a variable. As an example, the survey question on five innovative activities asks if each activity was "*linked* to new or significantly improved products" or production processes. For the question on the acquisition of machinery or equipment "linked" to innovation, the activity could be the innovation itself, such as when a firm acquires innovative production machinery. In this case, the activity question could be interpreted as an innovation indicator (or dependent variable), instead of as an input into innovation (or explanatory variable). Tang's work takes this approach, whereas the chapters by Le and Tang, and, Mohnen and Therrien assume that the five activity variables are innovation inputs.

3. RELEVANCE OF ECONOMETRICS

Before discussing the appropriate econometric techniques for ordinal and nominal variables from cross-sectional studies, it is worth briefly summarizing what we hope to learn from econometric analyses and how these techniques can support descriptive statistics. As noted above, a major advantage of the Canadian innovation survey is that it provides reliable prevalence data, due to a very high response rate. Simple cross-tabulation techniques, such as determining the percentage of innovative firms by sector and firm size, or the percentage of firms by size and sector that collaborate on R&D or use different IPR instruments, can tell us a lot about innovative behaviour.

There are two basic types of descriptive statistics that can be computed from innovation surveys. The most common statistic is to calculate frequencies for categories of interest at the level of the firm. For instance, Table 4 of the chapter by Tang shows that 36.8% of all firms active in high technology sectors used a patent to protect their intellectual property, while 82.1% acquired new technology. The second type of descriptive statistic aggregates results by an indicator of economic output, such as employment, value-added, or revenue. Tang also reports that 69.4% of high technology firms, after weighting by production revenue, used a patent to protect their intellectual property, and 79.9% acquired new technology. For patents, the correct interpretation is that 69.4% of all production revenue among high technology firms is from firms that used a patent.

When a variable is rare among small firms but common among large firms, the difference between firm level and aggregate descriptive results can be substantial, as is the case for the use of patents by high technology firms to protect intellectual property. This is because large firms account for only a small percentage of the number of firms but for a large percentage of economic output. Conversely, there is little difference between the firm-level and aggregate indicators for the acquisition of new technology, showing that this activity is common among all size classes of high technology firms.

Both firm-level and aggregate descriptive statistics are required to provide a thorough evaluation and understanding of innovation. For example, the firm-level indicator for patent use suggests that there is a serious need for policies to promote the greater use of patents among high technology firms. In contrast, the aggregate indicator suggests that the problem (if any) is not crucial and is limited to small firms.

The additional value provided by econometrics over descriptive analyses, particularly because we can rarely infer causality in a cross-sectional survey, is simply to control for possible confounding factors in firm-level analyses. For example, we know that both large firms and high technology firms are more likely to innovate than small firms and low technology firms. But, what if most large firms are also active in high technology sectors? If true, the apparent correlation between firm size and innovative status would be due to confounding between firm size and sector. Econometric or multivariate approaches control for the effects of other variables and therefore can provide a clearer picture of the impact of each variable.

Another advantage of econometrics is to simplify the interpretation of the raw statistics by replacing numerous cross-tabulations, which increase exponentially with the number of explanatory variables, by a single multivariate analysis. The econometric analysis is moreover guided by some sort of structural model that distinguishes between dependent and independent variables and which tries to correct for various types of estimation biases (errors in variables, simultaneity, sample selection, and omitted variables). Econometric analysis, if done properly, can provide greater clarity and an improvement in the precision of statistical correlations between variables of interest.

In some cases, the econometric techniques used in the seven chapters in this volume do not go far enough. In several chapters it would have been very useful for the econometric results to have been explored further, either through additional descriptive statistics, or by building econometric models through several stages, for instance by adding variables to explore their effect on other variables. As an example, most of the regressions use dummy variables to capture sector differences. However, the "gold standard" in econometrics is not to use sector dummies at all, but to identify the factors that create sector differences. These factors can be explored and possibly identified by carefully building a model in steps, to see if specific variables can capture the effect of sector dummies. Of course, if none of the available variables explain sector differences, it is advisable to use sector dummies as a proxy for more precise sector-specific influences. The need for further careful analysis is most pressing when evaluating some of the policy relevant conclusions. Several examples of these problems are given in Section 6.

4. ECONOMETRIC TECHNIQUES FOR QUALITATIVE DEPENDENT VARIABLES

As mentioned before, the Survey of Innovation 1999 data are mostly discrete: dichotomous (binary), ordered categorical (such as five-point importance scales) or unordered categorical (product or process innovation). Some variables are censored, i.e. collected only for a subset of the firms in the entire population. This section gives a brief overview of the estimation of regression models for discrete dependent and latent variables.[2] In fact, discrete dependent variables are defined in terms of latent variables that have a given probability distribution conditional on a set of explanatory variables.

4.1 Dichotomous Variables

Dichotomous or binary variables take only two values, y=1 and 0, corresponding in general to a yes/no answer. An example is whether or not a firm introduced at least one innovation between 1997 and 1999. The technique is used in the chapters by Therrien and Chang; Mohnen and Therrien; and, Le and Tang. The idea is to model the probability of reaching these two values, referring to a probability distribution for the random part of the model. Therefore we define a latent variable $y^*=X\beta+\varepsilon$ such that

$$p(y=1)=p(y^*>0)=p(\varepsilon>-X\beta)=1-F(-X\beta)$$

and

$$p(y=0)=p(y^*\leq 0)=p(\varepsilon\leq -X\beta)=F(-X\beta)$$

where F is the cumulative logistic distribution (in the case of the logit model) or the cumulative normal distribution (in the case of the probit model).[3] Parameters β are then estimated by maximizing, with respect to β, the likelihood function

$$L(y,X,\beta)=\Pi_1 F(X\beta)\Pi_0(1-F(X\beta)),$$

where index 1 corresponds to the observations where y equals 1 and index 0 corresponds to those where y equals 0. Both the logistic and the probit cumulative distribution functions are symmetric. The two distributions differ only by the thickness of the tails of the distribution. Therefore, they usually yield similar results. The logit model has a functional form that is easier to manipulate. It does not require an integration procedure.

4.2 Ordered Categorical Variables

The ordered categorical model differs from the dichotomous models by allowing multiple responses that correspond to an increasing order of importance. This kind of model is used in the chapters by Mohnen and Therrien to evaluate the intensity of innovation, and by Therrien and Chang to evaluate the novelty of the firm's most important innovation. An ordered model is simply an extension of the previous model for dichotomous dependent variables. Instead of having just two responses that divide the area under the cumulative distribution function into two parts, we now have multiple responses, each corresponding to successive areas under the cumulative distribution function. The dependent variable y can take values j = 0,1,..., m. Assume the same formulation for the latent variable y* as above. The probability that y = j can be written as

$$p(y=j) = p(c_j < y^* < c_{j+1}) = F(c_{j+1} - X\beta) - F(c_j - X\beta)$$

with $c_0 = -\infty$ and $c_m = \infty$. The likelihood function is

$$L(y, X, \beta) = \Pi_j [\, p(y=j)^{n_j} \,]$$

where nj is the number of observations where y=j.

The categories j must be mutually exclusive, i.e. $\sum_j p(y=j) = 1$. Various distribution functions can be chosen. The most common ones are the logistic and the normal distributions, corresponding to the ordered logit and the ordered probit respectively.

4.3 Unordered Categorical Variables

Suppose that we have several categories or outcomes, but that there is no logical order or ranking between them. This type of condition often occurs in consumer choice models, in which an individual decides to purchase one of several similarly priced items — a weekend vacation, a computer, or a college course. The chapter by Landry and Amara uses this model in an analysis of the novelty of the firm's most important innovation.

Assume that we have m categories, where the mth category is the reference category. Let P1, P2, ..., Pm be the respective probabilities associated with these categories. If we define

$$P_j / (P_j + P_m) = F(X\beta_j)$$

we obtain

$$P_j / P_m = F(X\beta_j) / (1 - F(X\beta_j)) = G(X\beta_j)$$

with $\beta_m = 0$ by normalization.

If we choose the logistic distribution for F(.), we obtain the most often used unordered categorical variable model, the multinomial logit model. In that case, $G(X\beta_j) = \exp(X\beta_j)$. We can now write

$$P_j = G(X\beta_j) / (1 + \sum_{k=1}^{m-1} G(X\beta_k)).$$

The likelihood function is as in Section 4.2. The coefficients can thus be interpreted in terms of the effects of explanatory variables on the relative odds of pairs of categories, and thereby the multinomial logit model is a natural extension of the bivariate logit model.

One advantage of the multinomial logit is its flexibility. It allows a variable such as firm size to have a positive effect on one pair of odds and a negative one on another pair of odds. To have this with an ordered logit would require introducing nonlinearities in the explanatory variables.

4.4 Limited Dependent Variables

It often happens in innovation surveys that data on some variables are only collected for innovating firms, either because these variables make no sense for non-innovating firms or because the questionnaire is designed so as not to ask unnecessary questions to non-innovating firms (e.g. R&D expenditures can also be incurred by non-innovating firms, but are typically only collected for innovators). In that case, data for these variables are said to be censored. As the dependent variable is only observed on a subset of the population, the probability of being in this subset has to be taken into account.

Suppose we have a binary variable y_1 that takes the values 0 and 1 depending on whether the latent variable y_1^* defined as $y_1^* = X_1\beta_1 + \varepsilon_1$ is negative or positive. This first variable is for instance the dichotomous variable to innovate or not. Then we have a second variable y_2 (for instance the intensity of innovation) which is only defined when $y_1^* > 0$, in which case it is defined in terms of the latent variable $y_2 = y_2^* = X_2\beta_2 + \varepsilon_2$, where X_1 and X_2 are explanatory variables, β_1 and β_2 are parameters to be estimated, and ε_1 and ε_2 are jointly normally distributed random variables. The conditional mean of y_2 is then given by

$$E(y_2 \mid y_1^* > 0) = X_2\beta_2 + \rho\sigma_2 \frac{\phi(X_1\beta_1)}{\Phi_1(X_1\beta_1)}$$

where ρ is the correlation coefficient between the two errors terms and σ_2 is the standard error of ε_2. Because it cannot separately be identified $\sigma_1 = 1$.

The regression line of the second equation gives only a biased representation of the conditional mean. Two possibilities exist to correct for this selection bias. One is to add the expression $\phi(X_1\beta_1)/\Phi_1(X_1\beta_1)$, known as the inverse Mills ratio, to the regressors, as in Therrien and Chang. The other one is to estimate the two equations and the joint distribution of their error terms simultaneously, as in Mohnen and Therrien. If the second dependent variable is continuous, we have the generalized tobit model. Mohnen and Therrien specify a probit equation (with explanatory variables X) and an ordered probit equation (with explanatory variables Z) defined on the censored data. The likelihood function underlying their estimation procedure is:

$$\begin{aligned}
\ln L = & \sum\nolimits_{INNO=0} \ln(\Phi_1[-X\beta]) + \sum\nolimits_{INNO=1, INNO_S=1} \ln(\Phi_2[t_1 - Z\gamma, X\beta, \rho]) + \\
& \sum\nolimits_{INNO=1, INNO_S=2} \ln(\Phi_2[t_2 - Z\gamma, X\beta, \rho] - \Phi_2[[t_1 - Z\gamma, X\beta, \rho]) + \\
& \sum\nolimits_{INNO=1, INNO_S=3} \ln(\Phi_2[t_3 - Z\gamma, X\beta, \rho] - \Phi_2[t_2 - Z\gamma, X\beta, \rho]) + \\
& \sum\nolimits_{INNO=1, INNO_S=4} \ln(\Phi_2[t_4 - Z\gamma, X\beta, \rho] - \Phi_2[t_3 - Z\gamma, X\beta, \rho]) + \\
& \sum\nolimits_{INNO=1, INNO_S=5} \ln(\Phi_2[t_5 - Z\gamma, X\beta, \rho] - \Phi_2[t_4 - Z\gamma, X\beta, \rho]) + \\
& \sum\nolimits_{INNO=1, INNO_S=6} \ln(1 - \Phi_2[t_5 - Z\gamma, X\beta, \rho])
\end{aligned}$$

where Φ_1 is the cumulative univariate normal distribution, Φ_2 is the cumulative bivariate normal distribution, the indices under the summation signs indicate the observations over which the sums are taken, and t_i (i=1 to 5) are the known categorical bounds.

4.5 Structural Equation Models With Latent Dependent and Independent Variables

Structural equation models constitute an integrated framework of linear equations that model the relationship between two sets of latent variables and between the latent variables and some observed characteristics. This model is used in the work by Tang. His structural equation model consists of three equations: a measurement equation for the dependent latent variable,

$$(1)\quad y = M\eta + \varepsilon .$$

a measurement equation for the independent latent variable,

$$(2)\quad x = \Lambda\xi + \delta .$$

and a structural equation linking the latent dependent and independent variables

$$(3)\quad \eta = \Gamma\xi + \varsigma .$$

The variables are defined as follows: y is a vector of observed variables for the dependent variables; x is a vector of observed variables for the independent variables; η is a random vector of latent dependent variables; ξ is a random vector of latent independent variables; M, Λ and Γ are matrices of coefficients; while ε, δ and ς are random errors, mutually uncorrelated.

As in traditional ordinary-least square regression models, it is assumed that the errors in (1), (2) and (3) are uncorrelated with the latent explanatory variables in these equations, i.e. that ε is uncorrelated with η, δ is uncorrelated with ξ, and ς is uncorrelated with ξ.

If an indicator is related to a latent variable, then the parameter between the indicator and the latent variable is free; otherwise, it is fixed to be zero. For the parameters to be identified, at least the number of free elements may not exceed the number of observed correlations.

This model is an extension of the multivariate linear regression model and of the factor analysis model. The parameters of the model can be estimated by a least squares criterion (minimizing the distance between the sample covariances of all indicators and the covariances predicted by the model) or by the maximum likelihood method (maximizing the likelihood of observing the data at hand).[4]

5. OVERVIEW OF THE ECONOMETRIC ANALYSES

Four of the seven chapters focus on different measures of the innovative status of the firm as the dependent variable, while Cozzarin evaluates the effect of innovation strategies on performance outcomes, Tang focuses on inventors and adopters as main innovation strategies, and Hanel evaluates the effect of several factors on the probability of using specific forms of intellectual property. We do not discuss Hanel's econometric analysis below because his work does not provide full regression results.

Several chapters use the same dependent variable but either different models or different sets of independent variables. Three chapters use the innovative status (yes or no) of the firm as the dependent variable, two use a measure of the novelty of the firm's most important innovation (world-first, Canada-first, or firm-first), and two use the innovative sales share, or the percentage of 1999 sales from new or significantly improved products. This variety allows us to assess the effect of model choice and explanatory variables on the results. This is done by comparing the results for a set of common independent variables and for those that are only included in a few models.

The choice of a common set of variables is based on several "stylized facts" that can be drawn from analyses of innovation surveys in Canada, the United States, and Europe. Almost all research finds that the probability of innovating increases with firm size, is higher for firms that perform R&D, is higher for firms active in medium and high technology sectors compared to firms active in low technology sectors, and is higher for firms that receive innovation subsidies.

5.1 Innovation Status

A firm is defined as an innovator if it has introduced one or more product or process innovations between 1997 and 1999. The dependent variable is dichotomous, taking the value of 1 for an innovator and 0 for a non-innovator. Either a probit or a logit model can be used, with little substantive difference between them.

The very high prevalence of innovative firms reduces interest in the factors that differentiate non-innovative from innovative firms — four out of five Canadian firms were innovators. Partly for this reason, all of the three chapters that use the innovative status of the firm as the dependent variable also develop a model for the intensity of the firm's innovative activities. The main motivation for Therrien and Chang and Mohnen and Therrien to develop a model for innovative status is to correct for censoring in a full model for innovation intensity.

Table 1 summarizes the model results for innovation status. Of interest, the "stylized fact" for an increase in the probability of innovating by firm size is not found in the study by Le and Tang. One explanation is that the variables for innovation activities capture firm size effects because all five of these activities are more prevalent among larger than smaller firms.

Therrien and Chang include a dummy variable that equals 1 when the firm strongly agrees that "production (or office) technologies change rapidly." This measure of the competitive environment is positively correlated with the probability of innovating. This should be reflected through process innovation, but the work does not provide a more detailed analysis of this.

5.2 Innovation Novelty

Table 2 summarizes the results of two chapters that use the novelty of the firm's most important innovation, introduced between 1997 and 1999, as the dependent variable. The novelty variable includes three categories: world-first, Canada-first, and firm-first innovators. An additional fourth category for non-innovators can be constructed from the results.[5] The choice of which econometric model to use depends on how the variable for novelty is interpreted. There are two main possibilities.

First, the responses to the novelty question can be used to rank the innovativeness of each firm along four levels. Non-innovators make up the lowest level of innovativeness, while world-first innovators have

Table 1: Studies Using Innovative Status (Yes or No) as the Dependent Variable

		Independent Variables						
Author	*Model*	*Firm Size*	*Sector*[1]	*R&D Status*	*Gov. Subs.*[2]	*Comp. Env.*[3]	*Collab.*[4]	*Other*
Therrien and Chang	Probit	+	+	+	+	+		Includes dummy variables for province. Size entered as two categorical variables plus a reference category of 20-49 employees.
Mohnen and Therrien	Probit	+	+					No constant, so no reference category for sector, but the coefficient increases by technology class. Size entered as the log of the number of employees.
Le and Tang	Logit	Ns	+	+			+	Includes dummy variable for innovative activities (acquisition of technology, engineering and design, tolling and start-up, training). Size entered as two categorical variables plus a reference category of 20-49 employees. Separate dummies included for 19 sectors with miscellaneous manufacturing as the reference category. The coefficients for high/medium technology sectors are generally positive while the coefficients for low technology sectors are generally negative.

Notes: + = variable statistically significant and positive, - = variable statistically significant and negative, Ns = not significant, ref = reference category. A blank cells means that the variable was not included.

[1]Positive sector effect, with activity in either high or medium technology sectors increasing the probability of being an innovator compared to low technology sectors.

[2]One or more measures of the use of government support programs included in the model, such as for R&D, venture capital, technology support, information or Internet services, or training.

[3]One or more measures of the competitive environment included in the model, such as rapid product obsolescence or change in production or office technologies.

[4]One or more measures of cooperative or collaborative arrangements included in the model.

Table 2: Studies Using Innovative Novelty (World-first, Canada-first, Firm-first) as the Dependent Variable

		Independent Variables						
Author	*Model*	*Firm Size*	*Sector*	*R&D Status*	*Gov. Subs.*	*Comp. Env.*	*Collab.*	*Other*
Therrien and Chang	Ordered probit (3 levels), plus inverse Mill ratio for non-innovators	+	+	(+)[1]	+		+	Includes dummies for 4 provincial regions, with the Atlantic provinces as the reference category.
Landry and Amara	Multinomial logit (World-first versus firm-first and Canada-first versus firm-first): no adjustment for non-innovators	+	+	+	+	-	+	Includes four sets of variables for information sources: internal, market sources, research sources, and generally available sources.

Notes: + = variable statistically significant and positive, - = variable statistically significant and negative, Ns = not significant, ref = reference category. A blank cells means that the variable was not included. See Table 1 for definitions and interpretation of sector, Government subsidies (Gov. subs.), competitive environment (Comp. env.), and collaboration (collab.).

[1]The authors do not include a direct variable for R&D performance, but R&D is included in a variable for the number of innovation activities.

the highest level of innovativeness. This ordinal progression assumes that creative innovation is of a higher order than innovation as an adoptive process. World-first innovators are assumed to have developed a completely new innovation themselves. In contrast, Canada-first and firm-first innovators are assumed to have adopted the innovation from another firm, such as a supplier or possibly their head office. Therefore, they have not acted to create an innovation themselves. The appropriate model is an ordered logit or probit. An acceptable alternative, although it results in some loss of information, is to divide the four categories into two and use a logit model.

Second, the novelty of the firm's most important innovation can be viewed as one of several optional choices. In this case, the decision to choose a "world-first" innovation as the most important to the firm is one of several alternatives, with the firm also able to choose a "Canada-first" or a "firm-first" innovation. The appropriate model is a multinomial logit (or probit) with each optional choice being compared against all other possibilities (as in Landry and Amara). A major justification for the choice of this model is that the novelty question only refers to the firm's *most important innovation*. It is entirely possible that a firm that decides to respond to the question with a Canada-first or firm-first innovation could have also developed a world-first innovation, but simply not found this innovation to be as important. One problem is that the question does not define "importance." We do not know if this is the firm's most technically advanced innovation or if it is the firm's most economically important innovation. The former would support the use of an ordered logit based on a ranking of the technical sophistication of each firm's creative innovative capabilities. The latter would tend to support the use of a multinomial model.

Unfortunately, none of the authors fully investigated these issues by comparing the responses to the novelty questions against several other questions that could provide clues to whether or not Canada-first or firm-first respondents also developed world-first innovations. These other questions include the number of innovations developed by the firm (Canada-first firms with a lot of innovations have probably also developed a world-first innovation) or whether or not the firm applied for a patent, which is a measure of innovation as a creative activity.[6]

An additional complication is that it is difficult to establish a meaningful difference in the level of innovativeness of Canada-first and firm-first innovators, since both types of firms probably adopted an innovation that was developed outside their firm.

Given these limitations, neither model is fully suited to the data. The choice of an ordered logit or probit model will contain error to the extent that some Canada-first and firm-first innovators also developed a world-first innovation. Conversely, the multinomial model will contain error to the extent that some firms did not introduce each of the three types of innovations and therefore had a restricted option set for their most important innovation.

Of interest, given these concerns, both models in Table 2 provide similar results for the main independent variables. The probability of introducing a higher-level innovation (or a world-first compared to a firm-first in Landry and Amara's multinomial logit model) increases with firm size, and is higher for high technology firms, R&D performers, firms that received government subsidies for innovation, and firms involved in collaboration. The results are stronger than those for the simple comparisons based on innovation status given in Table 1. The probable explanation is that the models in Table 2 contain more information for the 81% of firms that innovate, with three levels of innovativeness.

Landry and Amara's multinomial logit model focuses on the use of different types of information sources for innovation. Of interest are the types of information sources that are associated with a world-first innovation, which assumes creative innovative activities on the part of the firm. The use of "market" linked sources such as suppliers, clients and competitors reduces the probability of being a world-first innovator, compared to the reference category of firm-first innovators. In contrast, sourcing information from trade fairs and exhibitions, public research agencies, and from internal R&D and marketing staff increases the probability of being a world-first innovator. The negative effect of market-based sources is of particular interest because of the emphasis in the literature on the importance of "user-producer" relationships to innovation. Landry and Amara's results help to clarify some of the confusion over this issue. User-producer relationships, not surprisingly, are more important for adoptive innovation, such as acquiring new technology from a supplier. In contrast, world-first innovations require different knowledge sources (although not necessarily to the exclusion of market sources) that underpin technology invention, such as internal R&D or the public research infrastructure.

5.3 Sales from Innovative Products

The survey questionnaire asks innovative firms about the percentage of their total sales in 1999 that were due to new or significantly improved products, goods or services (the innovative sales share). Six ordinal categories are provided: 1% to 5%, 6% to 15%, and through to 76% to 100%. Le and Tang, and, Mohnen and Therrien used the innovative sales share as the dependent variable, as shown in Table 3.

Le and Tang use a simple logit model, with firms divided into two groups: those with an innovative sales share below 15% (approximately half of the firms) and those with a sales share above 15%. The analysis is limited to innovative firms that *only* introduced a product innovation and which answered all relevant questions, plus non-innovators. The reference group in the logit model consists of product innovators with a sales share below 15% plus the non-innovators. It is not clear why firms that were both product and process innovators are excluded from the model. In contrast, Mohnen and Therrien use an ordered probit with known thresholds (for instance the boundary of 5% between the first and second category) that maximizes the use of the available information. Firms that only introduced a process innovation are excluded, but firms that were both product and process innovators are included in the model.

Both Le and Tang, and, Mohnen and Therrien include explanatory variables for size, sector, R&D, and government subsidies, plus a dummy variable for participation in at least one cooperative or collaborative agreement for innovation. The model by Mohnen and Therrien corrects for sample selection by estimating the probit and ordered probit equations simultaneously, taking into account the joint distribution of the error terms of the two equations.

Both models find differences by sector and higher innovative shares among firms that perform R&D Conversely, Le and Tang do not find a significant effect for firm size or for collaboration, while Mohnen and

Table 3: Studies Using the Innovative Sales Share as the Dependent Variable

		Independent Variables						
Author	*Model*	*Firm Size*	*Sector*	*R&D Status*	*Gov. Subs.*	*Comp. Env.*	*Collab.*	*Other*
Le and Tang	Logit: above and below 15% innovative sales share. Limited to product innovators only plus non-innovators (1,851 firms). Non-innovators assigned to below 15% innovative sales share category.	Ns	+[1]	+			Ns	Includes separate variables for each of the five innovative activities (only R&D and training significant). Firm size is entered as two categorical variables with 20 – 49 employees as the reference category.
Mohnen and Therrien	Ordered probit with 5 known thresholds (6 categories). Excludes firms that only introduced a process innovation, leaving 4,404 firms in the model.	+[2]	+[3]	+	Ns	+	+	Includes a variable for more than the median number of the five innovation activities (+) and use of public science as an information source (ns). Competitive pressure is defined as a strategy of either "opening up new markets" or "increasing market share." Firm size is the log of employees.

Notes: + = variable statistically significant and positive, - = variable statistically significant and negative, Ns = not significant, ref = reference category. A blank cells means that the variable was not included. See Table 1 for definitions and interpretation of sector, Government subsidies (Gov. subs.), competitive environment (Comp. env.), and collaboration (collab.).

[1]Model uses 19 sector dummies, with miscellaneous manufacturing as the reference category. The only statistically significant differences are for four low technology sectors (food, wood, primary metals, fabricated metals), all of which have negative coefficients.

[2]Positive for Canada, but negative using the CIS results for Europe.

[3]Mohnen and Therrien do not use a constant and therefore lack a reference category for sector, with dummies for low, medium and high tech sectors. However, the coefficients increase from low to high tech sectors, indicating a positive sector effect.

Therrien find that both firm size and collaboration increases the probability of a higher innovative sales share. The latter difference is of interest, since innovation policy in several countries expressly encourages cooperation. So, which is more likely, does collaboration have no effect, or a positive effect?

There are three possible explanations of this difference in the results. The first explanation is due to differences in the two models, with the model by Mohnen and Therrien including six categories of innovative sales share versus the two categories in the logit model of Le and Tang. This would explain the results if the prevalence rate for cooperation is highest for firms with very high innovative product shares. The second explanation is due to differences in the choice of independent variables. In this respect, perhaps the greatest difference between the two models is in the treatment of the sector variables, which are based on three technology classes in the ordered probit model and on 19 sector dummies for the simple logit. It is possible that the latter are capturing the effect of cooperation, which could be strongly associated with specific sectors that do not correlate with the low, medium and high technology classification in the model by Mohnen and Therrien. The third explanation is due to differences in the firms that are included in the model. Le and Tang exclude a large number of firms that are involved in both product and process innovation.

The first and third possible explanations are due to greater information included in the model by Mohnen and Therrien, while the second explanation (the treatment of sectors) results in more information being included in the model by Le and Tang. On this basis, the model by Mohnen and Therrien is probably more accurate, but we do not know if a finer analysis of sectors would reduce the significant effect for collaboration and firm size in Mohnen and Therrien's model. Whichever is the correct explanation, further research on the effect of cooperation is warranted to determine if cooperation is or is not positively correlated with innovative product sales, and the factors that influence this relationship.

5.4 Innovation as an Adoptive or Creative Activity

Given its low overall R&D intensity, Canada is often accused of being more of a technology adopter than a technology inventor. Tang examines which firm objectives — developing products, improving production processes or reducing production costs — lead to more adoption or invention and whether the government's actions encourage one or the other. The three main objectives and the two innovation activities are not directly observed. What the questionnaire provides are six indicators on innovation activities (R&D, patent, trademark, copyright, industrial engineering and design, and acquisition of machinery, equipment and other technologies) and nine indicators of business objectives. The issue is then to translate the indicators into latent concept variables and to examine the link between the constructed latent variables.

Tang concludes that the objectives to develop products and to reduce production costs are the driving forces behind technology invention whereas improving the production process seems to be the driving force behind technology adoption. He also obtains some signs of substitutability between technology adoption and technology invention, insofar as factors that encourage one impede the other. Government support is correlated with both technology invention and technology adoption.

5.5 Effect of Innovation on Economic Performance

Cozzarin uses a GLS model to evaluate the effect of innovation activities on four performance measures: employment, labour productivity (average value added per employee), market share, and total value-added. These models include three dummies for the innovative novelty of the firm (world-first, Canada-first, and firm-first) as explanatory variables, with non-innovators as the reference category.

The model is essentially exploratory because of problems with the cause and effect relationship between innovation novelty and the performance measures. The latter are for 1997, while innovation novelty is based on the most important innovation between 1997 and 1999. Therefore, the model must assume that the variables for innovation novelty reflect long-standing differences in innovative capabilities, such that innovation novelty between 1997 and 1999 accurately reflects the firm's capabilities for several years before 1997. Furthermore, the model also assumes that the three dummy variables for innovative novelty are accurate, such that a firm that states that its most important innovation was a firm-first innovation did not also develop a world-first innovation.

The accuracy of the coefficients for innovative novelty will depend on the viability of these two main assumptions.

The most interesting performance measures are labour productivity and total value-added, since both can increase living standards. The labour productivity of both world-first and Canada-first innovators did not differ significantly from non-innovators, although firm-first innovators had higher labour productivity. This could be due to greater investment in labour-saving new production technology than in high value-added product innovation. Total value added showed a linear trend by innovative novelty, with the highest level for world-first innovators and the lowest level for the reference category of non-innovators.

6. DRAWING POLICY CONCLUSIONS

The results of innovation surveys can help inform the strategic and investment decisions of firms, test academic theories of innovation, or provide guidance to policy. Ironically, given that most innovation surveys are government-funded, the latter is possibly the most difficult application because it requires a careful sifting of the evidence. All of the seven chapters draw policy relevant conclusions, but many of these conclusions are only partly supported by the evidence. There is an enormous temptation to "stretch" the evidence or to fall back on common assumptions. To be fair, almost all innovation researchers (including the two authors of this chapter) have succumbed to these temptations.

As an example, several of the chapters draw policy implications for the lower rates of innovative activities among small firms. Hanel suggests that the lower use of patents by small firms should be remedied by policies to help educate them in how to use Intellectual Property Rights (IPRs) effectively. Hanel turns to the results of other research to suggest that the cause is due to small firms being less likely than large firms to develop a patentable invention, or to the high cost of patents to small firms. However, Hanel does not provide evidence from the Survey of Innovation 1999 to support these claims. It is also doubtful how educational programmes will help if small firms do not develop patentable inventions. In order to enhance Canada's innovative performance, Tang recommends that government policy should encourage innovation among small firms. This may be a worthwhile goal, but it could have little effect on Canada's overall innovative performance, since large firms account for most economic output. A complete evaluation of this policy prescription requires a careful evaluation of descriptive data at the aggregate level.

The solution to many of these problems is to conduct further in-depth analyses. For instance, we can't make an informed recommendation for patent policy unless we know if small *innovative* firms lag behind in their use of IPRs. Furthermore, our understanding of innovation among small firms would be enhanced by separate regressions limited to them. These could identify the factors that are specifically correlated with innovative activities among these firms.

Another example of the need for further analysis is from the study by Therrien and Chang. Based on the results for province level dummies in a model of innovation novelty, they conclude that Ontario and Quebec are good at promoting technology diffusion but the Prairie provinces are better at encouraging world-first innovations among the subset of innovative firms. This is a fascinating result that requires careful evaluation. One option is that the effect of Ontario and Quebec as a base for world-first innovation could be captured in the sector dummies — for example if more high technology and medium technology firms are located in Quebec and Ontario. As a first step, the authors could build the regression through several stages, for example by first excluding the dummies for sector and seeing if this has an effect on the dummy variables for Quebec and Ontario. Second, we need a careful analysis of other information in the survey. For example, are there provincial differences in the percentage of sales from innovative products, the number of innovations, or the number of patents? Answers to all of these questions would help to explain the apparent differences in the style of innovation between provinces.

7. CONCLUSIONS

The innovation surveys of Statistics Canada provide a wealth of information that can be used to identify current innovation rates and, potentially, to track changes in innovation activities over time. Econometric techniques are particularly useful for identifying patterns of activities and how these are correlated to firm characteristics such as their size or sector.

A major result of the Survey of Innovation 1999 is that a large majority — 81%, of manufacturing firms innovate. This reduces both academic and policy interest in the factors that are correlated with innovation. Instead, the focus of econometric research is shifting to an analysis of the factors that are correlated with the quality of innovation or the outputs of innovation. This includes the analysis by Tang of innovation as either an adoptive or creative activity, the many chapters that have looked at the novelty of innovation, and the two chapters on the sales share of innovative products. Cozzarin's analysis tries to go a step further by evaluating the effect of innovation on economic performance. (Surprisingly, none of the chapters examined the results to a question on the number of innovations introduced by the firm.)

Research on the quality of innovation is also of vital importance to the Canadian policy goal of substantially improving the innovative capacity of Canadian firms. This change must take place in an economy that has traditionally been based on innovation as an adoptive process — for instance through the purchase of new production technology from foreign firms or head offices. This strategy limits the need for creative innovation within Canada, but it also limits the economic benefits that can flow from creative innovation, such as higher economic growth rates and improved terms of trade. The challenge for Canadian policy makers is to encourage firms to invest more in creative innovation and to profit from these investments, as shown by improved economic performance. In this respect, Canadian policy makers need excellent data on the quality of innovation in Canada and the types of programs that can encourage successful investment in creative innovation. This is perhaps the key purpose of the Canadian Survey of Innovation 1999 and of econometric analyses of the data. Many of the chapters in this volume take up the challenge.

The evaluation of the econometric results in Tables 1 through 3 show that several variables, such as firm size and sector, are highly robust and remain statistically significant regardless of the model specification or the definition of innovation. However, these variables are of less interest to policy than the use of government subsidies for innovation or participation in collaboration. The significance of these two variables differs among the econometric models. The use of government subsidies increases the probability of being a world-class innovator in the work by Therrien and Chang but has no effect on the innovative sales share in Mohnen and Therrien. Collaboration increases the probability of being a world-class innovator in two chapters (Therrien and Chang, and, Landry and Amara) and the innovative sales share in Mohnen and Therrien, but has no effect on the latter in Le and Tang. These differences in key factors of relevance to policy need to be carefully evaluated to determine their cause. We need to know if the differences are due to the model specification, or perhaps the effect of collaboration or government subsidies depends on other factors, such as sector or firm size.

The analyses of the Canadian Survey of Innovation 1999 also point to several problems that could be overcome by "tweaking" the design of future innovation surveys. Essentially, we need to be able to distinguish between innovation as a creative, inventive activity and innovation as adoption. The current question on the novelty of innovation cannot fully meet this need because it is only asked for one innovation, albeit the most important. The survey needs to solve the definitional problem of which firms are world-first, Canada-first, or firm-first innovators by determining if each firm has introduced *any* world-first innovations. It would also be worthwhile to know the percentage of sales that are from world-first innovations. Alternatively, additional questions should focus on the most important innovation. At a minimum, it would be worthwhile to know if this innovation was patented, received government support for its development, or was partly developed by another firm. Finally, several of the chapters also noted that the ability to draw conclusions from the econometric results was hampered by the lack of a variable for firm ownership.

Notes

[1]The ability to redefine the target population as a smaller fraction of the original sample partly depends on the cells that define each sampling unit and the sampling fractions (the percent of firms that are surveyed) within each cell. The most common restriction in the analyses of the Statistics Canada survey is to limit the target population to innovative firms. This will introduce error in the econometrics because the sample was not selected with any prior knowledge on innovative status. However, the amount of error should be acceptable, since the proportion of innovative firms is very high (81% on average) and the sample fractions are comparatively large, with 5,220 useable responses from the entire population of 9,303 firms.

[2]For further discussion of these techniques, see Greene (2000), Maddala (1983), and Thomas (2000).

[3]For simplicity, the observations are not indexed.

[4]For more on the structural equation model and the software LISREL, the reader is referred to Jöreskog and Sörbom (1984).

[5]On the basis of non-parametric tests, Cozzarin concludes that a fifth group, the unknown innovators who do not report whether they are world-first, Canada-first and firm-first innovators, can be assimilated with the firm-first innovators.

[6]A logical choice would be to use a multinomial model after restricting the firms to those that have either applied for a patent or introduced an above average number of innovations.

The First to Market

Section Overview

Fred Gault

The goal of any entrepreneur is to profit from doing something new. This could be bringing an entirely new product to market, introducing a new production or delivery process, or a combination of the two. Understanding the kind of firm that can make such innovation happen, and the social and economic environment that nurtures it, is important for everyone, for the decision makers in firms that want to lead their markets, for the policy makers who want to build an economic and social environment conducive to sustainable growth, and for creators and transmitters of the knowledge that drives innovation.

Innovation is not yet as systematised as research and development became at the end of the 19th century and in the early 20th, but that systematisation is the logical next step which can build on what can be learned from firms that are engaging in the activity of innovation. Landry and Amara in the first chapter of this section make the point that the observed high level of innovation among the population of manufacturing firms does not help in understanding the phenomenon. The question of what constitutes innovation has to be made more restrictive.

To deal with this, they turn to the data on firms with world-first, Canada-first and firm-first innovations and describe the different characteristics of these firms. They do not deal with all characteristics, but with the sources of innovation, and they find that firms with world-first innovations turn more to internal R&D and to government laboratories for information than firms that claim only a firm-first innovation. Firm-first innovators make more use of clients and suppliers as sources of information.

There are two points that make these findings of interest. The first is that the Canadian Innovation Strategy promotes greater commercialization of intellectual property from government and university laboratories. Intellectual property can be licensed to existing firms, or used to spin-off new firms. However this is done, it appears from the work of Landry and Amara that this information is a potential source of world class innovation in manufacturing and that is a significant impact of the outcome of a successful commercialization activity. The second point is that this finding is not immediately obvious if the sources of innovation are examined without controlling for novelty of innovation. The tabulations in Schaan and Anderson (2001) show government laboratories quite low on the list of sources of information and there is a danger, were this data used uncritically, that government laboratories would be dismissed as ineffective players in the innovation system. This is not the case. Size of the firm is also a factor in the use of information from government laboratories, with larger firms making more use of government sources than those that are medium or small.

Therrien and Chang look at local collaboration as a factor influencing the activity of innovation and find that firms with world-first innovations are more likely to engage in collaboration and especially to have established a local collaboration. Collaboration is seen as an important factor associated with world-first innovation. These firms are more likely to use R&D, as well as the output from government laboratories, although these effects are not as significant as collaboration. The authors also look at size of firm and whether it is high, medium or low-tech, as factors, and find that world-first innovators are more likely to be in the high-tech sector and to be large firms.

Therrien and Chang introduce clusters as part of their discussion of collaboration. This is an important direction for future work, given the targets announced in the Innovation Strategy (Government of Canada 2002a) that Canada will have ten internationally recognized technology clusters by 2010 and will significantly improve the innovative performance of communities across Canada. Cluster analysis brings a local anchor to the discussion of innovation systems and, for that discussion to be informed, there have to be data on actors, activities, linkages and outcomes at the local level.

Cozzarin, in the final chapter of this section, finds that world-first innovators are large, make use of intellectual property, and are able to spend time on the development of innovations. Cozzarin looks at various measures of firm performance and finds that market share, total employment, labour productivity and price-cost margins are all correlated with world-first innovation.

What emerges from these three chapters is that there are significant differences between firms that produce world- or Canada-first innovations, compared with those that are firm-first innovators. The first to market firms are large, collaborative, engage in R&D and use government programmes, compared with the first to firm innovators that tend to be small, and are less inclined to collaborate, or to do R&D, or to use government programmes. However there are smaller firms that are first to market. As about nine out of ten firms in Canada have fewer than 50 employees, even a small percentage increase in the small first to market firms would make a significant economic difference.

4

Effects of Sources of Information on Novelty of Innovation in Canadian Manufacturing Firms

Réjean Landry and Nabil Amara

1. INTRODUCTION

Three questions are addressed in this chapter: What is the extent of novelty of innovation in Canadian manufacturing firms? Are there differences among provinces and industries in the matter of sources of information influencing novelty of innovation in manufacturing firms? What are the determinants of novelty of innovation in Canadian manufacturing firms?

The study presented in this chapter explored the effects of internal and external sources of information on novelty of innovation. The literature on innovation and social capital claims that information is embedded in networks and is derived from networks of relationships (Nahapiet and Ghoshal 1998). Moreover, certain networks would enhance a firm's ability to identify and develop innovation opportunities that could not be identified and developed otherwise (Powell 1990; Burt 2000). The overall thesis of the study reported here was that innovation opportunities exist because of information asymmetry, and that firms that have access to a greater variety of sources of information are in a better position to identify and develop innovation opportunities (Venkataraman 1997). Still more precisely, it was expected that firms introducing world-first innovations (innovations that have the highest degree of novelty) would be more likely to use a greater variety of sources of information to develop or improve their products or manufacturing processes than firms introducing Canada-first or firm-first innovations.

However, empirical studies supporting the suggestion that a greater variety of sources of information enhances the degree of novelty of innovation in manufacturing firms are few. Until now, the empirical literature has focused its attention on the relationship between the likelihood of innovation and sources of information, disregarding the relationship between the degree of novelty of innovation and sources of information influencing the development and improvement of products and manufacturing processes. An analysis of the United Kingdom's Innovation Survey, conducted by Tether (2000), shows that firms which have introduced innovations with higher degrees of novelty are more likely than other innovative firms to rely on external sources of information to develop or improve their products or processes. A still more recent analysis of the data from the UK Innovation Survey shows that small high-tech firms which have introduced major product and process innovations are more likely than others to rely on external sources of information, such as science laboratories, universities and public financial institutions (Romijn and Albu 2001).

The contribution of the study presented in this chapter to the debate derives from its use of a new dependent variable: the degree of novelty of innovation. Most studies based on innovation surveys operationalize innovation with the following two questions: "During the last three years, did your firm offer new or significantly improved products (goods or services) to your clients? Yes or No" and "During the last three years, did your firm introduce new or significantly improved production/manufacturing processes? Yes or No." As the percentage of innovative firms increases, especially in Canada where the Survey of Innovation 1999 indicated that 81.5% of manufacturing firms had been innovative during the three years preceding the survey, it becomes important to operationalize innovation in terms of the degree of novelty of the products or processes considered as innovations. This new operationalization of the variable "innovation" is likely to shed new light on innovation by showing that different factors explain different degrees of novelty of innovation in manufacturing firms. Furthermore, by taking into account the impact of provinces and industries, the empirical results of the study reported here will show that there are significant differences across provinces and industries in the matter of sources of information used to develop or improve products or processes. The use of the degree of novelty of innovation as a dependent variable and the tests of differences regarding the degree of novelty of innovation among regions and industries are likely to suggest that the differences in provincial and industrial contexts hinder the search for one optimum permanent policy valid for each and every province and industry.

2. RELATIONSHIPS BETWEEN NOVELTY OF INNOVATION AND SOURCES OF INFORMATION

Studies on innovation tend to suffer from three conceptual and methodological problems: composition of the study population; specification of the dependent variable "innovation"; and specification of the independent variables.

2.1 Composition of the Study Population

Almost all the studies on innovation tend to be based either on small samples or on samples of single industries. In addition, regional innovation surveys, such as those conducted in Quebec by Landry and colleagues (Landry 1996, 1998; Landry and Amara 1998; Landry and Frechette 1998; Landry, Lamari and Amara 2000a, 2000b) and in British Columbia by Holbrook and Hughes (2000), cover all manufacturing firms, including very small firms, whereas national innovation surveys, such as those of Statistics Canada, exclude firms with fewer than 20 employees and with annual revenue of less than $250,000. As for innovation surveys of single industries, almost all tend to cover science-based industries, especially those involving biotechnology and information technology, while neglecting those industries that are less science-based.

The sample for the study presented here, taken from the Statistics Canada Survey of Innovation 1999, was national and very large. It was especially appropriate for capturing the impact of the major sources of information explaining novelty of innovation in the Canadian manufacturing sector.

2.2 Specification of the Dependent Variable "Innovation"

There is as yet no agreement regarding the conceptualization and operationalization of the variable "innovation." There are two issues at stake in this regard: the first is related to the approaches to innovation surveys; and the second is concerned with the very definition of innovation and innovative firms.

2.2.1 Approaches to Innovation Surveys

Archibugi and Pianta (1996) have distinguished two approaches to innovation surveys, which have been adopted by many others (Evangelista 1997; Pianta and Sirilli 1998): the object approach and the subject approach.

In the object approach, the unit of observation is the individual innovation. Examples of such enumerations of individual innovations can be found in Townsend, Henwood, Thomas, Pavitt and Wyatt (1981), Acs and Audretsch (1990), Wallmark and McQueen (1991), and Coombs, Saviotti and Walsh (1996). The main shortcoming of this approach is that it rests on a definition of innovation that is arbitrary in the sense that the selection of individual innovations is based on the subjective decisions of experts. As a consequence, comparisons between studies and between countries are difficult to make.

In the subject approach, the unit of observation is the product/manufacturing process developed or improved at the level of the firm. This approach has been crystallized in the *Oslo Manual* (OECD 1992; OECD/Eurostat 1997). The main advantage of this approach is that it provides data on many explanatory variables of innovation. However, its major disadvantage is that, in most surveys using this approach, information regarding the degree of novelty of the innovations developed by firms is not collected. The subject approach has been adopted by the national statistical agencies of most European Union (EU) countries, Australia and Canada. Statistics Canada has conducted two innovation surveys of the manufacturing sector, the first in 1993 and the second in 1999. Guellec and Pattinson (2001) and Sandven and Smith (1998) provide a more detailed discussion of the lessons learned from OECD countries regarding the use of innovation surveys at the firm level.

2.2.2 Definition of Innovation

With regard to the definition of innovation, the two major issues discussed relate to: the criteria used to determine whether or not a firm is innovative; and the degree of novelty of innovation.

Development or improvement of products or processes: In accordance with the *Oslo Manual*, two questions are used to determine whether a firm is innovative: "During the last three years, did your firm offer new or significantly improved products (goods or services) to your clients? Yes or No" and "During the last three years, did your firm introduce new or significantly improved production/manufacturing processes? Yes or No." However, this measure of innovation (offering new or improved products or introducing new or improved processes) is becoming less and less useful for the explanation of the innovation process and as a policy tool, as the proportion of firms classified as innovative according to these criteria increases well above 50%. In their review of lessons learned from innovation surveys in EU countries, Guellec and Pattinson (2001) report the following results regarding the percentage of firms that developed or improved their products or processes during the period 1994-96: Ireland: 73%; Denmark: 71%; Germany: 69%; Austria: 67%; the Netherlands: 62%; Sweden: 54%; Italy: 48%; Norway: 48%; France: 43%; Luxembourg: 42%; Finland: 36%; Belgium: 34%; United Kingdom: 34%; Spain 29%; Portugal: 26%; all EU countries: 51%.

Most of the empirical studies based on innovation surveys conducted at the firm level use this definition of innovation. However, as the proportion of innovative firms increases well over 50%, it becomes important to conceptualize innovation in terms of the degree of novelty of the products or processes. The conceptualization of novelty has five complementary aspects: time; financial resources; frequency; creativity; and degree of novelty of innovation.

Time: One very simple way to conceptualize the degree of novelty of innovation is to ask firms to indicate how many months elapsed between their first investment in the development of their innovations and their first introduction in the firm or their first offer on the market. The assumption behind this question is that the degree of novelty of innovation increases as the number of months required for developing innovations increases. Landry, Lamari and Amara (2002) have used this conceptualization of the degree of novelty of innovation on the basis that it is easy for respondents to provide the answer to this question and that it constitutes fact, not perception.

Financial resources: Where the degree of novelty of innovation is measured in terms of financial investment in the development of products or processes (in terms of investments/sales), this figure tends to be inversely correlated with volume of sales, suggesting that small firms tend to invest relatively more in innovation than large firms do. This suggests that the ratio of investment in innovation to sales is not a valid measure of the degree of novelty of innovation.

Frequency: In innovation surveys, the frequency with which a firm introduces innovations is measured with a question like: "Please indicate how many new or significantly improved products were offered in the last three years." The answer to this question provides evidence on the culture of a firm by showing to what extent the development or improvement of products or processes constitutes either a regular recurrent pattern of behaviour or an exceptionally rare pattern of behaviour. However, it does not provide evidence on the degree of novelty of innovation.

Creativity: Another measure of novelty of innovation was developed by Eurostat for the second European Community Innovation Survey (March 1997). Immediately after the question on the development of products or processes came the following: "Who developed these products (or processes)? Mainly other enterprises or institutes? Your enterprise and other enterprises or institutes? Mainly your enterprise?" This question has been used to determine a firm's internal capability of developing innovations rather than to conceptualize novelty of innovation. However, Landry and Amara (2001) have recently used it for such a conceptualization.

Degree of novelty of innovation: Finally, a measure of the degree of novelty of innovation was used by Statistics Canada in its Survey of Innovation 1999. It was based on the following question: "Was this most important new or significantly improved product (good or service) or production/manufacturing process: a. a world first? b. a first in Canada? c. a first for your firm? Yes, No, Do not know." As far as can be ascertained, no empirical studies have used the degree of novelty of innovation as a dependent variable, as was the case in the study reported here.

2.3 Specification of the Independent Variables

In spite of a large body of empirical literature on the determinants of innovation, there is not yet a consensus regarding the categories of factors that explain innovation. The pioneering studies on innovation implicitly assumed that innovation was the result of events initiated by isolated entrepreneurs or inventors. The emerging theories of innovation lay stress on four categories of explanation:

(i) Innovations result not from discrete events, but primarily through problem-solving processes (Dosi 1982).

(ii) Innovations are determined not only by factors internal to firms, but also by interactive processes involving relationships between firms and the different actors in their environment (Kline and Rosenberg 1986).

(iii) Innovations are determined not only by an additional productivity of operations, but also by a systemic productivity of relations (Foray 1998, 2000; Lengrand and Chatrie 1999), generating diversified learning processes: learning-by-using, learning-by-doing, learning-by-sharing (Rosenberg 1982; Lundvall 1988, 1995; Malerba 1992). Learning may arise from internal or external sources of knowledge (Dodgson 1991). External learning refers to the absorption capacity of firms (Cohen and Levinthal 1990).

(iv) Innovations are determined not only by isolated learning, but also by social processes of exchange of knowledge generating a system — an innovative system (Johnson 1995; Lundvall 1992a; Acs 2000; Braczyk, Cooke and Heidenreich 1999; Cooke, Boekholt and Tödtling 2000; de la Mothe and Paquet 1998; Edquist 1997; Edquist and Hommen 1999; Holbrook and Wolfe 2000; Landry and Amara 1998; McFetridge 1993; Niosi 1993, 2000), a "système social d'innovation" (Amable, Barré and Boyer 1997; de la Mothe 2001), a "milieu innovateur" (Maillat 1995; Storper 1997) or an innovation cluster (Porter 1999, 2000).

These new conceptualizations of innovation have led to revisions in innovation theories. Traditional theoretical and empirical studies focused on firm-specific determinants (R&D activities, firm size, etc.). Over time, studies based on firm-specific determinants have tended to add an increasingly greater variety of determinants external to firms, especially with respect to external sources of information used by firms to develop or improve their products or processes.

3. DATA

The data used in the study presented here were taken from the responses of 5,455 manufacturing firms to the Survey of Innovation 1999. The survey was carried out according to the methodology developed in the *Oslo Manual*, and it covered both the manufacturing and the resource industries. The data analyzed in the study were for the manufacturing industries only. The unit of analysis was the provincial enterprise, which is defined by Statistics Canada as the combination of all legal units of a given enterprise group in the same industry within a province.

The survey was postal and mandatory. The sampling excluded firms with annual revenue of less than $250,000 and fewer than 20 employees. More than 9,000 firms met these criteria. The sampling was therefore biased against small firms with fewer than 20 employees, which make up a very large proportion of manufacturing firms in many non-metropolitan regions of Canada. Despite this bias, the survey provided rare and new evidence on innovation in all manufacturing industries and many explanatory variables of innovation.

Approximately 6,000 provincial enterprises in manufacturing industries were randomly selected for the survey, and a total of 5,455 participated. The questionnaire was completed by the firms' chief executive officers (CEOs) and the net response rate was over 90%. Of the

5,455 firms, 81% indicated that they had introduced new or significantly improved products or manufacturing processes during the period 1997-99. In the study reported here, the analysis was restricted to innovative firms that responded to the question on the degree of novelty of innovation: "Was this most important new or significantly improved product (good or service) or production/manufacturing process: a. a world first? b. a first in Canada? c. a first for your firm?"

Each firm in the Survey of Innovation 1999 database was assigned a weight, which allowed that firm to represent other firms in the population with similar characteristics.[1] The total population of manufacturing firms was 8,921, a number that was equal to the sum of the population weights of the in-sample firms.

4. FINDINGS

The findings of the study described in this chapter are presented in two parts. The first part presents descriptive statistics and tests of differences regarding various aspects of novelty of innovation and the sources of information used to develop or improve products or processes. The second part considers variables affecting the sources of information and other determinants of the novelty of innovations introduced on the market by Canadian manufacturing firms.

4.1 Descriptive Statistics

Internal, market, research and generally available sources of information used by Canadian manufacturing firms to develop or improve their products or processes were examined. Each source of information was initially considered independently; then, in a second stage, the sources were used to create four indices of variety of use: the index of internal sources; the index of market sources; the index of research sources; and the index of generally available sources. The sources of information used to develop or improve products or processes were considered in relation to the following factors: degree of novelty of innovation; number of employees; number of months required to develop innovations; level of firms' technological intensity; and province of operation.

4.1.1 Sources of Information and Degree of Novelty of Innovation

The results with respect to the sources of information used by firms to develop innovation according to the degree of novelty of their most important innovation are summarized in Table 1. Overall, the higher the degree of novelty of innovation introduced by firms, the more likely they were to use diverse internal and external sources of information. More specifically, of the firms that had introduced world-first innovations on the market, 77.8% relied on their R&D staff compared to 67.7% of the firms that had introduced Canada-first innovations and only 45.7% of the firms that had introduced firm-first innovations.

As for market sources, Table 1 indicates that clients were used more frequently as a source of information for innovation by firms that had introduced world-first (70.4%) or Canada-first (69.4%) innovations than by firms that had developed firm-first innovations (63.9%).

As expected, research sources were used much less frequently as a source of information to develop or improve products or processes, although firms developing innovations with higher degrees of novelty tended to rely on research sources more than other firms.

Finally, the generally available sources of information used most frequently by firms as a source of ideas to develop or improve products or processes were trade fairs and exhibitions. Again, the frequency of participation at trade fairs and exhibitions was higher for firms that had introduced world-first (72.5%) or Canada-first (71.7%) innovations than for firms that had developed firm-first innovations (68%).

Following the hypotheses of network theories of innovation, it was assumed that it was insufficient to consider the various sources of information only independently; this was because the development and improvement of products and processes cannot depend only on one source of information, but must rely on recourse to a variety of sources. As a result, the four indices of variety of use — the index of internal sources, the index of market sources, the index of research sources and the index of generally available sources — were developed. The index of internal sources was created by adding together the number of different internal sources of information used by firms. This index of variety could range from 0 to 4 because a firm could use up to four of the following internal sources: R&D staff, marketing staff, production

Table 1: Sources of Information Used by Firms to Develop Innovation, by Degree of Novelty of Their Most Important Innovation, 1997-99

Sources of Information	*Degree of Novelty of Innovation*						*Pearson Chi-Square Test*[a]
	World-First		*Canada-First*		*Firm-First*		
	Number	*% Within World-First*	*Number*	*% Within Canada-First*	*Number*	*% Within Firm-First*	
Internal Sources							
• R&D Staff	592	77.8	867	67.7	1,472	45.7	355.9*
• Marketing Staff	599	78.8	969	75.6	2,002	62.1	125.8*
• Production Staff	556	73.2	977	76.3	2,312	71.8	2.29
• Management Staff	575	75.6	1,005	78.4	2,475	76.8	2.34
Market Sources							
• Related Firms in Same Corporate Group	321	42.2	571	44.6	974	30.2	100.1*
• Suppliers of Equipment, Material and Components	442	58.2	853	66.5	2,215	68.8	31.2*
• Clients	535	70.4	889	69.4	2,057	63.9	19.7*
• Competitors	287	37.8	465	36.3	1,212	37.6	0.8
• Consultancy Firms	168	22.1	288	22.5	567	17.6	17.9*
Research Sources							
• Universities and Colleges	117	15.4	158	12.3	226	7.0	65.5*
• Federal Agencies and Research Laboratories	137	18.0	147	11.5	211	6.6	103.2*
• Provincial Agencies and Research Laboratories	86	11.3	91	7.1	144	4.5	53.3*
Generally Available Sources							
• Trade Fairs and Exhibitions	551	72.5	919	71.7	2,190	68.0	9.7**
• Internet- or Computer-Based Information Networks	358	47.1	525	41.0	1,169	36.3	32.9*
• Professional Conferences, Meetings and Publications	441	58.0	722	56.3	1,626	50.5	21.3*

Source: Statistics Canada, Survey of Innovation 1999. Tabulated by authors.

[a]This tested the independence of the frequency of use of a particular source of information from the variable indicating the degree of novelty of the most important innovation (world-first, Canada-first, firm-first).

*, ** and *** indicate that the null hypothesis (independence of the frequency of use of a particular source of information from the degree of novelty of the most important innovation) could be rejected at the 1%, 5% and 10% level of significance, respectively.

staff and management staff. The construction of the three other indices was based on the same logic. To compare these indices with regard to the different degrees of novelty of innovation, a one-way ANOVA was used, more specifically Duncan's multiple-range test, which compares the means for groups in homogeneous subsets. This test was appropriate for grouping the different degrees of novelty of innovation into homogeneous subsets — that is, degrees of novelty between which the mean differences were not statistically significant — and hence for comparing the means of the different subsets. The null hypothesis tested was the equality of means for the variable index of variety of sources of information between the different degrees of novelty of innovation (world-first, Canada-first and firm-first).

The results of Duncan's test are reported in Table 2. The means of the index of internal sources range from 2.56 to 3.05. Duncan's test shows that there were two homogeneous subsets of degrees of novelty of innovation in relation to the variety of internal sources used by firms. As shown in Table 2, there were no significant statistical differences between firms that had introduced world-first or Canada-first innovations with respect to the variety of internal sources used to develop or improve products or processes. These two groups were more likely to use a greater variety of internal sources of information than firms that had introduced firm-first innovations. As for the index of market sources, Duncan's test shows that there were three homogeneous subsets of firms: firms developing Canada-first innovations, which used the greatest variety of market sources; firms developing world-first innovations; and, lastly, firms developing firm-first innovations, which used the smallest variety of market sources.

As for the indices of research sources and of generally available sources of information, the results of Duncan's test reported in Table 2 show that there were three homogeneous subsets of firms: firms developing world-first innovations, which used the greatest variety of research sources and of generally available sources; firms developing Canada-first innovations; and, lastly, firms developing firm-first innovations, which used the smallest variety of research sources and of generally available sources.

Table 2: Means of the Indices of the Sources of Information Used to Develop Innovation for Groups of Firms in Homogeneous Subsets, by Degree of Novelty of Their Most Important Innovation (Duncan's Test), 1997-99

Index of Internal Sources

Degree of Novelty of Innovation	*Number of Observations*	*Subset for α = 0.05* 1	2
Firm-First	3,221	2.56	
Canada-First	1,281		2.98
World-First	760		3.05
Significance[a]		1.000	0.071

Index of Research Sources

Degree of Novelty of Innovation	*Number of Observations*	*Subset for α = 0.05* 1	2	3
Firm-First	3,221	0.18		
Canada-First	1,281		0.31	
World-First	760			0.45
Significance		1.000	1.000	1.000

Index of Market Sources

Degree of Novelty of Innovation	*Number of Observations*	*Subset for α = 0.05* 1	2	3
Firm-First	3,221	2.18		
World-First	760		2.31	
Canada-First	1,281			2.39
Significance		1.000	1.000	1.000

Index of Generally Available Sources

Degree of Novelty of Innovation	*Number of Observations*	*Subset for α = 0.05* 1	2	3
Firm-First	3,221	1.55		
Canada-First	1,281		1.69	
World-First	760			1.77
Significance		1.000	1.000	1.000

Source: Statistics Canada, Survey of Innovation 1999. Tabulated by authors.

[a]Duncan's test compares means for groups in homogeneous subsets. When the significance test is above the threshold $\alpha = 0.05$, the null hypothesis (non-differences of means) cannot be rejected.

4.1.2 Sources of Information and Number of Employees

When the sources of information used by firms to develop or improve their products or processes were considered in relation to size of firm, it was clear that the frequency of utilization of the various internal and external sources of information tended to increase gradually as the number of employees increased. This result could be expected with respect to internal R&D and external research sources. However, Tables 3a and 3b show that this result held even in the case of market sources with regard to suppliers as sources of information, but not for clients.

The results of Duncan's test for this variable are reported in Tables 4a and 4b. They show that the variety of sources of information increased as the number of employees increased — from small (1-99 or 1-49 employees) to medium-sized (100-499 or 50-249 employees) and to large firms (500+ or 250+ employees). The only exception to this trend was that small and medium-sized firms were not significantly different regarding the variety of research sources.

4.1.3 Sources of Information and Number of Months Required to Develop Innovations

A consideration of the number of months that firms needed to develop their most important innovation in relation to the sources of information they used reveals the following: the frequency of utilization of internal and external research sources, marketing staff, related firms in the same corporate group, clients and consultancy firms, participation in professional conferences and meetings, and use of publications increased in a statistically different manner for innovations that required less than one year, innovations that required between one and two years, and innovations that required more than two years. The results reported in Table 5a indicate that there were no significant differences regarding the relationship between the number of months required to develop innovations and the use of the other sources of information. Another taxonomy of the number of months required to develop innovations is used in Table 5b, and it shows results similar to those of the taxonomy used in Table 5a.

Table 3a: Sources of Information Used by Firms to Develop Innovation, by Number of Employees, 1997-99

Sources of Information	*Number of Employees*						*Pearson Chi-Square Test*[a]
	1-99		*100-499*		*500+*		
	Number	*% Within 1-99*	*Number*	*% Within 100-499*	*Number*	*% Within 500+*	
Internal Sources							
• R&D Staff	1,588	50.1	1,126	63.3	217	68.9	104.28*
• Marketing Staff	2,027	63.9	1,315	74.0	226	72.0	55.42*
• Production Staff	2,252	71.0	1,341	75.5	252	80.3	20.05*
• Management Staff	2,422	76.4	1,395	78.5	238	75.8	3.04
Market Sources							
• Related Firms in Same Corporate Group	902	28.5	787	44.3	178	56.5	189.13*
• Suppliers of Equipment, Material and Components	2,073	65.4	1,211	68.1	224	71.3	6.98**
• Clients	2,096	66.1	1,182	66.5	203	64.4	0.49
• Competitors	1,178	37.2	654	36.8	132	41.9	3.07
• Consultancy Firms	515	16.3	428	24.1	80	25.4	52.19*
Research Sources							
• Universities and Colleges	274	8.6	183	10.3	43	13.7	10.29*
• Federal Agencies and Research Laboratories	270	8.5	177	10.0	48	15.2	16.12*
• Provincial Agencies and Research Laboratories	186	5.9	109	6.1	26	8.3	2.84
Generally Available Sources							
• Trade Fairs and Exhibitions	2,204	69.5	1,247	70.1	208	66.2	1.91
• Internet- or Computer-Based Information Networks	1,182	37.3	748	42.1	121	38.4	10.94*
• Professional Conferences, Meetings and Publications	1,584	50.0	1,006	56.6	197	62.7	32.96*

Source: Statistics Canada, Survey of Innovation 1999. Tabulated by authors.

[a]This tested the independence of the frequency of use of a particular source of information from the variable indicating size of firm.

*, ** and *** indicate that the null hypothesis (independence of the frequency of use of a particular source of information from size of firm) could be rejected at the 1%, 5% and 10% level of significance, respectively.

Table 3b: Sources of Information Used by Firms to Develop Innovation, by Number of Employees, 1997-99

Sources of Information	Number of Employees: 1-49 Number	1-49 % Within 1-49	50-249 Number	50-249 % Within 50-249	250+ Number	250+ % Within 250+	Pearson Chi-Square Test[a]
Internal Sources							
• R&D Staff	696	45.5	1,689	57.6	545	67.9	116.40*
• Marketing Staff	929	60.8	2,053	70.0	587	73.1	51.43*
• Production Staff	1,038	67.9	2,183	74.5	625	77.8	32.73*
• Management Staff	1,155	75.6	2,293	78.2	607	75.6	5.03***
Market Sources							
• Related Firms in Same Corporate Group	450	29.5	983	33.5	434	54.0	150.03*
• Suppliers of Equipment, Material and Components	954	62.5	1,981	67.6	573	71.4	21.14*
• Clients	966	63.2	1,985	67.7	530	66.0	9.11**
• Competitors	567	37.1	1,090	37.2	307	38.2	0.34
• Consultancy Firms	234	15.3	564	19.2	225	28.0	54.32*
Research Sources							
• Universities and Colleges	119	7.8	263	9.0	118	14.7	31.50*
• Federal Agencies and Research Laboratories	128	8.4	264	9.0	104	13.0	14.26*
• Provincial Agencies and Research Laboratories	99	6.5	151	5.2	71	8.8	15.54*
Generally Available Sources							
• Trade Fairs and Exhibitions	1,015	66.4	2,099	71.6	546	68.0	13.85*
• Internet- or Computer-Based Information Networks	555	36.3	1,175	40.1	321	40.0	6.39**
• Professional Conferences, Meetings and Publications	725	47.5	1,572	53.6	490	61.0	39.90*

Source: Statistics Canada, Survey of Innovation 1999. Tabulated by authors.

[a]This tested the independence of the frequency of use of a particular source of information from the variable indicating size of firm.

*, ** and *** indicate that the null hypothesis (independence of the frequency of use of a particular source of information from size of firm) could be rejected at the 1%, 5% and 10% level of significance, respectively.

Table 4a: Means of the Indices of the Sources of Information Used to Develop Innovation for Groups of Firms in Homogeneous Subsets, by Size of Firm (Duncan's Test), 1997-99

Index of Internal Sources			
Number of Employees	*Number of Observations*	*Subset for $\alpha = 0.05$* 1	2
1-99	3,169	2.61	
100-499	1,777		2.91
500+	314		2.97
Significance[a]		1.000	0.288

Index of Research Sources			
Number of Employees	*Number of Observations*	*Subset for $\alpha = 0.05$* 1	2
1-99	3,169	0.23	
100-499	1,777	0.26	
500+	314		0.37
Significance		0.306	1.000

Index of Market Sources				
Number of Employees	*Number of Observations*	*Subset for $\alpha = 0.05$* 1	2	3
1-99	3,169	2.13		
100-499	1,777		2.40	
500+	314			2.59
Significance		1.000	1.000	1.000

Index of Generally Available Sources			
Number of Employees	*Number of Observations*	*Subset for $\alpha = 0.05$* 1	2
1-99	3,169	1.57	
500+	314		1.67
100-499	1,777		1.69
Significance		1.000	0.768

Source: Statistics Canada, Survey of Innovation 1999. Tabulated by authors.

[a]Duncan's test compares means for groups in homogeneous subsets. When the significance test is above the threshold $\alpha = 0.05$, the null hypothesis (non-differences of means) cannot be rejected.

Table 4b: Means of the Indices of the Sources of Information Used to Develop Innovation for Groups of Firms in Homogeneous Subsets, by Size of Firm (Duncan's Test), 1997-99

Index of Internal Sources					*Index of Research Sources*			
Number of Employees	*Number of Observations*	*Subset for α = 0.05* 1	2	3	*Number of Employees*	*Number of Observations*	*Subset for α = 0.05* 1	2
1-49	1,527	2.50			1-49	1,527	0.23	
50-249	2,931		2.80		50-249	2,931	0.23	
250+	802			2.94	250+	802		0.36
Significance[a]		1.000	1.000	1.000	Significance		0.832	1.000
Index of Market Sources					*Index of Generally Available Sources*			
Number of Employees	*Number of Observations*	*Subset for α = 0.05* 1	2	3	*Number of Employees*	*Number of Observations*	*Subset for α = 0.05* 1	2
1-49	1,527	2.08			1-49	1,527	1.50	
50-249	2,931		2.25		50-249	2,931		1.65
250+	802			2.58	250+	802		1.69
Significance		1.000	1.000	1.000	Significance		1.000	0.363

Source: Statistics Canada, Survey of Innovation 1999. Tabulated by authors.

[a]Duncan's test compares means for groups in homogeneous subsets. When the significance test is above the threshold α = 0.05, the null hypothesis (non-differences of means) cannot be rejected.

Table 5a: Sources of Information Used by Firms to Develop Innovation, by Number of Months Required to Develop Their Most Important Innovation, 1997-99

Sources of Information	Number of Months Required to Develop Most Important Innovation: 1-11 Number	1-11 % Within 1-11	12-24 Number	12-24 % Within 12-24	24+ Number	24+ % Within 24+	Pearson Chi-Square Test[a]
Internal Sources							
• R&D Staff	629	46.1	1,535	55.9	746	66.0	99.05*
• Marketing Staff	841	61.7	1,908	69.5	808	71.5	34.16*
• Production Staff	1,022	74.9	2,012	73.3	792	70.1	7.53**
• Management Staff	1,020	74.8	2,162	78.7	852	75.4	9.74*
Market Sources							
• Related Firms in Same Corporate Group	450	33.0	951	34.6	460	40.7	17.99*
• Suppliers of Equipment, Material and Components	906	66.4	1,834	66.8	754	66.7	0.06
• Clients	859	63.0	1,819	66.2	792	70.1	13.97*
• Competitors	491	36.0	1,046	38.1	422	37.3	1.71
• Consultancy Firms	229	16.8	539	19.6	251	22.2	11.66*
Research Sources							
• Universities and Colleges	87	6.4	276	10.0	135	11.9	24.22*
• Federal Agencies and Research Laboratories	99	7.3	246	9.0	146	12.9	24.50*
• Provincial Agencies and Research Laboratories	70	5.1	156	5.7	92	8.1	11.35*
Generally Available Sources							
• Trade Fairs and Exhibitions	940	68.9	1,929	70.2	775	68.6	1.38
• Internet- or Computer-Based Information Networks	534	39.1	1,067	38.8	441	39.0	0.04
• Professional Conferences, Meetings and Publications	660	48.4	1,440	52.4	673	59.6	31.50*

Source: Statistics Canada, Survey of Innovation 1999. Tabulated by authors.

[a]This tested the independence of the frequency of use of a particular source of information from the variable indicating the number of months required to develop the most important innovation. *, ** and *** indicate that the null hypothesis (independence of the frequency of use of a particular source of information from the number of months required to develop the most important innovation) could be rejected at the 1%, 5% and 10% level of significance, respectively.

Table 5b: Sources of Information Used by Firms to Develop Innovation, by Number of Months Required to Develop Their Most Important Innovation, 1997-99

Sources of Information	*Number of Months Required to Develop Most Important Innovation*						Pearson Chi-Square Test[a]
	1-5		*6-24*		*24+*		
	Number	*% Within 1-5*	*Number*	*% Within 6-24*	*Number*	*% Within 24+*	
Internal Sources							
• R&D Staff	156	42.7	2,008	53.6	746	66.0	80.06*
• Marketing Staff	208	57.0	2,540	67.8	808	71.5	26.67*
• Production Staff	271	74.2	2,763	73.8	792	70.1	6.30**
• Management Staff	275	75.3	2,907	77.6	852	75.4	3.02
Market Sources							
• Related Firms in Same Corporate Group	100	27.4	1,301	34.7	460	40.7	24.79*
• Suppliers of Equipment, Material and Components	235	64.4	2,505	66.9	754	66.7	0.94
• Clients	241	66.0	2,437	65.1	792	70.1	9.83*
• Competitors	134	36.7	1,403	37.5	422	37.3	0.08
• Consultancy Firms	65	17.8	703	18.8	251	22.2	7.23**
Research Sources							
• Universities and Colleges	21	5.8	342	9.1	135	11.9	14.41*
• Federal Agencies and Research Laboratories	24	6.6	321	8.6	146	12.9	22.94*
• Provincial Agencies and Research Laboratories	16	4.4	209	5.6	92	8.1	11.93*
Generally Available Sources							
• Trade Fairs and Exhibitions	240	65.8	2,630	70.2	775	68.6	3.75
• Internet- or Computer-Based Information Networks	142	38.8	1,459	39.0	441	39.0	0.006
• Professional Conferences, Meetings and Publications	146	40.0	1,954	52.2	673	59.6	45.30*

Source: Statistics Canada, Survey of Innovation 1999. Tabulated by authors.

[a]This tested the independence of the frequency of use of a particular source of information from the variable indicating the number of months required to develop the most important innovation. *, ** and *** indicate that the null hypothesis (independence of the frequency of use of a particular source of information from the number of months required to develop the most important innovation) could be rejected at the 1%, 5% and 10% level of significance, respectively.

The results of Duncan's test indicate that the most important innovations that were developed in less than 12 months had involved a significantly lower number of internal sources and market sources of information than innovations developed over longer periods of time. As for the index of research sources, Table 6a shows that there were three homogeneous groups of firms: firms developing their most important innovation within 12 months using a smaller variety of research sources than the two other categories of firms; firms developing their most important innovation within 12-24 months using a greater variety of research sources of information; and firms developing their most important innovation in more than 24 months using the greatest variety of research sources. Another taxonomy of the number of months required to develop innovations is used in Table 6b, and it shows results similar to those of the taxonomy used in Table 6a.

4.1.4 Sources of Information and Level of Firms' Technological Intensity

To study differences in the use of 15 sources of information by 44 industries, the Industry Canada classification of industries into high-, medium- and low-tech categories was used (see Appendix 1 for the full classification). This classification is itself based on indicators of R&D intensity developed by the OECD. The classification was used to facilitate comparisons regarding relationships between industries and across sources of information. In a progression from the low-tech through the medium-tech to the high-tech category, firms showed a significant propensity to use internal and external research sources, marketing staff, related firms in the same corporate group, clients, competitors, Internet- or computer-based information networks, and professional conferences, meetings and publications more frequently (Table 7). The only exception was provincial agencies and research laboratories.

The results of Duncan's test, reported in Table 8, suggest that firms in the high-tech category tended to use a greater variety of internal, market, research and generally available sources of information to develop their innovations. By comparison, there were generally no significant differences between firms in the low- and medium-tech

Table 6a: Means of the Indices of the Sources of Information Used to Develop Innovation for Groups of Firms in Homogeneous Subsets, by Number of Months Required to Develop Their Most Important Innovation (Duncan's Test), 1997-99

Index of Internal Sources

Number of Months	*Number of Observations*	*Subset for α = 0.05*	
		1	*2*
1-11	1,363	2.57	
12-24	2,746		2.77
24+	1,130		2.83
Significance[a]		1.000	0.136

Index of Research Sources

Number of Months	*Number of Observations*	*Subset for α = 0.05*		
		1	*2*	*3*
1-11	1,363	0.19		
12-24	2,746		0.25	
24+	1,130			0.33
Significance		1.000	1.000	1.000

Index of Market Sources

Number of Months	*Number of Observations*	*Subset for α = 0.05*		
		1	*2*	*3*
1-11	1,363	2.15		
12-24	2,746		2.25	
24+	1,130			2.37
Significance		1.000	1.000	1.000

Index of Generally Available Sources

Number of Months	*Number of Observations*	*Subset for α = 0.05*	
		1	*2*
1-11	1,363	1.56	
12-24	2,746	1.61	1.61
24+	1,130		1.67
Significance		0.173	0.124

Source: Statistics Canada, Survey of Innovation 1999. Tabulated by authors.

[a]Duncan's test compares means for groups in homogeneous subsets. When the significance test is above the threshold α = 0.05, the null hypothesis (non-differences of means) cannot be rejected.

Table 6b: Means of the Indices of the Sources of Information Used to Develop Innovation for Groups of Firms in Homogeneous Subsets, by Number of Months Required to Develop Their Most Important Innovation (Duncan's Test), 1997-99

Index of Internal Sources				*Index of Research Sources*				
Number of Months	*Number of Observations*	*Subset for α = 0.05* 1	2	*Number of Months*	*Number of Observations*	*Subset for α = 0.05* 1	2	3
1-5	365	2.49		1-5	365	0.17		
6-24	3,745		2.73	6-24	3,745		0.23	
24+	1,130		2.83	24+	1,130			0.33
Significance[a]		1.000	0.056	Significance		1.000	1.000	1.000

Index of Market Sources				*Index of Generally Available Sources*			
Number of Months	*Number of Observations*	*Subset for α = 0.05* 1	2	*Number of Months*	*Number of Observations*	*Subset for α = 0.05* 1	2
1-5	365	2.12		1-5	365	1.44	
6-24	3,745	2.23		6-24	3,745		1.61
24+	1,130		2.37	24+	1,130		1.67
Significance		0.051	1.000	Significance		1.000	0.257

Source: Statistics Canada, Survey of Innovation 1999. Tabulated by authors.

[a]Duncan's test compares means for groups in homogeneous subsets. When the significance test is above the threshold α = 0.05, the null hypothesis (non-differences of means) cannot be rejected.

Table 7: Sources of Information Used by Firms to Develop Innovation, by Level of Technological Intensity, 1997-99

Sources of Information	*Level of Technological Intensity*						*Pearson Chi-Square Test*[a]
	High-Tech		*Medium-Tech*		*Low-Tech*		
	Number	*% Within High-Tech*	*Number*	*% Within Medium-Tech*	*Number*	*% Within Low-Tech*	
Internal Sources							
• R&D Staff	144	77.0	1,778	60.7	1,008	47.0	129.17*
• Marketing Staff	144	77.0	2,031	69.3	1,395	65.0	17.67*
• Production Staff	123	65.8	2,122	72.4	1,601	74.7	8.58**
• Management Staff	146	77.7	2,254	76.9	1,656	77.2	0.12
Market Sources							
• Related Firms in Same Corporate Group	84	44.9	1,095	37.3	688	32.1	22.61*
• Suppliers of Equipment, Material and Components	112	59.9	1,807	61.6	1,591	74.2	91.72*
• Clients	148	79.1	2,003	68.3	1,331	62.1	36.10*
• Competitors	87	46.5	1,145	39.1	732	34.1	19.79*
• Consultancy Firms	58	31.0	539	18.4	426	19.9	18.34*
Research Sources							
• Universities and Colleges	33	17.6	319	10.9	147	6.9	38.45*
• Federal Agencies and Research Laboratories	28	15.0	294	10.0	173	8.1	12.67*
• Provincial Agencies and Research Laboratories	24	12.8	161	5.5	135	6.3	16.90*
Generally Available Sources							
• Trade Fairs and Exhibitions	130	69.1	1,972	67.3	1,559	72.7	17.07*
• Internet- or Computer-Based Information Networks	119	63.6	1,214	41.4	719	33.5	82.01*
• Professional Conferences, Meetings and Publications	128	68.4	1,540	52.5	1,121	52.3	18.65*

Source: Statistics Canada, Survey of Innovation 1999. Tabulated by authors.

[a]This tested the independence of the frequency of use of a particular source of information from the variable indicating level of technological intensity.

*, ** and *** indicate that the null hypothesis (independence of the frequency of use of a particular source of information from level of technological intensity) could be rejected at the 1%, 5% and 10% level of significance, respectively.

Table 8: Means of the Indices of the Sources of Information Used to Develop Innovation for Groups of Firms in Homogeneous Subsets, by Level of Technological Intensity (Duncan's Test), 1997-99

Index of Internal Sources					*Index of Research Sources*			
Level of Techno-logical Intensity	*Number of Observations*	*Subset for α = 0.05* 1	2	3	*Level of Techno-logical Intensity*	*Number of Observations*	*Subset for α = 0.05* 1	23
Low-Tech	2,144	2.64			Low-Tech	2,144	0.21	
Medium Tech	2,931		2.79		Medium-Tech	2,931	0.26	
High-Tech	187			2.97	High-Tech	187		0.46
Significance[a]		1.000	1.000	1.000	Significance		0.187	1.000

Index of Market Sources				*Index of Generally Available Sources*			
Level of Techno-logical Intensity	*Number of Observations*	*Subset for α = 0.05* 1	2	*Level of Techno-logical Intensity*	*Number of Observations*	*Subset for α = 0.05* 1	2
Low-Tech	2,144	2.22		Low-Tech	2,144	1.58	
Medium-Tech	2,931	2.25		Medium-Tech	2,931	1.61	
High-Tech	187		2.61	High-Tech	187		2.01
Significance		0.729	1.000	Significance		0.670	1.00

Source: Statistics Canada, Survey of Innovation 1999. Tabulated by authors.

[a]Duncan's test compares means for groups in homogeneous subsets. When the significance test is above the threshold $\alpha = 0.05$, the null hypothesis (non-differences of means) cannot be rejected.

categories with respect to the variety of sources of information used to develop or improve products or processes, except for use of internal sources: here, the low-, medium- and high-tech firms represented three homogeneous subsets.

4.1.5 Sources of Information and Province of Operation

Were there differences in the use of sources of information among provinces? The results reported in Table 9 suggest that there were significant differences from province to province with respect to the frequency of use of the various sources of information for the development or improvement of products or processes. The results of Duncan's test, reported in Table 10, suggest that there were two, four, five and three homogeneous subsets of provinces with respect to the indices of internal sources, market sources, research sources and generally available sources of information, respectively. Overall, the manufacturing firms in British Columbia, Manitoba and Alberta tended to use a smaller variety of sources of information, whereas firms located in Prince Edward Island, Newfoundland and Nova Scotia generally used a greater variety of sources of information to develop or improve their products or processes.

4.2 Model

4.2.1 Dependent Variable

The qualitative dependent variable used in the study presented here was the degree of novelty of innovation. It took a value of 1 when the most important product or process innovation developed by a firm between 1997 and 1999 was considered by the firm to be a world-first innovation; it took a value of 2 when the most important product or process innovation developed by a firm between 1997 and 1999 was considered by the firm to be a first in Canada; and it took a value of 3 when the most important product or process innovation developed by a firm between 1997 and 1999 was considered a first for the firm. The latter was identified as the reference category.

Table 9: Sources of Information Used by Firms to Develop Innovation, by Province of Operation, 1997-99

Sources of Information	Province of Operation[b]																				
	Newfoundland		*P.E.I.*		*Nova Scotia*		*New Brunswick*		*Quebec*		*Ontario*		*Manitoba*		*Sask.*		*Alberta*		*B.C.*		*Chi-Square Test*[a]
	N	*%*	*N*	*%*	*N*	*%*	*N*	*%*	*N*	*%*	*N*	*%*	*N*	*%*	*N*	*%*	*N*	*%*	*N*	*%*	
Internal Sources																					
R&D Staff	19	51.4	18	62.1	68	59.6	47	47.0	1,131	59.9	1,100	55.2	98	59.8	47	58.8	180	51.6	223	43.5	53.33*
Marketing Staff	23	63.9	22	75.9	81	71.1	59	59.0	1,391	73.7	1,334	67.0	95	57.9	49	61.3	207	59.3	308	60.2	70.14*
Production Staff	31	86.1	21	72.4	95	83.3	68	68.0	1,412	74.8	1,443	72.5	119	72.6	58	72.5	238	68.2	361	70.4	19.96**
Management Staff	34	94.4	28	96.6	98	86.0	84	84.8	1,222	64.8	1,662	83.5	143	87.2	71	87.7	283	81.1	429	83.6	259.62*
Market Sources																					
Related Firms in Same Corporate Group	19	52.8	12	42.9	45	39.5	38	38.0	606	32.1	819	41.1	49	29.7	30	37.5	102	29.2	147	28.7	62.55*
Suppliers of Equipment, Material and Components	31	86.1	19	65.5	87	76.3	69	69.0	1,240	65.7	1,388	69.7	103	62.8	48	60.0	220	63.0	304	59.3	37.69*
Clients	20	55.6	25	86.2	74	64.9	65	65.0	1,318	69.9	1,282	64.4	107	65.2	58	71.6	216	61.7	318	62.0	29.73*
Competitors	10	27.8	11	39.3	49	43.0	39	39.0	685	36.3	745	37.4	65	39.6	32	40.0	132	37.7	196	38.2	4.79
Consultancy Firms	9	25.0	7	24.1	25	21.9	28	28.0	369	19.6	379	19.0	40	24.4	20	25.0	68	19.4	78	15.2	16.39***
Research Sources																					
Universities and Colleges	4	11.1	6	20.7	11	9.6	12	12.0	201	10.7	188	9.4	13	7.9	14	17.3	21	6.0	31	6.6	26.24*
Federal Agencies and Research Laboratories	6	16.2	6	21.4	21	18.4	17	17.0	226	12.0	132	6.6	13	7.9	12	15.0	27	7.7	35	6.8	65.59*
Provincial Agencies and Research Laboratories	12	33.3	7	24.1	21	18.4	16	16.0	138	7.3	64	3.2	13	7.9	7	8.8	18	5.1	25	4.9	148.09*
Generally Available Sources																					
Trade Fairs and Exhibitions	24	66.7	16	57.1	89	78.1	72	72.0	1,379	73.1	1,390	69.8	98	59.8	54	67.5	193	55.1	345	67.3	60.72*
Internet- or Computer-Based Information Networks	19	51.4	10	34.5	56	49.1	33	33.0	722	38.3	799	40.2	45	27.4	41	51.3	124	35.5	203	39.6	26.68*
Professional Conferences, Meetings and Publications	24	66.7	14	50.0	72	63.2	50	50.0	1,012	53.6	1,029	51.7	89	54.3	52	65.0	186	53.1	261	50.9	15.18***

Source: Statistics Canada, Survey of Innovation 1999. Tabulated by authors.

[a]This tested the independence of the frequency of use of a particular source of information from the variable indicating province of operation.

[b]The Territories were not included in this analysis.

*, ** and *** indicate that the null hypothesis (independence of the frequency of use of a particular source of information from province of operation) could be rejected at the 1%, 5% and 10% level of significance, respectively.

Table 10: Means of the Indices of the Sources of Information Used to Develop Innovation for Groups of Firms in Homogeneous Subsets, by Province of Operation (Duncan's Test), 1997-99

Index of Internal Sources

Province of Operation	*Number of Observations*	*Subset for α = 0.05* 1	2
B.C.	512	2.57	
N.B.	99	2.58	
Alberta	349	2.60	
Quebec	1,886	2.73	2.73
Manitoba	164	2.77	2.77
Ontario	1,990	2.78	2.78
Saskatchewan	80	2.80	2.80
Nfld.	36		2.97
Nova Scotia	114		3.01
P.E.I.	28		3.04
Significance[a]		0.206	0.074

Index of Research Sources

Province of Operation	*Number of Observations*	*Subset for α = 0.05* 1	2	3	4	5
B.C.	512	0.18				
Alberta	349	0.19				
Ontario	1,990	0.19				
Manitoba	164	0.24	0.24			
Quebec	1,886	0.30	0.30	0.30		
Saskatchewan	80		0.41	0.41	0.41	
N.B.	99			0.46	0.46	
Nova Scotia	114			0.47	0.47	
Nfld.	36				0.59	0.59
P.E.I.	28			0.68		
Significance		0.231	0.071	0.086	0.063	0.301

Index of Market Sources

Province of Operation	*Number of Observations*	*Subset for α = 0.05* 1	2	3	4
B.C.	512	2.03			
Alberta	349	2.11	2.11		
Manitoba	164	2.21	2.21	2.21	
Quebec	1,886	2.23	2.23	2.23	2.23
Ontario	1,990	2.32	2.32	2.32	2.32
Saskatchewan	80	2.34	2.34	2.34	2.34
N.B.	99	2.38	2.38	2.38	2.38
Nova Scotia	114		2.45	2.45	2.45
Nfld.	36			2.48	2.48
P.E.I.	28				2.57
Significance		0.056	0.059	0.143	0.063

Index of Generally Available Sources

Province of Operation	*Number of Observations*	*Subset for α = 0.05* 1	2	3
Manitoba	164	1.41		
P.E.I.	28	1.43		
Alberta	349	1.44		
N.B.	99	1.55	1.55	
B.C.	512	1.57	1.57	1.57
Ontario	1,990	1.62	1.62	1.62
Quebec	1,886	1.65	1.65	1.65
Saskatchewan	80		1.83	1.83
Nfld.	36		1.84	1.84
Nova Scotia	114			1.90
Significance		0.176	0.087	0.053

Source: Statistics Canada, Survey of Innovation 1999. Tabulated by authors.

[a]Duncan's test compares means for groups in homogeneous subsets. When the significance test is above the threshold α = 0.05, the null hypothesis (non-differences of means) cannot be rejected.

The probability of choosing novelty category k (k = 1,2,3) was given by:

$$\text{Prob}_{ik} = \frac{e^{\beta_k X_i}}{1+\sum_{k=1}^{3} e^{\beta_k X_i}} \qquad (1)$$

where X_i was the matrix of novelty attributes and β_k was m x 1 vector of parameters.

As in the case of bivariate logit models, coefficients for reference choice were set equal to zero. Such normalization was taken into account when the rest of the model coefficients were interpreted. Since the development of innovations considered as a first for a firm was taken as the reference category, the estimated parameters were interpreted in the following way:

$$\frac{Prob_{i1}}{Prob_{i3}} = \frac{e^{\beta_1 X_i}}{e^{\beta_3 X_i}} = e^{(\beta_1 - \beta_3)Xi} = e^{\beta_1 X_i}$$

and (2)

$$\frac{Prob_{i2}}{Prob_{i3}} = \frac{e^{\beta_2 X_i}}{e^{\beta_3 X_i}} = e^{(\beta_2 - \beta_3)Xi} = e^{\beta_2 X_i}$$

or

$$\text{Ln}\left(\frac{Prob_{i1}}{Prob_{i3}}\right) = (\beta_1 - \beta_3)X_i = \beta_1 X_i$$

and (3)

$$\text{Ln}\left(\frac{Prob_{i2}}{Prob_{i3}}\right) = (\beta_2 - \beta_3)X_i = \beta_2 X_i$$

From (3), the estimated coefficients — for instance, βlj (j = 1,....,m) — were interpreted as the marginal change in the logarithm of the odds of the assessment by firms of the most important innovation they developed between 1997 and 1999 as a world-first introduction over the category assessing the most important innovation as a firm-first introduction due to a marginal change in the attribute j. However, while marginal changes in the logarithm of the odds were not always intuitively understandable, the exponential of parameters also referred to as odds ratios could be used. They offered a straightforward model interpretation. Indeed, exp(βlj) was the factor by which the odds changed when the jth independent variable increased by one unit. If βlj was positive, this factor — i.e., exp(βlj) — would be higher than 1, meaning that the odds were increased. Conversely, if βlj was negative, exp(βlj) was less than 1, implying that the odds were decreased. And if βlj was 0, exp(βlj) was equal to 1, which left the odds unchanged. For example, for a continuous variable such as the number of a firm's employees, exp(βlj) measured the factor by which the odds of the firm being a world-first innovator rather than a firm-first innovator changed when the number of employees was increased by one unit between 1997 and 1999. Analogously, if attribute j was a dummy variable, the exponential of parameters — i.e., exp(βlj) — measured the factor of change in the odds with respect to the reference variable.

4.2.2 Independent Variables

To identify the explanatory variables of the novelty of innovations developed by firms, two multinomial logit regressions were estimated by maximum likelihood using the different alternatives of novelty of innovation as dependent variables, and as independent variables the list of variables discussed in the second section of this paper. In the first model, the sources of information were included one by one as dichotomic explanatory variables. In the second model, the sources of information were modelled as indices of variety of sources.

Table 11 provides an overview of the operationalization for the explanatory variables. Cronbach's alpha for all the independent variables based on multiple-item scales is shown in Table 12. The values of the alpha coefficients, reported in Table 12, indicate that all the multiple-item scales employed in the study were reliable.

Table 11: Definition of the Independent Variables

Sources of Information	*The sum of the number of different sources of information used to develop new products and processes*
Internal Sources	• R&D staff • Marketing staff • Production staff • Management staff
Market Sources	• Related firms in same corporate group • Suppliers of equipment, material and components • Clients • Competitors • Consultancy firms
Research Sources	• Universities and colleges • Federal agencies and research laboratories • Provincial agencies and research laboratories
Generally Available Sources	• Trade fairs and exhibitions • Internet- or computer-based information networks • Professional conferences, meetings and publications
Government Support Programs	*The sum of the number of different federal and provincial government programs used by a firm between 1997 and 1999*
	• R&D tax credits • Government R&D grants • Government venture capital support • Government technology support and assistance programs • Government information or Internet services • Government support for training
Competitive Environment	*Level of agreement with the following statements (1 = strongly disagree to 5 = strongly agree)*
	• My client's demands are easy to predict • My clients can easily substitute my products for the products of my competitors • My competitors' actions are easy to predict • The arrival of new competitors is a constant threat • The arrival of competing products is a constant threat • My firm can easily replace its current suppliers • It is difficult to hire qualified staff and workers • It is difficult to retain qualified staff and workers • My products quickly become obsolete • Production technologies change rapidly • Office technologies change rapidly
Number of Employees	*Total number of a firm's employees*
R&D Activities	*Binary variable coded 1 if, between 1997 and 1999, a firm undertook R&D activities, and 0 otherwise*

Source: Statistics Canada, Survey of Innovation 1999. Tabulated by authors.

Table 11: Definition of the Independent Variables – Concluded

Innovative Practices	*The sum of the number of different innovative practices linked to offering new or significantly improved products or to introducing new or significantly improved production/manufacturing processes*
	• R&D linked to new or significantly improved products or production/ /manufacturing processes • Acquisition of machinery, equipment or other technology linked to new or significantly improved products or production/manufacturing processes • Industrial engineering and industrial design linked to new or significantly improved products or production/manufacturing processes • Tooling up and production start-up linked to new or significantly improved products or production/manufacturing processes • Training linked to the introduction of new or significantly improved products or production/manufacturing processes
Barriers to Knowledge Exchange	*The sum of different obstacles encountered by a firm to the access of information regarding the development or improvement of products or processes*
	• High cost of developing new or significantly improved products or processes • Inability to devote staff to projects to develop new or significantly improved products or processes on an ongoing basis because of production requirements • Inability to qualify for government assistance programs or R&D tax credits • Lack of skilled personnel to develop or introduce new or significantly improved products or processes • Lack of financing for the development or introduction of new or significantly improved products or processes • Lack of marketing capability to market new or significantly improved products • Lack of information on technology relevant to the development or introduction of new or significantly improved products or processes • Lack of external technical support services required to develop or introduce new or significantly improved products or processes • Lack of access to expertise in universities that could assist in developing or introducing new or significantly improved products or processes • Lack of access to expertise in government laboratories that could assist in developing or introducing new or significantly improved products and/or processes • Lack of co-operation with other firms
Technological Intensity	*Binary variables indicating degree of technological intensity*
	• **High:** Binary variable coded 1 if a firm operated in a high-tech sector, and 0 otherwise • **Medium:** Binary variable coded 1 if a firm operated in a medium-tech sector, and 0 otherwise • **Low:** Binary variable coded 1 if a firm operated in a low-tech sector, and 0 otherwise
Collaborative Arrangements	*Binary variable coded 1 if, between 1997 and 1999, a firm was involved in co-operative and collaborative arrangements with other firms or organizations to develop new or significantly improved products or processes, and 0 otherwise*

Source: Statistics Canada, Survey of Innovation 1999. Tabulated by authors.

Table 12: Coefficients of Reliability (Cronbach's Alpha) for Variables, Including Multiple-Item Scales

Names of Variables	*Number of Cases*	*Number of Items in Scales*	*a*
Internal Sources of Information (INTSOUR)	5,263	4	0.66
Market Sources of Information (MARSOUR)	5,263	5	0.48
Research Sources of Information (RESSOUR)	5,263	3	0.60
Generally Available Sources of Information (INFSOUR)	5,263	3	0.63
Competitive Environment (COMPET)	5,263	11	0.62
Innovative Practices (INNOPRA)	5,263	5	0.81
Government Support Programs (GVTSUPP)	5,263	12	0.67
Barriers to Knowledge Exchange (OBSTACLE)	5,263	14	0.62

Source: Statistics Canada, Survey of Innovation 1999. Tabulated by authors.

5. RESULTS

The results of the regression of the degree of novelty of the most important innovation developed or improved by firms — that is, a world-first or Canada-first introduction, or a firm-first introduction, identified as the reference category in the model — are summarized in Tables 13a and 13b, respectively, for sources of information included one by one in the model and for sources of information included as indices of variety. The two equations had acceptable predictive power, with 61.9% and 61.5% of correct predictions, respectively. The value of the Cox and Snell R^2 was 0.17 and 0.14, respectively, which is quite reasonable for qualitative dependent variable models. Furthermore, the computed values of the likelihood ratio (i.e., 876.7 and 747.3) were much larger than the critical value of the chi-squared statistic at the 1% level, with 48 and 26 degrees of freedom, respectively. This suggests that the null hypothesis — that all the parameter coefficients (except the intercept) were zeros — was strongly rejected. Consequently, the two models were significant at the 1% level.

5.1 Impact of the Individual Sources of Information

The results reported in Table 13a show that, in comparison to firms introducing firm-first innovations, firms developing world-first or Canada-first innovations were more likely to use their R&D and marketing staff to develop or improve their products or processes. No significant impacts were found regarding the use of production and management staff.

As for market sources of information, the results of the estimation show that firms that had introduced world-first or Canada-first innovations were more likely than firms that had introduced firm-first innovations to use related firms in the same corporate group as sources of information for innovation. The negative relationship between competitors as a source of information and the introduction of world-first or Canada-first innovations means that the use of competitors as a source of information decreased the likelihood of developing such innovations. Likewise, the use of suppliers of equipment, material and components as a source of information decreased the likelihood of developing world-first or Canada-first innovations. Firms that had introduced world-first innovations were more likely than firms that had introduced firm-first innovations to use clients as a source of information for innovation. Consultancy firms were found to exert no significant impact on the degree of novelty of innovation.

The Survey of Innovation 1999 identified three research sources that firms could use to develop or improve their products or processes. Of these sources, only federal agencies and research laboratories were found to influence the degree of novelty of innovation. More specifically, the results reported in Table 13a show that, in comparison to firms introducing firm-first innovations, firms introducing world-first innovations were more likely to use federal and provincial agencies and research laboratories to develop their innovations. No significant impacts were found regarding the use of universities and colleges as research sources.

Of the three sources of generally available information that firms could use to develop or improve their products or processes, two were found to exert no impact on the degree of novelty of innovation. The results show that firms introducing world-first innovations were more likely than other firms to use trade fairs and exhibitions as sources of information.

The column headed Exp (β) in Table 13a shows that, compared to firms that did not use R&D and marketing staff as sources of information, firms that did use these sources were, respectively, 2.11 and 1.42 times more likely to develop world-first innovations and 1.52 and 1.35 times more likely to introduce Canada-first innovations than firm-first innovations.

Compared to firms that did not use related firms in the same corporate group as a source of information for innovation, firms that did use this source of information were 1.26 times more likely to introduce world-first innovations and 1.49 times more likely to introduce Canada-first innovations than firm-first innovations.

Compared to firms that did not use competitors as a source of information for innovation, firms that did use this source were 0.78 times less likely to introduce Canada-first innovations than firm-first innovations. Again, in comparison to firms that did not use clients as a source of information for innovation, firms that did use this source were 0.81 times less likely to introduce world-first innovations than firm-first innovations.

As for research sources, Table 13a shows that, compared to firms that did not use federal and provincial agencies and research laboratories as sources of information for innovation, firms using these sources were 1.59 and 1.44 times more likely, respectively, to introduce world-first innovations than firm-first innovations.

Finally, firms using trade fairs and exhibitions as sources of information for innovation were 1.25 times more likely than firms that did not use these sources to introduce world-first innovations than firm-first innovations.

5.2 Impact of the Sources of Information as Indices of Variety of Sources

In the second model, the sources of information used by firms to develop or improve their products or processes were introduced as indices of variety of sources. As indicated above (see Table 11), these indices resulted from the addition of the number of different sources of information used by firms within four broad categories: internal sources; market sources; research sources; and generally available sources of information. These four indices were indicators used to measure the variable network that was hypothesized as one of the crucial determinants of innovation in network theories of innovation. Three out of the four forms of networks identified have been found to influence novelty of innovation significantly. Hence, firms using a greater variety of internal sources of information to develop innovations were more likely to develop world-first or Canada-first innovations. Moreover, firms using a greater variety of research sources to develop their innovations were more likely to develop world-first innovations. Finally, the negative relationship between the variety of market sources and novelty of innovation means that, as firms increased the number of different market sources they used to develop their innovations, the likelihood that they would develop world-first innovations rather than simply firm-first innovations decreased (Table 13b).

The column headed Exp (β) in Table 13b shows that an increase by one unit in the index of internal sources increased by 1.30 times the likelihood that firms would develop world-first innovations, and by 1.25 times the likelihood that they would develop Canada-first innovations, rather than firm-first innovations.

An increase by one unit in the index of research sources increased by 1.32 times the likelihood that firms would introduce world-first innovations rather than firm-first innovations.

Finally, an increase by one unit in the index of market sources decreased by 0.82 times the likelihood that firms would introduce world-first innovations rather than firm-first innovations.

5.3 Impact of the Other Explanatory Variables

The results reported in Tables 13a and 13b indicate that, in comparison to firms introducing firm-first innovations, firms introducing world-first or Canada-first innovations were more likely to carry out R&D activities, operate in industries in the medium-tech sector, engage in a greater variety of innovative practices, use a greater variety of

Table 13a: Multinomial Logit Estimation for the Factors Influencing Novelty of Innovation in the Development or Improvement of Products or Processes by Canadian Manufacturing Firms (Sources of Information Introduced Separately)

Independent Variables[a]	*Dependent Variable (Degree of Novelty of Innovation)*			
	World-First/First-First		*Canada-First/Firm-First*	
	Coefficient (β)	*Exp* (β)	*Coefficient* (β)	*Exp* (β)
Intercept	-2.611		-2.177	
Sources of Information (Binary Variables:1 = YES; 0 = NO)				
Internal Sources				
• R&D Staff	0.748*	2.113	0.418*	1.519
• Marketing Staff	0.352*	1.422	0.299*	1.349
• Production Staff	0.055	1.057	0.058	1.060
• Management Staff	-0.055	0.946	0.063	1.065
Market Sources				
• Related Firms in Same Corporate Group	0.234**	1.264	0.401*	1.493
• Suppliers of Equipment, Material and Components	-0.558*	0.572	-0.193**	0.825
• Clients	-0.209**	0.811	-0.017	0.983
• Competitors	-0.298*	0.742	-0.252*	0.777
• Consultancy Firms	-0.172	0.842	0.002	1.002
Research Sources				
• Universities and Colleges	-0.050	0.951	0.119	1.127
• Federal Agencies and Research Laboratories	0.467*	1.595	0.076	1.079
• Provincial Agencies and Research Laboratories	0.364**	1.440	0.035	1.036
Generally Available Sources				
• Trade Fairs and Exhibitions	0.225**	1.252	0.138	1.148
• Internet- or Computer-Based Information Networks	0.017	1.017	-0.161	0.851
• Professional Conferences, Meetings and Publications	-0.105	0.900	0.019	1.019
Competitive Environment (COMPET)	-0.046*	0.955	-0.024*	0.976
Barriers to Knowledge Exchange (OBSTACLE)	0.035	1.035	0.019	1.020
Government Support Programs (GVTSUPP)	0.105*	1.111	0.061**	1.063
Number of Employees (TOTEMPL)	0.0003*	1.000	0.0001	1.000
Innovative Practices (INNOPRA)	0.193*	1.213	0.243*	1.275
Collaborative Arrangements (USECOLL)	0.593*	1.809	0.231*	1.260
High-Tech Sector (BINHIGH)	0.787*	2.197	0.226	1.253
Medium-Tech Sector (BINMED)	0.598*	1 818	0.248*	1.282
R&D Activities (BINR&D)	0.840*	2.317	0.314*	1.369
Number of Cases (Total = 4,826)	717/2,927		1,182/2,927	

Cox & Snell R^2: 0.166
Chi-square (df): 876.75 (48)
Overall percentage of correct predictions: 61.9%

Source: Statistics Canada, Survey of Innovation 1999. Tabulated by authors.

[a]See Table 11 for the definition of the independent variables.

*, ** and *** indicate that the variable was significant at the 1%, 5% and 10% level, respectively.

Table 13b: Multinomial Logit Estimation for the Factors Influencing Novelty of Innovation in the Development or Improvement of Products or Processes by Canadian Manufacturing Firms (Sources of Information Aggregated as Indices)

Independent Variables[a]	*Dependent Variable (Degree of Novelty of Innovation)*			
	World-First/First-First		*Canada-First/Firm-First*	
	Coefficient (β)	*Exp (β)*	*Coefficient (β)*	*Exp (β)*
Intercept	-2.938		-2.298	
Sources of Information (Indices of Variety)				
• Internal Sources (INTSOUR)	0.266*	1.304	0.224*	1.251
• Market Sources (MARSOUR)	-0.193*	0.825	-0.007	0.993
• Research Sources (RESSOUR)	0.275*	1.317	0.093	1.098
• Generally Available Sources (INFSOUR)	0.013	1.013	-0.027	0.973
Competitive Environment (COMPET)	-0.045*	0.956	-0.025*	0.975
Barriers to Knowledge Exchange (OBSTACLE)	0.032	1.033	0.016	1.016
Government Support Programs (GVTSUPP)	0.120*	1.127	0.066*	1.068
Number of Employees (TOTEMPL)	0.0003*	1.000	0.0002	1.000
Innovative Practices (INNOPRA)	0.186*	1.204	0.238*	1.269
Collaborative Arrangements (USECOLL)	0.610*	1.840	0.273*	1.314
High-Tech Sector (BINHIGH)	0.904*	2.470	0.250	1.284
Medium-Tech Sector (BINMED)	0.654*	1.923	0.264*	1.302
R&D Activities (BINR&D)	1.093*	2.983	0.416*	1.516
Number of Cases (Total = 4,826)	717/2,927		1,182/2,927	

Cox & Snell R^2: 0.143
Chi-square (df): 747.31 (26)
Overall percentage of correct predictions: 61.5%

Source: Statistics Canada, Survey of Innovation 1999. Tabulated by authors.

[a]See Table 11 for the definition of the independent variables.

*, ** and *** indicate that the variable was significant at the 1%, 5% and 10% level, respectively.

government support programs, and be involved in collaborative arrangements with other firms or organizations to develop or improve their products or processes.

Firms introducing world-first innovations were more likely than other firms to have a larger number of employees and to operate in industries in the high-tech sector.

Finally, the negative relationship between the competitive environment index and the degree of novelty of innovation suggests that an increase in competitive pressure decreased the likelihood of the development of world-first and Canada-first innovations.

The columns headed Exp (β) in Tables 13a and 13b show that, compared to firms that did not carry out R&D activities, firms that did carry out such activities were 2.32 (2.98 in Table 13b) times more likely to introduce world-first innovations and 1.37 (1.52 in Table 13b) times more likely to introduce Canada-first innovations than firm-first innovations.

Table 13a shows that an increase by one unit in the index of government support programs increased by 1.11 times the likelihood that firms would introduce world-first innovations and by 1.06 times the likelihood that they would introduce Canada-first innovations rather than firm-first innovations. Similarly, the results of the second model, presented in Table 13b, indicate that the marginal impacts of these factors were 1.13 and 1.07, respectively.

The columns headed Exp (β) in Tables 13a and 13b indicate that an increase by one unit in the index of innovative practices increased by 1.21 times the likelihood of developing world-first innovations and by 1.27 times the likelihood of developing Canada-first innovations.

Finally, the column headed Exp (β) in Table 13a shows that firms operating in the high-tech and medium-tech sectors were, respectively, 2.20 and 1.82 times more likely to introduce world-first innovations than firms in the low-tech sector. In the second model (Table13b), firms in the high-tech and medium-tech sectors were, respectively, 2.47 and 1.92 times more likely to develop world-first innovations than firms in the low-tech sector.

6. DISCUSSION AND POLICY IMPLICATIONS

Some of the results of the study presented here shed new light on innovation, suggesting that the fact that most prior studies of innovation have focused their attention on the likelihood (the presence or absence) of innovation rather than the degree of novelty of innovation needs discussion. The selection of the degree of novelty of innovation instead of likelihood of innovation as a dependent variable was a fruitful choice, given that the results of the estimations of the two multinomial logit models revealed that different degrees of novelty of innovation were predicted by different explanatory variables. The use of a new dependent variable has become a pressing need in a context where, as in Canada, more than 80% of manufacturing firms are innovative, and more than 60% of firms develop innovations that are firsts for them. Now that a very large proportion of firms are innovative, the next policy challenge is to find ways to increase the degree of novelty of innovation, in order to have more firms developing world-first or country-first innovations.

The impact of the sources of information used by firms to develop or improve their most important innovation was estimated in two different ways: at the level of the individual sources and at the network level. When the impact of the various individual sources of information on innovation is considered, the results of the study described here confirm the results of prior innovation studies in showing the influence on innovation of sources such as the following: R&D staff; marketing staff; related firms in the same corporate group; suppliers of equipment, material and components; clients; competitors; federal and provincial agencies and research laboratories; and trade fairs and exhibitions. However, contrary to prior studies based on the client-supplier network theories of innovation, this study suggests that the use of suppliers of equipment, material and components, clients and competitors as sources of information to develop or improve products or processes exerts a negative impact on the novelty of innovation. This means that firms using these sources are less likely to introduce world-first or Canada-first innovations than firm-first innovations. Compared to firms that introduce firm-first innovations, firms introducing world-first or Canada-first innovations are more likely to use R&D staff, marketing staff and federal agencies and research laboratories.

In the second model, the sources of information were introduced as indices of variety in order to capture a network effect. Following Granovetter (1973), it was assumed that "weak ties" linking firms to the many different categories of sources of information are more useful for developing or improving products or processes than "strong ties" such as those linking firms only to their clients or suppliers. In theory, firms strongly tied to a small variety of sources of information are more likely to come across the same information than firms weakly tied to a large variety of sources of information; firms with "weak ties" are more likely to hear of opportunities that are different from those encountered by firms with "strong ties." According to Burt (1992), access to a greater variety of sources of information increases non-redundancy of information while providing firms with the chance to receive more timely information about opportunities for the development or improvement of products or processes. A large variety of sources of information provides benefits that are additive rather than redundant. According to Burt (2000: 256), "Certain network forms deemed social capital can enhance the manager's ability to identify and develop opportunities."

This idea of variety of sources of information was operationalized through four forms of network: variety of internal sources of information; variety of market sources of information; variety of research sources of information; and variety of generally available sources of information. The impact of the variety of generally available sources of information was found to have no significant relationship to novelty of innovation in firms. However, firms that could count on a greater variety of internal sources of information were found to be more likely to introduce world-first or Canada-first innovations than other innovative firms. Likewise, firms that used a greater variety of research sources were more likely to develop world-first innovations than other innovative firms. However, the results of the second model sound a note of caution regarding the impact of market sources of information on innovation. The fact that the results of the study presented here suggest that the use of market sources of information decreased the likelihood of developing world-first innovations does not mean that market sources do not influence the likelihood (the presence or absence) of innovation. However, the results of the study suggest that the impact of market sources of information on innovation found in other studies does not hold for world-first innovations. Indeed, these results simply suggest that innovations embodying more radical changes in products or processes require more research-based information than incremental changes that can be implemented with market sources of information, which are much less research-based. The policy implications deriving from these results are very important: radical innovations could be increased by developing policies promoting stronger linkages between firms and government laboratories and universities, whereas incremental changes might be more efficiently influenced by stronger linkages between firms and market sources of information.

As expected, R&D activities, collaborative arrangements, a large number of employees, use of a great variety of government support programs, and operation in the high-tech sector had a positive effect on novelty of innovation. These results suggest that, in order to foster novelty of innovation in manufacturing firms, policy-makers should continue to offer incentives regarding R&D and collaborative arrangements, and, more generally, provide a large variety of government support programs.

The absence of a significant relationship between barriers to knowledge exchange and novelty of innovation provides a very important policy lesson: policy-makers should focus attention on missed opportunities, on policy interventions that would induce firms to seize opportunities for innovation that they would not otherwise seize. The results of the study, showing that higher degrees of novelty of innovation are related to recourse to a greater variety of sources of information, suggest that policy-makers should develop measures that would broaden the variety of sources of information used by firms in the development or improvement of their products or processes.

Appendix 1: Distribution of Enterprises in Manufacturing Industries, by Technological Intensity

Description of Industries	*Total*
1. **High-Tech**	
Pharmaceutical and Medicine Manufacturing	62
Computer and Peripheral Equipment Manufacturing	45
Communications Equipment Manufacturing	66
Audio and Video Equipment Manufacturing	5
Semiconductor and Other Electronic Component Manufacturing	49
Aerospace Product and Parts Manufacturing	71
Total	**298**
2. **Medium-Tech**	
Petroleum and Coal Products Manufacturing	54
Basic Chemical Manufacturing	111
Resin, Synthetic Rubber, and Artificial and Synthetic Fibres and Filaments Manufacturing	42
Pesticide, Fertilizer and Other Agricultural Chemical Manufacturing	40
Paint, Coating and Adhesive Manufacturing	72
Soap, Cleaning Compound and Toilet Preparation Manufacturing	55
Other Chemical Product Manufacturing	83
Plastics and Rubber Products Manufacturing	546
Non-Metallic Mineral Product Manufacturing	370
Primary Metal Manufacturing	217
Fabricated Metal Product Manufacturing	1,231
Agricultural, Construction and Mining Machinery Manufacturing	147
Industrial Machinery Manufacturing	89
Commercial and Service Industry Machinery Manufacturing	41
Ventilation, Heating, Air-Conditioning and Commercial Refrigeration Equipment Manufacturing	122
Metalworking Machinery Manufacturing	175
Engine, Turbine, and Power Transmission Equipment Manufacturing	52
Other General Purpose Machinery Manufacturing	210
Navigational, Measuring, Medical and Control Instruments Manufacturing	118
Manufacturing and Reproducing Magnetic and Optical Media	15
Electrical Equipment, Appliance and Component Manufacturing	214
Motor Vehicle Manufacturing	27
Motor Vehicle Body and Trailer Manufacturing	98
Motor Vehicle Parts Manufacturing	183
Railroad Rolling Stock Manufacturing	10
Ship and Boat Building	58
Other Transportation Equipment Manufacturing	9
Miscellaneous Manufacturing	404
Total	**4,790**
3. **Low-Tech**	
Food Manufacturing	821
Beverage and Tobacco Product Manufacturing	94
Textile Mills	164
Textile Product Mills	134
Clothing Manufacturing	517
Leather and Allied Product Manufacturing	78
Sawmills and Wood Preservation	282
Veneer, Plywood and Engineered Wood Product Manufacturing	114
Other Wood Product Manufacturing	292
Paper Manufacturing	291
Printing and Related Support Activities	560
Furniture and Related Product Manufacturing	485
Total	**3,833**
TOTAL	**8,921**

Source: Innovation Policy Branch, Industry Canada, based on Statistics Canada's Survey of Innovation 1999.

Notes

The authors thank Pierre Therrien from Industry Canada, and Fred Gault, Brian Nemes, Frances Anderson and Susan Schaan from Statistics Canada for their support and advice. The authors alone are responsible for the opinions expressed in this chapter and for any errors and omissions that may remain.

[1]For methodological issues and an overall description of the survey, see Schaan and Anderson (2001).

5

Impact of Local Collaboration on Firms' Innovation Performance

Pierre Therrien and Vanessa Chang

1. INTRODUCTION

Competition has intensified over the last few decades with both the growing openness of the global economy and the reduced time to market demanded by new and increasingly complex technologies. Producing, or at least adopting, innovations is now recognized as crucial for firms wishing to remain at the forefront of competition. As competition now arises from foreign as well as from domestic sources, firms must keep abreast of new technologies available worldwide.

While the demand for speed to market increases, this new generation of technologies, which fuse recent advances from formerly distinct fields into new ones, such as mechatronics and photonics, has made the development of novel ideas more and more complex. This shift from a single to a multidisciplinary approach to the creation and application of knowledge has made it much more difficult for a single company to cover all disciplines, as was done in the past, and vertical disintegration has been observed (Wintjes and Cobbenhagen 2000).

As a result, firms — even large ones — have become more specialized in their core competencies, and consequently they have had to outsource or collaborate with others in order to acquire complementary knowledge and technologies. Technological collaboration allows firms to access new knowledge and markets, as well as to share costs and risks (OECD 2000a). Two distinct types of collaboration — international and local — are emerging, each with different intentions and results.

The first, international collaboration, is spurred on by the proliferation of faster/cheaper transportation and communication systems. Global communication networks and markets enable a firm to source information efficiently from a distance. For example, in 1995 Texas Instruments' high-speed telecommunications chip was created through an international effort. It was conceived by engineers from the Ericsson Telephone Company in Sweden, and designed in Nice with software tools developed by the company in Houston. The chip was then rolled off production lines in Japan and Dallas, tested in Taiwan, and wired into Ericsson line-cards which monitor telephone systems in Sweden, the United States, Mexico and Australia (Quah 1995).

The second type, local collaboration, is driven by a growing awareness of the benefits derived from tacit knowledge and exchange of ideas, which are promoted through face-to-face contact and close geographical proximity. To flourish, local collaboration on innovation needs strong local knowledge infrastructures, such as universities, research centres and geographically concentrated interdependent firms.

This chapter presents the results of a study that investigated the impact of collaboration on the innovation performance of firms in Canadian manufacturing industries. It distinguished between different aspects of collaboration, based on whether participation was local or not, and examined the potentially different impacts of this on firms' innovation performance.

2. LITERATURE SURVEY

2.1 Collaborative Efforts

Several reasons for firms to participate in collaborative efforts on innovation emerge from a literature survey. The most frequently mentioned benefit derived from collaboration is the mutual exchange of ideas and thoughts, which leads to novel ideas and discoveries. Studies from Sweden (Melin 2000; Okubo and Sjoberg 2000) show that collaborations lead to work of higher scientific quality, and that interactions among researchers tend to open up new ways of thinking and produce new ideas. They also point out that firms can continue to specialize in specific areas, knowing that other competencies that they lack are readily available when needed.

The last point, which describes the merging of outside competencies with a firm's own abilities, has gained in importance in the last few years as an explanatory factor for innovation performance. In a literature review, the OECD (2000a) concluded that collaboration is now becoming an essential part of innovation activities because the increasing complexity of technologies and the need for speed to market make it impossible for a single firm to master all the specific fields required to innovate. Leiponen (2000a) points out that, nowadays, performing R&D alone is insufficient for producing profitable innovations. But combining a firm's own R&D capabilities with those of other firms can result in improved innovation performance. Firms with strong internal capabilities have been observed to outpace other firms (in terms of innovative and financial performance) when they engage in collaboration. A strong innovative capacity can be developed through seeking and identifying complementarities between employees' skills, internal R&D investment and other essential innovation activities, such as training, acquisition of machinery and equipment, and tooling up. Therefore, the choices made as to how innovation activities are organized, in order to attain the greatest benefit, may strongly influence the success or failure of a firm's innovation performance.[1]

The main reason that firms choose to be involved in collaboration arrangements is to capture "missing" knowledge that is embodied in other workers, firms or organizations. Drastic improvements in telecommunications tools have facilitated information flows, and nowadays a firm can choose its best partner, regardless of the distance separating them. However, if the shared knowledge cannot be easily transferred, distance can still matter, because the flow of ideas and the transmission of knowledge between different groups tend to decrease as the geographical distance between them increases. Therefore, the nature and complexity of the knowledge to be shared can modify a firm's behaviour regarding the choice of its partners. Evidence of this behaviour is seen in a study done by the Conference Board of Canada (2000), showing that collaboration with public organizations, such as universities and government laboratories, tends to decrease as distance increases. However, this trend is not perceived when firms collaborate with other private firms, such as competitors or clients.

To understand the behaviour of collaborating firms, a distinction needs to be made between information and (tacit) knowledge. Unlike information, which is highly codified and easily acquired, tacit knowledge is disembodied, informal and needs "constant effort" to acquire.[2] For example, Nauwelaers and Wintjes (2000: 5) state that "knowledge differs from innovation. ... distance does not seem to be a barrier to the transmission of information, but in the transmission of knowledge, it does." This is in line with the assumption that transmission of tacit knowledge (which is ubiquitous in institutions such as universities and government laboratories) decreases with distance. Baptista and Swann (1998: 528) expand this argument:

> So long as much technological knowledge has a tacit nature and cannot be codified through plans, instructions or scientific articles, it seems reasonable to expect a greater concentration of innovators. This type of knowledge can only be learned through everyday practice and use of a technology, and its transmission relies for the most part on informal personal contact.

Geographical proximity matters to knowledge spillovers and the exchange of ideas, and tacit visions, leading to coordinated investment decisions that provide a strong knowledge base for that particular region. Local collaboration — considered as a formal type of network — contributes to the transfer of information and knowledge generated by the close proximity of interrelated firms and organizations.

2.2 Concentration of Innovators and Technological Infrastructure

In contrast to a simplistic linear model, in which the origin of an innovation can always be traced back to basic research that drives scientific discoveries to product development, a more complex and dynamic model has been emerging. Kline and Rosenberg (1986) introduced the "chain-link" model, in which research, product testing and development, and market research are all interdependent, and dynamic feedback among all actors takes place at various stages in the innovation process.

According to this model, innovation may be initiated by any actors at any stage, and it tends to be circular rather than sequential.

The "chain-link" model not only recognizes the importance of a firm's R&D capabilities, but also stresses the significance of other sources of knowledge in the innovation process. Feldman (1994) has identified four sources of knowledge in the innovation process: university research, industrial R&D, related industry presence and business services. University research produces basic knowledge, which can then be developed and refined through industrial research, while business services provide financial and commercial knowledge.[3] Together, these resources and the interactions they generate constitute a technological infrastructure that supports innovative activity.

Because of the cumulative nature of the knowledge required to innovate, and the way in which it is possible to accumulate this knowledge (according to the "chain-link" model), innovation activities tend to be geographically concentrated. A region with strong innovation assets has a comparative advantage, and, since knowledge is cumulative, this advantage is self-reinforced and leads to a geographical concentration of innovative activity (Arthur 1990). Therefore, the cumulativeness of innovation is also highly relevant in geographical patterns of innovative activities and performance.

Local collaboration can be viewed as a formal attempt to take advantage of the technological infrastructure of a region. Being engaged in collaborative agreements with local partners then allows the optimal transfer of tacit knowledge, leading to innovative breakthroughs. Also, local collaboration is not the only means by which to appropriate the knowledge generated by technological infrastructure: seminars, professional conferences and chamber of commerce meetings can also be important sources of knowledge. However, a formal partnership — such as local collaboration — implies a strong commitment on the part of both partners, and it allows disclosure of sensitive information that would not be diffused through informal networks.

The literature also identifies problems associated with collaborative arrangements. For example, "groupthink" can slow down the rate of innovation, as closely related firms start to share a uniform approach to innovation and competition. "Groupthink" tends to reinforce old behaviours, suppress new ideas and create rigidities that may hinder improvements. Truly radical innovation may be discouraged under these conditions (Horibe 2001). Another problem connected with collaboration is that, under some conditions, there may be opportunities and incentives for free riders (information asymmetry associated with any type of knowledge-sharing), or for one party to extract more of, or even expropriate all of, the gain (Nooteboom 1999). Similarly, the concentration of firms in one location may have the effect of rendering human resources scarce and "bid up" their costs. There may also be other costs associated with the increased congestion created within a growing cluster, which may have a negative effect on innovation (Porter 1998).

3. DATA[4]

To assess the impact of different types of collaboration on a firm's innovation performance, the study reported in this chapter used data from the Survey of Innovation 1999. The sample consisted of approximately 6,000 provincial enterprises from 31 manufacturing industries. The questionnaire was designed to identify innovative firms from the period 1997-99, as well as to analyze the innovation process itself. Respondents were asked to define the novelty of their most important innovation as (a) a world first, (b) a first in Canada or (c) a first for the firm, as well as to identify the percentage of sales derived from their new or significantly improved products.

The use of direct innovation measures, such as the importance of innovation or the percentage of increased sales from innovation, presents some advantages over such indirect measures as R&D expenditures or patents. As it is now well recognized that R&D expenditure is not the sole path to innovation, using this measure alone would draw an incomplete picture of innovation activities in the economy. Similarly, the propensity to patent varies across industries (possibly denoting different technological opportunities among industries, as well as different strategies used to protect intellectual property), and therefore data on patents cover only a portion of innovative activities in the economy.

However, the direct measures used by the survey to identify innovators and the importance of innovations were susceptible to the subjective biases of chief executive officers (CEOs).[5] A further examination would be useful to determine whether the new or significantly improved products/processes reported were technically innovations or not. Moreover, depending on the size of a firm, the perception of an innovation's importance could be biased (larger firms are expected to devote more resources to strategic intelligence[6] than smaller firms and therefore would be better informed about the true importance of their own innovation). However, tests for the reliability of the variables related to novelty of innovation revealed that no specific bias should be expected.[7]

Innovative firms were asked if they were actively involved in collaboration/co-operation agreements with other firms (such as suppliers, customers or competitors) or with public-sector organizations (such as universities or provincial or federal laboratories) to develop new or significantly improved products/processes. Moreover, each firm involved in collaboration was asked whether its collaborators were located in Canada, the United States, Europe or the Pacific Rim. It should be noted that sub-questions regarding the location of collaborators were problematic when the firm had more than one Canadian location. For single-location firms, collaboration in Canada was broken down as follows: within 100 kilometres; in the same province; in the rest of Canada; and outside Canada. Local collaboration was defined as collaboration with another firm/organization located within 100 kilometres of the firm. However, this breakdown of Canadian collaboration was not possible for multi-location firms.

Therefore, first, the complete subsample of innovative firms was used to assess the impact of collaboration (as a whole) on performance. Second, the subsample of multi-location collaborative firms was removed so that the specific effect of local collaboration on an innovative firm's performance could be estimated.[8]

Each manufacturing firm was assigned to a specific region and industry. In the end, there were 31 industries (in accordance with the 1997 North American Industry Classification System (NAICS)) in five regions: the Atlantic, Quebec, Ontario, the Prairies and British Columbia. The size of each firm, defined by the number of employees, was also available. See Table A1.1 in Appendix 1 for more details on the region and industry definitions used.

4. ECONOMETRIC APPROACH

In the study reported in this chapter, the effect of collaboration on firms' innovation performance was analyzed for the period 1997-99. Performance can be defined in several ways. It can be measured by innovativeness or the number of innovations that a particular firm brings to the market; by the increase in sales that result from an innovation; by the importance of an innovation; or by other direct and indirect impacts of innovation, such as increased productivity, profitability, or greater international market share.[9]

Using the Survey of Innovation 1999, the variables defined above were all categorized. For instance, the categories "world-first," "Canada-first" and "firm-first" were used to define novelty of innovation. The share in sales of new or significantly improved products was categorized as "1-5%," "6-25%," "26-50%," "51-75%" and "76-100%."[10]

Working with such discrete (or categorized) dependent variables required a specific econometric method. The appropriate technique was to use either the ordered logit or the probit model, depending on the assumption about the distribution of the error term.[11]

Moreover, as noted in section 3, only innovative firms (or, at least, firms that tried to innovate) answered questions regarding the importance of innovation, and sales resulting from innovation and collaboration. To obtain unbiased estimates, the two-stage estimator method designed by Heckman was used (see Greene 1997, chapter 20). In the first stage, a probit regression was estimated to obtain the probability of being innovative; in the second stage, the inverse Mills' ratio was computed and inserted into the regression on the importance of innovation. The inverse Mills' ratio took into account the fact that only a non-random subsample (of innovators) in the whole population was being analyzed.

Therefore, the two equations were:

$$(1)\quad innov^* = \Psi' W + \mu$$

$$innov = 1 \text{ if } innov^* > 0 \text{ which implies that } \mu > -\Psi' W$$
$$= 0 \text{ otherwise}$$

where *innov** is a latent variable for the observed variable *innov*, *W* is a set of explanatory variables and the error term is $\mu \sim N(0, \sigma u^2)$, and

$$(2)\ importance\_inn^* = \beta' X + \varepsilon$$

where *importance_inn** is observed only if *innov* = 1, *X* is a set of explanatory variables, and the error term ε is normally distributed with zero mean and variance σe^2. Both μ and ε follow a bivariate normal distribution with zero means, variance of 1 and correlation ρ. Correction for the bias generated was obtained by computing the inverse Mills' ratio from equation (1) and inserting it into equation (2). Equation (2) becomes:

$$(2')\ importance\_inn^* = \beta' X + \beta_\lambda \lambda(\Psi' W) + v$$

where $\lambda(\ )$ is the inverse Mills' ratio computed from equation (1) and β_λ estimates ρ, which is the correlation between μ and ε.

4.1 Explanatory Variables

The explanatory variables for the model were: industrial sectors; technological infrastructure; size of firm; innovation activities; and collaboration.

4.1.1 Industrial Sectors

The technological environment influences a firm's innovation performance. Even though the capability to transform technological opportunities into successful innovations is intrinsic to a firm, it is well understood that each firm in a specific industry faces a similar environment in terms of opportunity and appropriability, degree of cumulativeness of knowledge and type of knowledge required to run a business successfully. However, because this technological environment varies across industries, its effects on firms' innovation performance will also vary across industries. Dummies representing industrial sectors — low-, medium- and high-tech intensity — were used to capture the industrial effect.[12]

4.1.2 Technological Infrastructure

Ideally, the impact of technological infrastructure on firms' innovation performance would be measured by an index using the aforementioned four sources of knowledge: university research, industrial R&D, related industry presence and business services. However, the location of a firm (dummies by cities, regions or provinces) can be a proxy for technological infrastructure. The finest measure for the location of a firm used by the Survey of Innovation 1999 was the province. Even though a more precise location would have been preferable, the province in which a firm was located was still valuable for an understanding of the innovative process and the influence of location on firms' innovation performance. Dummies representing the Atlantic region (New Brunswick, Nova Scotia, Prince Edward Island and Newfoundland), Quebec, Ontario, the Prairies (Manitoba, Saskatchewan and Alberta) and British Columbia were added in the model.

4.1.3 Size of Firm

Attempts to innovate are not without cost. Results from the Survey of Innovation 1999 show that the two major impediments to innovation were the high cost of developing innovation and the inability to devote enough staff to the development of new products/processes because of current production demands (Statistics Canada 2001). This, in combination with the observation that larger firms (with greater financial and human capacities) can fund research more easily, indicates that size matters in innovation. Size of firm was represented by three categories, according to the number of employees: small (20-49 employees), medium (50-249 employees) and large (more than 250 employees).

4.1.4 Innovation Activities

Combining several innovation activities would increase a firm's innovation performance if those activities were complementary. Firms that are able to combine these activities — innovation management, R&D, training, acquisition of machinery and equipment, collaboration with private or public partners, and use of government support programs — optimally are expected to outperform their counterparts in the industry. Innovation activities included in the model were R&D, acquisition of machinery, industrial engineering and industrial design, tooling up, and training linked to innovation.

4.1.5 Collaboration

Collaboration can be viewed as another innovation activity. Because of the increasing complexity of technologies and the need for speed to market, collaboration is becoming an essential part of a firm's strategy for innovation and, therefore, it was studied separately from other innovation activities. A firm reporting active participation in joint projects to develop new or significantly improved products/processes was defined as collaborative.

4.2 Basic Models

The basic models (in two stages) to be estimated were as follows.

Stage 1 — Probit on innovation:[13]

$$Innovation = \alpha_0 + \sum_j \alpha_j Ind_j + \sum_m \alpha_m Region_m + \pi_S \; Succes\_factors + \pi_e \; Environment + \pi_g Gvt\_inn + \mu$$

and

Stage 2 — Ordered probit on importance of innovation:

$$Novelty\_inn = \beta_0 + \beta_1 Coll + \beta_2 Act\_inn + \beta_3 Gvt\_inn + \sum_j \delta_j Ind_j + \sum_m \delta_m Province_m + \beta_\lambda \lambda + v$$

Innovation is whether or not a firm introduced a new or significantly improved product/process, and *Novelty_inn* represents a firm's most important innovation. *Novelty_inn* can take three values: world-first, Canada-first or firm-first. *Coll* is a dummy indicating whether or not a firm was engaged in collaboration, *Gvt_inn* represents government support programs used by a firm and *Act_inn* represents a set of variables related to activities linked to innovation. *Succes_factors* and *Environment* refer respectively to a set of selected success factors of firms and selected variables indicating the competitive environment faced by a firm. *Province* is a proxy for technological infrastructure. $\lambda$is the inverse Mills' ratio computed from the first-step regression, and μ and v are the error terms.

An alternate model testing the effects of different kinds of collaboration (such as collaboration with partners within 100 kilometres and non-local collaboration) was also considered.

5. RESULTS

Table 1 shows that more than 80% of firms introduced a new or significantly improved product/process in the period 1997-99. The regional distribution of innovators shows that the percentage of innovative firms in each province was close to the Canadian average. Ontario and Quebec were slightly above the Canadian average with 83%. The Prairies and British Columbia were the least innovative provinces, but three-quarters of firms in these provinces were still innovative.

Table 1 also reveals that more than half of the entire population of firms came from the medium-tech sector and over 40% came from the low-tech sector, while only 4% came from the high-tech sector. The high-tech sector in Canada is growing, but it is still a small part of the entire Canadian economy.[14] As expected, firms from the high-tech sector were more likely to be innovative: 90% of the firms in this group were innovative, well above the Canadian average (80.7%).

The proportion of small firms that had introduced a new or significantly improved product/ process (75%) was slightly below the overall average, while the proportion of medium-sized firms that had brought an innovation to the market (82%) was approximately the same as the overall average. Large firms were the most innovative (88%).

Table 1 also presents some selected variables reflecting CEOs' perceptions (with regard to such factors as competitive environment and firms' success), as well as firms' behaviour (use of government support programs for innovation). Out of the 16 factors that potentially enhance or explain a firm's success, only two factors were retained: performing R&D within the firm, and developing new products/processes. These two factors have already been identified as the most important for distinguishing innovative from non-innovative firms.[15] Table 1 shows that, for innovative firms, performing R&D within the firm and developing

Table 1: Descriptive Data for Selected Variables – Innovators and Non-Innovators, 1997-99

Variables	*Status of Firm*		*All Firms*	
	Innovators	*Non-Innovators*	*%*	*N*
	80.7%	19.3%		
Region				
Atlantic	77%	23%	100	451
Quebec	83%	17%	100	3,125
Ontario	83%	17%	100	2,931
Prairies	74%	26%	100	1,106
British Columbia	74%	26%	100	896
Industry				
Low-Tech industries	79%	21%	100	3,670
Medium-Tech	82%	18%	100	4,552
High-Tech	90%	10%	100	287
Size of Firm				
20-49 Employees	75%	25%	100	2,598
50-249 Employees	82%	18%	100	4,738
≥ 250 Employees	88%	12%	100	1,173
Activity Related to Innovation				
Use of Gvt. Pgm.	59%	32%	54	4,558
Selected Firm Success Factors				
Performing R&D	59%	24%	52	4,464
Developing New Product/Process	72%	33%	65	5,500
Selected Competitive Environment Factor				
Production Technology Changes Rapidly	43%	23%	39	3,330
Total (weighted)	6,866	1,643		8,509

Source: Statistics Canada, Survey of Innovation 1999. Tabulated by authors.

new products/processes were far more important for success than they were for non-innovative firms. Similarly, using CEOs' perceptions about the competitive environment, the only significant factor that emerged in distinguishing an innovator from a non-innovator was the recognition that production technology changes rapidly.

Table 2 presents descriptive data qualifying the novelty of innovations brought to the market. Fourteen per cent of innovative firms introduced a new or significantly improved product/process that was a world-first, while 25% introduced a Canada-first innovation and 61% introduced a firm-first innovation.[16]

Disaggregating innovative firms by novelty of innovation reveals more internal variance among all variables (such as size, province, etc.) than comparing innovative firms (as a whole) with non-innovative firms. Firms from the Atlantic region, Quebec and British Columbia had fewer world-first innovators than the Canadian average (14%), while Ontario and the Prairies were above the Canadian average. Similarly, small firms (low-tech industries) tended to be below the Canadian average of world-first innovators, while large firms (high-tech industries) were far above the Canadian average.

A clear pattern emerged for activities related to innovation. The following activities were cited more often by world-first innovators than by other innovators: use of government support programs; collaborative agreements; R&D; and industrial engineering and industrial design. However, although several world-first innovators frequently mentioned acquisition of machinery and equipment as an innovation activity, it was not an activity that could be used to distinguish world-first innovators from other types of innovators. It is also worthwhile mentioning that almost all world-first innovators (96%) were engaged in R&D activities.

In summary, it is evident that, on average, world-first innovators were engaged in more activities (4.33) than firm-first innovators (3.72). This corroborates the assumption that complementarity of activities linked to innovation leads to better innovation performance.

These results also suggest that world-first innovators shared particular characteristics. Therefore, it proved possible to construct a profile and identify the best practices of these firms and, more precisely, the impact of engagement in collaboration on firms' innovation performance.

Table 2: Descriptive Data for Selected Variables, by Novelty of Innovation, 1997-99

Variables	*Importance of Innovation*			*All Innovative Firms**	
	World-First	*Canada-First*	*Firm-First*	*%*	*N*
Region					
Atlantic	12%	21%	67%	100	256
Quebec	13%	23%	64%	100	1,772
Ontario	16%	27%	57%	100	1,962
Prairies	16%	24%	60%	100	583
British Columbia	12%	21%	67%	100	485
Industry					
Low-Tech	9%	22%	69%	100	2,063
Medium-Tech	18%	26%	56%	100	2,819
High-Tech	27%	27%	46%	100	177
Size of Firm					
20-49 Employees	10%	20%	70%	100	1,325
50-249 Employees	15%	24%	61%	100	2,930
≥ 250 Employees	19%	34%	47%	100	803
Activities Related to Innovation					
Use of Gvt. Pgm.	77%	69%	55%	62	3,127
Collaboration	54%	41%	29%	36	1,804
Local Collaboration	16%	12%	9%	11	556
R&D	96%	89%	73%	80	4,062
M&E	88%	90%	88%	88	4,461
Engineering & Design	83%	80%	59%	68	3,424
Tooling Up	79%	81%	70%	74	3,740
Training	87%	89%	82%	84	4,270
No. of Innovation Activities	4.33	4.28	3.72		3.95
Total (weighted)	727	1,241	3,090		5,058

Source: Statistics Canada, Survey of Innovation 1999. Tabulated by authors.

*This is a subsample of innovative firms which gave an unambiguous answer regarding the importance of innovation.

5.1 Technological Infrastructure and Collaboration

The first part of Table 3 presents estimates from the first-stage probit on innovation. Being located in Quebec or Ontario increased the probability of innovating (with the Atlantic region as the baseline). Technological infrastructure in these two provinces could make diffusion of technology more effective. This finding is in line with the distribution of research chairs and other federal and provincial programs designed to improve the diffusion and creation of innovation within Canada (such as Networks of Centres of Excellence programs and public laboratories). Quebec and Ontario (primarily Montreal and Toronto) received the lion's share of these programs.[17]

Size of firm mattered given that medium-sized and large firms were more likely to innovate (with small firms as the baseline). As expected, selected variables reflecting CEOs' perceptions were all positive and significant, showing that CEOs who agreed with the statements on success factors were also more likely to belong to innovative firms. Finally, firms that used government support programs for innovation were also more likely to be innovative.

Surprisingly, being a high-tech firm did not have a significant effect on the probability of being an innovative firm (using the low-tech sector as the baseline). However, it should be noted that, by definition, a firm did not have to introduce a new or significantly improved product and/or process to the market to be considered innovative, but had only to introduce it into the firm.[18] This definition encompassed the adopter of technology as well as the creator. Fortunately, the questionnaire differentiated between adopters and creators of technology by distinguishing world-first innovators (creators) from firm-first innovators (adopters).

Using the information gathered by the first probit on innovation and correcting the bias generated by the use of a non-random subsample, it was possible to interpret the results from the second-stage ordered probit on the importance of innovation (see Table 3).[19] The model predicted that the probabilities of being world-first and Canada-first innovators were 12% and 25.5%, respectively. These results are in line with the actual frequency for the whole sample (14% and 25%, respectively, for world-first and Canada-first innovators), which validates the model used.

Table 3: Probit Model, Using Collaboration (as a Whole)

	Probit on Innovation		*Ordered Probit on Importance of Innovation*	
	Coefficient	*S.E.*	*Coefficient*	*S.E.*
Intercept	-0.231*	(0.08)	-1.879*	(0.13)
Region				
Quebec	0.136**	(0.08)	-0.037	(0.09)
Ontario	0.203*	(0.08)	0.120	(0.09)
Prairies	-0.071	(0.08)	0.226*	(0.10)
British Columbia	-0.014	(0.09)	0.067	(0.10)
Industry				
Medium-Tech	0.090*	(0.04)	0.288*	(0.04)
High-Tech	0.136	(0.11)	0.299*	(0.10)
Size of Firm				
50-249 Employees	0.108*	(0.04)	0.128*	(0.04)
≥ 250 Employees	0.217*	(0.06)	0.258*	(0.06)
Selected Competitive Environment Factor				
Production Technology Changes Rapidly	0.402*	(0.04)	—	
Selected Firm Success Factors				
Performing R&D	0.373*	(0.04)	—	
Developing New Product/Process	0.681*	(0.04)	—	
Activities Related to Innovation				
Use of Gvt. Pgm.	0.406*	(0.04)	0.136*	(0.04)
No. of Innovation Activities	—		0.115*	(0.02)
Collaboration	—		0.279*	(0.04)
Inverse Mills' Ratio	—		-1.055*	(0.11)
Prob (Y = World-First)	**—**		**0.12**	
Prob (Y = Canada-First)	**—**		**0.255**	
-2 log L	-3,476.75		-4,345.506	
No. Obs. (weighted)	8,509		5,058	

Source: Statistics Canada, Survey of Innovation 1999. Tabulated by authors.

*Significant at the 5% level.

**Significant at the 10% level.

As expected, firms that belonged to medium- and high-tech industrial sectors, as well as large firms, were more likely to produce world-first innovations. These results are not surprising given that, in the high-tech sector, for instance, opportunities to make an innovative breakthrough are more prevalent than in low-tech industries. Moreover, these firms by definition rely more on advances in technology for success and are driven almost exclusively by constant technological change.

Regarding size of firm, the literature points out that the emergence of information technologies has reduced the cost of codifying and diffusing information and, as a result, has lowered cost barriers to innovation (OECD 2000b). Moreover, the complexity of technology would favour highly specialized firms that can move quickly to new technologies. These new opportunities should leave more room for small firms to innovate. However, the results presented in Table 4 show that smaller firms had yet to benefit from these new opportunities, as they were still less likely (regardless of industrial sector) to introduce innovative breakthroughs.

The coefficients for firms located in Quebec and Ontario were no longer significant, while firms from the Prairies were the only ones with a significant (and positive) coefficient. The Prairies constitute a puzzling case because, even though they were one of the least innovative regions (see Table 1), firms in the Prairies, together with those in Ontario, were more likely to introduce a world-first innovation than firms in the other provinces. This result indicates that firms and organizations in the Prairies had been successful in building up technological infrastructures, leading to few but important innovations. However, using provinces or regions as proxies for assessing the impact of technological infrastructure remains hazardous. Provinces are only crude estimators; smaller geographical areas of study are preferable for determining the impact of proximity on the innovation performance of firms. However, at the time of the analysis, more disaggregated measures were unavailable.

Estimates of activities linked to innovation revealed, again, a clear pattern, in which the use of innovative tools positively affected the probability of a firm becoming a world-first innovator. Using government support programs was helpful in introducing world-first innovations. Similarly, increasing the number of activities linked to innovation (such as performing R&D, acquiring machinery and equipment, and training) increased the probability of being a world-first innovator.

Finally, being involved in collaborative agreements was positively related to the probability of introducing an innovative breakthrough to the market. Calculation of the marginal effect of each explanatory variable shows that being involved in collaboration increased the probability of being a world-first innovator by more than 5.8 percentage points.[20] In other words, the average Canadian firm would see its probability of being a world-first innovator increased from 11% to 16.8% if it engaged in collaboration. (It should be recalled that the probability of being a world-first innovator was about 12% with all variables at their mean values.) This marginal effect was larger than other marginal changes involving such variables as using government support programs (a change of 2.7 percentage points), size of firm (2.8 and 3.3, respectively, for medium-sized and large firms), or the number of activities linked to innovation (a change of 2.4 percentage points).[21] One interpretation of these results would be that, because technology is becoming more and more complex, it is unlikely that a firm on its own could easily introduce an innovative breakthrough without external expertise.

5.2 Local Collaboration

Disaggregating collaboration into two parts — namely, local collaboration and other types of collaboration — is not a simple task.[22] First, only firms that already designated themselves as collaborators were asked to describe their collaborative agreements. Second, collaborative firms were split into two categories: firms with only one Canadian location; and firms with more than one Canadian location. Only firms in the former category answered questions about local collaboration.

One solution to assessing the impact of local collaboration on a firm's innovation performance was to limit the sample to single-location firms involved in collaborative agreements. Such a subsample required corrections, because it was not a randomly selected subsample of the whole population. Tests on this restricted subsample were done, and the result was that all regressors used were non-significant. This suggests either that local collaboration had no marginal effect compared to the overall effect of collaboration (whether the collaboration was local, national or international), or that this model suffered from strong collinearity.

Another solution was to use a subsample in which collaborative firms with more than one Canadian location had been removed. Firms in the subsample were regrouped into single-location firms (for which information about local collaboration was available) and all other non-collaborative firms. The model described in Table 3 was recomputed using the latter subsample, and the results show that the effect of each variable remained unchanged (signs of all variables remained the same, and all variables that were significant remained statistically significant) (see Table A1.2 in Appendix 1).[23] This result was sufficiently reassuring for an investigation of the effect of local collaboration on a firm's innovation performance to be continued (see Table A1.3 in Appendix 1 for descriptive data, by different types of collaboration, using the subsample of single-location firms and other non-collaborative firms).

Estimates from Table 4 using the subsample of single-location firms and non-collaborative but innovative firms were very similar to those from Table 3. Collaboration was split into two components: local collaboration and other types of collaboration. Both variables were statistically significant and positive, implying that local collaboration, as well as other types of collaboration, positively affected the probability of a firm being a world-first innovator. Calculation of the marginal effect of local collaboration shows that involvement in local collaboration increased the probability of a firm being a world-first innovator by 4.5 percentage points, while the marginal effect of other types of collaboration increased the probability by 3.9 percentage points. For comparison, using government support programs increased the probability of introducing a world-first innovation by only 2.9 percentage points.

The marginal effects for local collaboration and other types of collaboration were very similar. Each one contributed to an increase in the probability of being a world-first innovator to almost the same extent. Local collaboration per se did not seem to leverage a higher amount of

Table 4: Probit Model, Distinguishing Local Collaboration from Other Types of Collaboration

	Ordered Probit on Importance of Innovation	
	Coefficient	*S.E.*
Intercept	-1.996*	(0.14)
Region		
Quebec	0.013	(0.10)
Ontario	0.195**	(0.11)
Prairies	0.297*	(0.11)
British Columbia	0.187	(0.11)
Industry		
Medium-Tech	0.329*	(0.04)
High-Tech	0.322*	(0.12)
Size of Firm		
50-249 Employees	0.169*	(0.05)
≥ 250 Employees	0.263*	(0.07)
Activities Related to Innovation		
Use of Gvt. Pgm.	0.166*	(0.05)
No. of Innovation Activities	0.106*	(0.02)
Collaboration		
Local Collaboration	0.225*	(0.06)
Other Types of Collaboration	0.174*	(0.06)
Inverse Mills' Ratio	-1.045*	(0.12)
Prob (Y = World-First)	**0.103**	
Prob (Y = Canada-First)	**0.242**	
-2 log L	-3,487.411	
No. Obs. (weighted)	4,224	

Source: Statistics Canada, Survey of Innovation 1999. Tabulated by authors.

Note: A first-stage probit on innovation was done. No significant change (when compared with Table 3) occurred in the results of the first stage and, therefore, only estimates from the second stage are presented.

*Significant at the 5% level.

**Significant at the 10% level.

knowledge into bringing an innovative breakthrough to the market (at least, not a significantly greater amount than collaboration as a whole). Therefore, for Canadian firms, being involved in local collaboration did have a positive and significant effect on the probability of being a world-first innovator, but its effect did not outweigh the impact of collaboration as a whole.

6. CONCLUSION

The study reported in this chapter sought to determine the effect of the new innovative paradigm, in which collaboration and proximity are now seen as fundamental to innovation performance. On the basis of data from the Survey of Innovation 1999 on Canadian manufacturing firms, a positive correlation was found between collaboration for innovation and world-first innovators. Moreover, being involved in collaboration had a greater effect on the probability of a firm being a world-first innovator than other factors, such as size of firm. This result supports the assumption that, due to the increasing complexity of technology, a firm on its own cannot easily introduce world-first innovations.

The importance of geographical proximity was examined from two different, but linked, perspectives: local collaboration and technological infrastructure. Local collaboration was shown to have an impact on the probability of a firm being a world-first innovator, but its effect was no greater than the impact of collaboration as a whole. An attempt to assess the impact of technological infrastructure on firms' innovation performance was made, but the sample design imposed additional constraints and a much larger geographical area was used than was desirable. However, the results show that Quebec and Ontario had been successful in providing a flourishing environment that promoted the diffusion of technology, and the Prairies had been successful in providing technological infrastructure for innovative breakthroughs.

Appendix 1: Definitions, Probit Model and Descriptive Data

Table A1.1: Region and Industry Definitions

Region	*Definition*	
Atlantic	Provinces of New Brunswick, Nova Scotia, PEI and Newfoundland	
Quebec	Province of Quebec	
Ontario	Province of Ontario	
Prairies	Provinces of Manitoba, Saskatchewan and Alberta	
British Columbia	Province of British Columbia	
Industry	*NAICS*	*Definition*
Low-Tech	311	Food
	312	Beverage and Tobacco
	313	Textile Mills
	314	Textile Product
	315	Apparel
	316	Leather
	321	Wood
	322	Paper
	323	Printing
	337	Furniture
Medium-Tech	324	Petroleum and Coal
	325	Chemical (except Pharmaceutical & Medicine)
	326	Plastics and Rubber
	327	Non-metallic Mineral
	332	Fabricated Metal
	333	Machinery
	3345	Navigational, Measuring, Medical & Control Instruments
	3346	Magnetic & Optical Media
	335	Electrical Equipment
	336	Transportation Equipment (except Aerospace)
	339	Miscellaneous Manufacturing
High-Tech	3254	Pharmaceutical & Medicine Manufacturing
	334	Computer and Electronics
	3364	Aerospace Product & Parts Manufacturing

Source: Industry Canada, from Statistics Canada, Survey of Innovation 1999. The industrial classification is adapted from Hatzichronoglou (1997).

Table A1.2: Probit Model, Using Subsample of Single-Location Firms

	Ordered Probit on Importance of Innovation	
	Coefficient	*S.E.*
Intercept	-1.997*	(0.13)
Region		
Quebec	0.014	(0.10)
Ontario	0.197**	(0.11)
Prairies	0.298*	(0.11)
British Columbia	0.189	(0.11)
Industry		
Medium-Tech	0.329*	(0.04)
High-Tech	0.326*	(0.12)
Size of Firm		
50-249 Employees	0.17*	(0.05)
≥ 250 Employees	0.261*	(0.07)
Activities Related to Innovation		
Use of Gvt. Pgm.	0.166*	(0.05)
No. of Innovation Activities	0.106*	(0.02)
Collaboration		
Collaboration	0.203*	(0.05)
Inverse Mills' Ratio	-1.048*	(0.12)
Prob (Y = World-First)	**0.103**	
Prob (Y = Canada-First)	**0.243**	
-2 log L	-3,487.635	
No. Obs. (weighted)	4,224	

Source: Statistics Canada, Survey of Innovation 1999. Tabulated by authors.

Note: A first-stage probit on innovation was done. No significant change (when compared with Table 3) occurred in the results of the first stage and, therefore, only estimates from the second stage are presented.
*Significant at the 5% level.
**Significant at the 10% level.

Table A1.3: Descriptive Data, by Different Types of Collaboration

	Type of Collaboration			*Innovative Firms*	
	Local	*Other Types*	*No Collab.*	*%*	*N*
	13.2%	**9.7%**	**77.1%**	**100**	
Region					
Atlantic	10%	12%	78%	100	208
Quebec	13%	9%	78%	100	1,584
Ontario	14%	11%	76%	100	1,538
Prairies	9%	7%	84%	100	485
British Columbia	18%	11%	72%	100	410
Industry					
Low-Tech	10%	8%	82%	100	1,763
Medium-Tech	14%	11%	74%	100	2,345
High-Tech	31%	12%	57%	100	115
Size of Firm					
20-49 Employees	13%	10%	77%	100	1,146
50-249 Employees	13%	8%	79%	100	2,511
≥ 250 Employees	13%	16%	71%	100	566
Activities Related to Innovation					
Use of Gvt. Pgm.	76%	76%	56%	60	2,545
Collaboration	100%	100%	0%	23	969
Local Collaboration	100%	0%	0%	13	557
R&D	90%	87%	75%	78	3,306
M&E	90%	89%	87%	88	3,704
Engineering & Design	76%	79%	62%	65	2,761
Tooling Up	77%	80%	71%	72	3,057
Training	87%	87%	82%	83	3,499
No. of Innovation Activities	4.21	4.21	3.76	3.87	
Total (weighted)	557	412	3,255		4,224

Source: Statistics Canada, Survey of Innovation 1999. Tabulated by authors.

Note: This is a subsample of single-location firms and non-collaborative firms.

Notes

[1]A previous study on Canadian firms shows that firms will improve their probability of innovating if they are engaged in several activities linked to innovation, such as R&D, training, acquisition of machinery and equipment, co-operation and tooling up (Therrien 2002). For more on complementarity, see, for instance, Athey and Stern (1998) and Topkis (1998).

[2]Information and knowledge are two distinct concepts, and the difference between them is well documented. See, for instance, Dosi (1996) and Lavoie and Roy (1998).

[3]The role played by business services must not be underestimated. In a note comparing Silicon Valley, California with Cambridge, UK, an emerging high-tech innovation economy, Kenney (2000) remarks that it is the lack of a business sector that prevents Cambridge from becoming the Silicon Valley of Europe. Kenney mentions that what makes Silicon Valley unique is the massive presence of specialized organizations, which assist in the establishment of new firms and attract resources (human and financial) to accelerate their growth.

[4]The target population of this survey was a provincial enterprise with a gross business income of at least $250,000 and more than 19 employees. However, due to a problem of data reconciliation, some firms with fewer than 20 employees were included in the sample. These firms were removed from the whole sample. A provincial enterprise included all establishments of a given enterprise in the same industry within a province. For the remaining analysis here, provincial enterprise and firm will refer to the same concept.

[5]See Hamdani (2000).

[6]Strategic intelligence is a service that allows managers to be kept informed of what competitors or other organizations are doing in relation to their industrial sector. For more information on strategic intelligence, see, for instance, http://www.glsreseaux.com/EN/Veille.html or http://www.fuld.com/.

[7]The correlation between the importance of innovation and the length of time between the initial conception of the idea and the introduction of the innovation to the market was significant and strong. Both variables could face the same bias (as they were measured by the same person — the CEO), but the length of time from inception to the introduction of an innovation was less likely to be subject to a CEO's perception.

[8]These multi-location collaborative firms had to be eliminated because questions regarding local collaboration were less relevant to them (for these firms, there were several local areas) and therefore they could not answer those particular questions. However, tests were done to assess the impact of removing these observations from the original sample and the results on collaboration (as a whole) remained unchanged. For more details, see section 5 of this chapter.

[9]It is important to note that each variable measures a different aspect of a firm's performance. An imitation (a firm-first innovation) can be financially more profitable (by increasing sales) than a world-first innovation.

[10]In this chapter, only novelty of innovation is analyzed.

[11]Usually, results from these two models are almost identical. Change occurs only if there is an important proportion of observations at the tails of the distribution.

[12]The aggregation of industries into low-, medium- and high-tech sectors follows the OECD's taxonomy and is based on the intensity of R&D (Hatzichronoglou 1997).

[13]The choice of regressors used in the first probit equation (being innovators or not) came from the results obtained by Therrien (2002). The regressors were a combination of industry and regional dummies, use of government programs, size of firm and some qualitative variables concerning the CEOs' perceptions with regard to competitive environment and firms' success factors. Results from regressions using the correction for selection bias were very similar to those without correction.

[14]See Lavoie and Therrien (1999). Employment in high-tech industries grows steadily, but it still remains the smallest employment group in manufacturing.

[15]See Therrien (2002) for more details. In the questionnaire, factors in a firm's success were split into three categories: markets and products (such as seeking new markets), human resources (such as hiring experienced employees) and other (which included performing R&D and developing new products/processes).

[16]It should be noted that not all innovative firms answered the question regarding the novelty of their innovation. Some firms did not answer at all, while others gave ambiguous answers (for example, some did not know how to rank their most important innovation). Therefore, the subsample of innovative firms was restricted to firms that described their most important innovation and gave unambiguous answers regarding the novelty of their innovation. However, tests showed that using the assumption that firms that gave ambiguous answers were firm-first innovators would give similar results.

[17]Quebec and Ontario are the most populous provinces in Canada. Nevertheless, the percentage of research chairs for Quebec and Ontario together from 2000 to 2005 is projected to be about 66%, while the joint population of these two provinces in 2000 was 62% of the total population of Canada (Industry Canada (Knowledge Infrastructure Directorate) and Statistics Canada, CANSIM, matrices 6367-6378 and 6408-6409).

[18]It should be noted that the definition used was the internationally accepted definition of innovation developed by the OECD (see OECD/Eurostat 1997).

[19]There is a proportional test related to the ordered probit concerning the assumption of constancy of effects across categories (testing whether the coefficients remain the same if the novelty of innovation moves from firm-first to Canada-first and from Canada-first to world-first). The proportional test was rejected, but if the model were recomputed using a dependent variable such as world-first versus all other innovators, the results would remain comparable.

[20]Marginal effect was computed by the difference in predicted probabilities (at the mean values). Predicted probabilities were computed by giving the variable of interest a value of 1 and 0 (for binomial variables) or by effecting a change of one unit for count variables (such as the number of activities linked to innovation).

[21]It should be pointed out that, given the cross-sectional nature of the data, the causality link cannot be clearly established. A firm can be asked to collaborate because it is a world-first innovator, and it can become a world-first innovator because it is engaged in collaboration. The high marginal impact of collaboration shows the important correlation (two-way causal effect) that that variable has with novelty of innovation.

[22]As stated earlier, local collaboration was defined as collaboration with another firm/organization located within 100 kilometres, while other types of collaboration were defined as collaboration that was not local (collaboration could be provincial, national or international).

[23]Computing the marginal effect of each regressor revealed only minor change, except for the variable "collaboration," whose effect decreased significantly. This was expected, since removing all collaborative firms with more than one Canadian location significantly decreased the number of collaborative firms in the subsample, while means for all other variables remained practically unchanged.

6

World-First Innovation and Firm Performance

Brian P. Cozzarin

1. INTRODUCTION

The overall objective of the research reported in this chapter was to determine the impact of producing a world-first innovation, a Canada-first innovation and a firm-first innovation on firms' economic performance. The research also aimed to shed light on firm-level characteristics that influence innovation activity. In this chapter, industrial organization and management literature serve as the basis for theoretical predictions relating to technological change and innovation at the firm level. (Schumpeter 1942) saw the importance of innovation for the continued success of firms (and capitalism itself). His assertions were the motivation for a large body of empirical research.

Schumpeter put forth two main hypotheses that related growth resulting from innovation to the size and market conduct of firms. Schumpeter's first hypothesis stated that larger firms are more likely to innovate (the hypothesis has been tested in another form as well: as firm size increases, there is more than a proportionate increase in innovation activity). Three reasons have been proposed in the literature to explain Schumpeter's first hypothesis. The first is that large firms have access to capital markets and can find financing for R&D (risk spreading), and they are therefore able to engage in more R&D than small firms. The second reason is that economies of scale inherent in the

R&D process give large firms a natural advantage over small firms. The third reason is that complementarities exist between R&D and other non-manufacturing-based functions of the firm (marketing and sales, logistics, etc.), which allow large firms to take on more R&D projects. A number of cross-sectional studies have found a positive monotonic relationship between the size of a firm and the amount of R&D undertaken (Soete 1979; Link 1980; Loeb 1983; Meisel and Lin 1983).[1]

Schumpeter's second hypothesis stated that monopolies are more likely to engage in innovation activities and to take advantage of technological growth opportunities because of their ability to fund in-house R&D. Nelson's and Winter's (1982) evolutionary models of economic growth support the notion that fast-paced innovation activity within an industry can cause an increase in the number of firms exiting, which in turn causes an increasingly concentrated industry. Empirical studies of market concentration and R&D (Scott 1984; Levin, Cohen and Mowery 1985) have found that relations between them depend on other industry conditions (such as the type of technology used within the industry, and the type of products — whether durables or consumables — produced within the industry).

Other studies have examined the effects of cash flow on R&D (Armour and Teece 1980; Link 1981). It has been shown that cash flow is usually correlated with higher R&D within a firm. High cash flow within a firm may be the result of a succession of profitable R&D programs. Characteristics other than those internal to a firm may be of importance for the success and use of R&D programs. For instance, companies may use suppliers' knowledge to supplement their R&D efforts (von Hippel 1988).

A number of recent studies have tested either Schumpeter's first or second hypothesis. A notable characteristic of the studies is that they often use the same measure of innovation. This common measure is the number of innovations recorded by the US Small Business Administration database. The database was compiled from trade literature in 1982 by a private firm (The Futures Group). According to Audretsch (1995), the database has some bias towards product innovation.

Results of tests of the first hypothesis — that large firms are more innovative than small firms —are contradictory. Brouwer and Kleinknecht (1996) have found support for the hypothesis, while Audretsch (1995), Holmes, Hutton and Weber (1991), and Dixon and Seddighi (1996) have found evidence against it. However, a possible confounding factor is that the rate of innovation does not depend only on size of firm. While firm size is important, consideration of other variables, such as the environment (low- or high-tech), the rate of obsolescence of products or processes, the industry, and the type of innovation (product or process), can help to establish a clearer relationship between firm size and innovation.

Similar results have emerged from tests of the second hypothesis. Acs and Audretsch (1988), and Brouwer and Kleinknecht (1996) provide evidence that rejects the hypothesis. Brouwer and Kleinknecht claim that lower concentration leads to faster diffusion. Acs and Gifford (1996) suggest that the rate of obsolescence determines the impact of market structure on innovation. Their argument is that, if firms are confronted with a low rate of obsolescence, innovation decreases with monopoly profits. If the rate of obsolescence is high, innovation increases with monopoly profits.

Because of the lack of data, only a handful of studies has been able to use an innovation variable (Acs and Audretsch (1988) and Audretsch (1995) being exceptions). Measurements of the importance and/or novelty of innovations, and hence their expected profitability for a firm, are even more difficult to come by.[2] Some researchers have used expert opinions to rank firms' innovations in order of importance. Then, on the basis of the expert ranking, analysis could proceed to determine whether the most important innovations were associated with the firm's performance. Another, more common, method involves counting the number of patents attributable to each firm within an industry and then performing tests (univariate or multivariate) to see whether higher patent counts correlate with higher firm performance. The problem with using patent data is that a significant number of innovations are never patented because of the nature of the industry. (Pharmaceuticals are often patented because it is easy to identify the chemical composition of drugs and reverse-engineer them; but the sword cuts both ways — legal challenges are easy to make when it is simple to prove that a competitor has copied a chemical formula. New process technologies are relatively easy to copy, and many firms would rather keep them secret than patent them.) Furthermore, patent data favour product innovations since many process innovations are kept secret. The Survey of Innovation 1999 attempted to measure the importance of innovations by asking firms directly what their most important innovation was over a three-year period (1997-99). Chief executive officers (CEOs) were then asked whether that innovation was a world-first, a Canada-first or a firm-first innovation.

2. STYLIZED FACTS

The study of innovation is not a new topic; there is a massive amount of literature dealing with issues ranging from the diffusion of hybrid corn to the effects that innovation within a firm will have on profits. Within this body of research, Cohen and Klepper (1996) have pointed out that there are four robust empirical patterns concerning R&D, innovation and firm size:

Stylized Fact 1: The likelihood of performing R&D rises with firm size.

Stylized Fact 2: R&D and firm size are closely and positively related within industries.

Stylized Fact 3: R&D rises proportionately with firm size in most industries.

Stylized Fact 4: The number of patents or innovations generated per dollar of R&D declines with firm size.

The major conclusion of the study is that large firms are able to spread the costs of R&D over a larger output, and thus realize a greater return for their R&D than small firms. Larger firms thus have an incentive to undertake more R&D projects than smaller firms, and they will also undertake riskier (in terms of lower mean payoffs) projects than small firms at the margin. Since the ability to capitalize on innovations depends to a large degree on intellectual property rights and other appropriation mechanisms, firms will exploit innovations with their own output. One consequence of cost spreading is that technical progress within an industry depends not only on the industry's total R&D spending, but also on market structure.

3. THE RELATIONSHIP BETWEEN INNOVATION NOVELTY AND PERFORMANCE

Despite the stylized facts about R&D, innovation and firm size, Audretsch (1995: 579) has lamented that "...very little is known about the relationship between firm performance and subsequent innovation activity." This is primarily because of the paucity of data that relate innovation inputs (expenditure on R&D, breadth of R&D) and output — namely, innovation itself. Audretsch's study made use of the US Small Business Administration database, which has recorded 8,074 innovations for service and manufacturing firms, and the *Business Week* survey of company-financed R&D expenditures. The two surveys were combined, and the total matched sample size was 631 firms. Audretsch used the firms' innovation rate (the number of innovations divided by the number of employees) as the dependent variable in his regressions. He found a positive relationship between the innovation rate and the ratio of R&D to sales, and a negative relationship between the innovation rate and sales, indicating that the innovation rate falls as firm size increases. He did not find a statistically valid relationship between the innovation rate and profits or growth in sales.

In an earlier study, Cohen and Levinthal (1989) constructed a model that characterized a firm's stock of technological and scientific knowledge as conditional on the firm's expenditure on R&D. The stock of knowledge had a direct effect on the firm's gross earnings (profits). Cohen and Klepper (1996) developed another model in which R&D expenditure had a direct effect on a firm's price-cost margin. Solving the first-order conditions of the price-cost margin equation yielded each firm's optimal expenditure on R&D within a business unit (or specific product line). Conditioning variables for R&D in both studies included the demand characteristics, technological opportunity (the ability to acquire knowledge and expertise outside the firm and/or the firm's industry) and the appropriation regime (the filing of patents, the use of licensing agreements) faced by the firm.

In the study presented in this chapter, two measures of innovation were of importance. The first was the number of new products and/or new processes that a firm introduced over the three-year period 1997-99. It was expected that, as the number of product and process innovations increased, a firm's price-cost margin would also increase (process innovations lower unit costs; product innovations increase the quality of existing products), or, in the case of totally new products, sales would increase. The second measure was the "novelty" of each firm's innovation(s): world-first, Canada-first or firm-first. It was expected that world-first innovations would yield higher returns (in terms of price-cost margins or sales) than Canada-first and firm-first innovations. Likewise, Canada-first innovations would yield higher returns than firm-first innovations. Firm-first innovations would most likely represent the diffusion of new innovations both from within the industry and external to it.

The novelty of innovation was the primary focus of the study, because it is a unique variable that should provide some insight into the reward for being "first." The conditioning variables for the number and novelty of innovations included: past expenditure on R&D; past profits; a firm's strategy profile; demand conditions within the industry; the type and use of intellectual property rights; and firm size.

The study examined four hypotheses. The first hypothesis — that innovative firms (world-first, Canada-first, firm-first) would have higher performance (in terms of the performance measures defined in the next section) than non-innovative firms — was tested empirically. It was hoped that the data would provide some insight into the veracity of the hypothesis with the construction of simple bivariate tables. Second, it was expected that the dichotomous innovation variables would be statistically different from zero (reject H_0) in the multivariate analysis. Third, it was also expected that the parameter estimates would be greater for world-first innovations compared to firm-first innovations or in terms of the estimated coefficients themselves. The fourth hypothesis was that firms in high-tech industries would perform better in terms of both the number of world-first innovations and overall performance than firms in medium- or low-tech industries.

4. DATA

Statistics Canada's Survey of Innovation 1999 provided the primary data set for the analysis. The collection unit for the survey was at the provincial enterprise level (the firm proper, not plants), using a mix of three- and four-digit NAICS (North American Industry Classification System) codes. The CEO of each firm was contacted to fill out the survey. If the firm operated in all 10 provinces, the CEO would have received 10 questionnaires. The survey probed the innovation capabilities and performance of firms in the Canadian manufacturing sector, as well as their R&D activity. Of a total population of provincial enterprises with a gross business income of at least $250,000 and 20 or more employees (9,303 firms), 5,944 were sampled. The response rate was 95%.

Economic performance data came from the 1997 Annual Survey of Manufacturers (ASM), which has been linked to the Survey of Innovation 1999. The linked data set contains the following firm-level variables from the ASM: number of production workers; number of salaried employees; person hours for production workers; dollar value of wages paid; dollar value of salaries paid; cost of fuel and electricity; cost of materials and supplies used; total value of shipments; and total value added.

The distribution of firms in the manufacturing sector by three-digit NAICS code is shown in Table 1. Table 2 shows the size distribution of firms in the linked data set. Most firms were medium-sized, with 50 to 249 employees. Fifty-nine per cent of firms undertook R&D. However, 19% were non-innovators (the definition of innovation used in the Survey of Innovation required a firm to have an innovation in the period 1997-99; some firms indicated that they had innovations still under development or that they had ultimately been unsuccessful). Few firms had product (14%) or process (12%) innovations alone; the majority of innovations (54%) were both product and process.

The value of shipments data and cost data were used to calculate the well-known Lerner index (otherwise known as a price-cost margin). The Lerner index for firm i was defined as ouput price minus marginal cost divided by output price:

$$\text{Lerner}_i = \frac{\text{Price}_i - C_i'(q)}{\text{Price}_i}$$

or, using the variables from the ASM:

$$\text{PriceCost}_i = \frac{\text{Shipments}_i - (\text{Wages}_i + \text{Salaries}_i + \text{Fuel \& Electric}_i + \text{Materials}_i)}{\text{Shipments}_i}$$

where:

PriceCost_i = price-cost margin for firm i
Shipments_i = value of shipments
Wages_i = wages of production workers
Salaries_i = earnings of salaried employees

$Fuel\&Electric_i$ = cost of fuel and electricity
$Materials_i$ = cost of materials and supplies used

Since each firm is assigned to an industry, it was possible to create domestic market share variables. The domestic market share of a firm was defined as the shipments value of the firm divided by the sum of shipments for all firms within the industry:

$$MarketShare_i = \frac{Shipments_i}{\sum_{m=1}^{N} Shipments_m}$$

Table 1: Linked Survey of Innovation – ASM Data Set, by Industry Characteristics

Industry	*Three-Digit NAICS*	*Proportion of Population Sampled*
		(% of Firms)
Food	311	9.1
Beverage & Tobacco	312	1.5
Textile Mills	313	2.4
Textile Product	314	1.6
Apparel	315	5.8
Leather	316	1.0
Wood	321	9.2
Paper	322	4.4
Printing	323	4.5
Petroleum & Coal	324	0.9
Chemical	325	6.3
Plastics & Rubber	326	6.4
Non-metallic Mineral	327	4.3
Primary Metal	331	3.4
Fabricated Metal	332	9.1
Machinery	333	8.9
Computer & Electronics	334	4.4
Electrical Equipment	335	2.9
Transportation Equipment	336	5.5
Furniture	337	4.8
Miscellaneous Manufacturing	339	3.5

Source: Statistics Canada, Survey of Innovation 1999. Tabulated by author.

where:

$MarketShare_i$	=	market share for firm i in the relevant three-digit NAICS industry
N	=	total number of firms within the relevant industry

Labour productivity was defined as a firm's manufacturing value added divided by the total number of employees (production and salaried).

To obtain meaningful comparisons between the performance variables, relative measures were constructed to control for the absolute size of each industry. To construct relative measures, NAICS industry-level variables were used. The 1997 industry-level ASM data were taken from CANSIM. For the tabular analysis, nine performance variables were used for each of the 21 NAICS industries. The means for the performance variables by industry are summarized in Table A1.1 in Appendix 1. The ranking of each industry for each performance variable is shown in Table A1.2 in Appendix 1. In terms of the mean

Table 2: Linked Survey of Innovation – ASM Data Set, by Firm Characteristics

Characteristics	*Population*
	(% of Firms)
Size	
Small (20-49 Employees)	27
Medium (50-249 Employees)	56
Large (250+ Employees)	17
R&D Activity	
R&D Undertaken	59
No R&D	41
Type of Innovation	
Product Only	14
Process Only	12
Product-Process Combined	54
Non-Innovator	19

Source: Statistics Canada, Survey of Innovation 1999. Tabulated by author.

performance ranking in 1997, Beverage and Tobacco was in the top four industries in six of the nine performance categories. Petroleum and Coal was in the top four industries five times. Leather, Plastics and Rubber, and, Primary Metal were among the top four industries three times each. The industry with the highest mean market share per firm was Petroleum and Coal (2.064%). Petroleum and Coal also had the highest labour productivity. Chemical Manufacturing had the highest price-cost margin (28.3%).

5. DEVELOPING A TAXONOMY FOR INNOVATION NOVELTY

Question 18 of the survey asked respondents to rate their firm's "most important new or significantly improved product (good or service) or production/manufacturing process." CEOs had three choices: a world-first, a first in Canada or a first for their firm. Of the 5,220 firms[3] included in the dataset, 976 were non-innovators (those which, according to their CEOs, had no product, process or product-process innovations from 1997 to 1999, or those which had unsuccessful or uncompleted projects).[4] Four hundred and eighty firms in the sample were world-first innovators, and 789 were Canada-first innovators.

At first glance, 2,975 firms remained to be classified as firm-first innovators. However, nearly 40% of these 2,975 firms claimed that they had a product, process or combined product-process innovation but were unable or unwilling to identify and describe the innovation (question 17), while others stated that they did not know if it was a world-first, Canada-first or firm-first innovation (question 18). In fact, 518 CEOs said that their innovation was not a world-first, Canada-first or firm-first. One hundred and fourteen said that their innovation was neither a world-first nor a Canada-first, but they did not know if it was a firm-first. Only one CEO admitted to not knowing whether the innovation was a world-, Canada- or firm-first. Five hundred and eleven respondents would not divulge the type, name and nature of the innovation in response to question 17 and were excluded from answering question 18. As a result of these factors, there were 1,144 firms that had innovations, but for some reason their description and/or importance was unknown or not reported. The question was, where did this rather substantial number of firms belong — with the world-first innovators, with the Canada-first group or with the firm-first group?

The "unknown" group of firms had to be compared to firms in the three innovator categories. A logical set of choice variables for comparison would have been the performance measures; however, they were unlikely to include all the relevant information that could differentiate firms. In the survey, there were only two questions relating to firm-specific characteristics that all CEOs were required to answer. Question 1 ("Competitive Environment") consisted of 11 non-forced Likert statements (the scale was from 1 to 5) concerning the competitive environment facing a firm. Question 2 (also consisting of non-forced Likert statements), called "Firm Success Factors," had three sections. The first section, called "Markets and Products," included six statements. The second section, "Human Resources," also had six statements. The last section, "Other," had four statements. The 27 variables were unwieldy, so principal components analysis was performed (Appendix 2). The first four principal components of the competitive environment variables had eigenvalues greater than 1, accounting for 53% of the total variation in the variables. The first four principal components of the firm success variables had eigenvalues greater than 1, which accounted for 53% of the total variation.

The objective was to differentiate firms in each of the innovator categories. The method used was to compare the means of the performance variables (market share — MSHR, relative employment — R_EMP, relative labour productivity — R_LPROD, and, relative profit — R_PRFT) and firm characteristic variables (competitive environment principal components — DEMAN_1 to DEMAN_4, firm success principal components — STRAT_1 to STRAT_4) for the "unknown" group to those for the three others. The standard procedure is to use a T-test for testing the hypothesis that the two means are equal (negation of H_0 indicates that the means are statistically different). However, if the two random variables are not normally distributed, the T-test cannot be used and a non-parametric test must be substituted. So, before any tests on the means could be performed, the variables had to be checked for normality using goodness-of-fit tests. The relevant goodness-of-fit test statistics include: the Kolmogorov-Smirnov (D) statistic; the Cramer-von Mises (W^2) statistic; and the Anderson-Darling (A^2) statistic (Mardia

1980). As shown in Appendix 3, all variables did not fit the normal distribution for each of the three tests. The only exception was the third principal component for demand characteristics, which was normally distributed.

The variables were then subjected to a Wilcoxon rank score test for the difference in means of the 12 variables across world-first, Canada-first, firm-first, non-innovator and "unknown" groups. The similarity in means between the "unknown" group, the three innovator groups and non-innovators determined where the 1,144 firms would be categorized. The Wilcoxon rank score tests found that H_0 ($\mu_1 = \mu_2$) was not rejected for all 11 variables for the firm-first innovators versus the "unknown" group (see Appendix 3 for details).[5] Thus, the firms in the "unknown" group could not be differentiated from the firm-first innovators and were placed with them. The innovator taxonomy now included four categories: world-first, Canada-first, firm-first and non-innovators. Over the period 1997-99, there were 727 world-first innovators or 8.55% of the population, and 1,241 Canada-first innovators and 4,898 firm-first innovators in the population (Table 3). The 1,643 non-innovating firms represented almost 20% of the population.

Table 3: Frequency of World-, Canada- and Firm-First Innovations, 1997-99

Innovation	*Frequency*	*%*	*Cumulative Frequency*	*Cumulative %*
			(Population Weighted)	
World-First	727	8.55	727	8.55
Canada-First	1,241	14.58	1,968	23.13
Firm-First	4,898	57.56	6,866	80.69
Non-Innovator	1,643	19.31	8,510	100.00

Source: Statistics Canada, Survey of Innovation 1999. Tabulated by author.

6. OVERVIEW OF SURVEY RESULTS IN TERMS OF THE TAXONOMY FOR INNOVATION NOVELTY

6.1 Competitive Environment and Firm Success Factors

Table 4 summarizes the extreme scores on a five-point Likert scale (scores of 4 or 5 indicated strong agreement) for the competitive environment of a firm in relation to the taxonomy for innovation novelty developed in the previous section. It is evident that there were no differentiating characteristics for world-first innovators. However, 63.6% of non-innovators strongly agreed that their products were easily substitutable, only 9.6% agreed that their products became obsolete quickly, and 23.4% agreed that production technology changed rapidly. The three responses indicate that non-innovators produced commodity-type products. With regard to the strategy profile of a firm (Table 5), innovators overall liked to seek new markets and satisfy existing clients. They

Table 4: Competitive Environment of a Firm and Innovation Novelty, 1997-99

Competitive Environment	*World-First*	*Canada-First*	*Firm-First*	*Non-Innovator*
	(% of Firms in Strong Agreement)			
Demands easy to predict	32.1	29.1	30.7	32.9
Easy product substitution	46.4	55.4	61.0	63.6
Competitors' actions easy to predict	25.0	21.4	22.4	22.1
Arrival of new competitors a constant threat	45.3	48.4	50.0	45.4
Arrival of competing products a constant threat	55.8	56.0	54.7	44.3
Easy to replace current suppliers	23.8	22.6	24.0	24.6
Difficult to hire qualified staff and workers	61.0	63.3	64.2	55.7
Difficult to retain qualified staff and workers	33.6	33.7	35.4	28.1
Products become obsolete quickly	12.9	11.0	11.9	9.6
Production technologies change rapidly	43.4	40.7	43.4	23.4
Office technologies change rapidly	66.2	67.3	64.2	56.3

Source: Statistics Canada, Survey of Innovation 1999. Tabulated by author.

Table 5: Strategy Profile of a Firm and Innovation Novelty, 1997-99

Strategy Profile	*World-First*	*Canada-First*	*Firm-First*	*Non-Innovator*
	(% of Firms in Strong Agreement)			
Markets and Products				
Seek new markets	81.5	79.5	76.0	58.5
Satisfy existing clients	95.9	96.8	97.5	92.7
Develop niche or specialized markets	79.0	74.8	70.3	52.5
Develop export markets	73.7	67.3	60.0	43.2
Promote firm or product reputation	86.3	85.3	83.1	75.2
Provide after-hour client support services	47.6	45.4	40.6	29.5
Human Resources				
Hire new graduates from universities	36.4	29.7	19.1	11.2
Hire new graduates from technical schools and colleges	50.8	46.2	38.3	24.7
Hire experienced employees	72.5	67.8	70.4	59.3
Recruit skilled people from outside Canada	14.4	12.6	7.2	6.3
Train employees	85.5	86.4	80.7	69.5
Use teams within the firm to bring together people with different skills	73.8	65.9	53.8	37.9
Other				
Perform R&D within the firm	81.6	70.3	53.0	24.4
Collaborate and cooperate with other firms	47.7	37.7	29.8	15.1
Develop new products and processes	88.0	82.7	67.3	32.7
Develop new industry-wide standards	39.5	37.7	32.6	20.8

Source: Statistics Canada, Survey of Innovation 1999. Tabulated by author.

also liked to hire experienced employees, and training seemed to be important as well. Seventy-nine per cent of world-first innovators were interested in developing niche markets, and 73.7% wanted to develop export markets. This would seem to fit with a firm's desire to produce a world-first innovation. Development of export markets was of much less interest to Canada-first and firm-first innovators and to non-innovators (67.3%, 60% and 43.2%, respectively). With respect to human resources, world-first innovators considered hiring new university and college graduates to be more important than did Canada-first and firm-first innovators. They were also more likely to use teams within their organizations to coordinate work.

More telling is the fact that 81.6% of world-first innovators considered R&D to be very important, while 70.3% of Canada-first innovators considered it important, and only 53.0% of firm-first innovators did so. Not surprisingly, R&D was considered important by only 24.4% of non-innovating firms. Almost 48% of world-first innovators were very interested in collaborative arrangements with other firms, and 88% wanted to develop new products and new processes. In other words, innovation was seen as a key priority within world-first innovating firms, but was less of a priority among Canada-first innovators (82.7%) and firm-first innovators (67.3%). Innovation was not important to non-innovators (only 32.7% of non-innovators strongly agreed that innovation was important to them).

6.2 Innovation Novelty and Performance

One quick glance at Table 6 shows that there was a monotonically increasing relationship between the performance variables and innovation novelty with the exception of the wage-related variables. Non-innovators were smaller (in terms of employment and market share) and their price-cost margins (absolute and relative) were lower than those of the innovators. It is clear that world-first innovators were larger both in absolute terms and in relation to firms in their own industries. Labour productivity and price-cost margins were larger for the world-first group in both absolute and relative terms.

Firms were divided into three equal-sized groups on the basis of performance (low, medium and high performers) (Table 7). A comparison of the percentage of firms in the high-performance category of world-first innovators with the remaining innovators and the non-innovating firms gives an idea of whether world-first innovators had an edge. Forty-one per cent of world-first innovators were classified as "high" in terms of market share and 38.9% as "high" in terms of relative employment. World-first innovators were more apt to be high performers in terms of labour productivity and price-cost margins than the other

Table 6: Innovation Novelty and Performance, 1997-99

Variable	Innovation Novelty: World-First	Canada-First	Firm-First	Non-Innovator
	(Mean of Performance Variable)			
Market Share	0.42%	0.30%	0.19%	0.17%
Total Employment	263	200	156	134
Wage Rate	15.67	15.88	14.87	15.16
Labour Productivity	109,728	98,422	83,033	86,841
Price-Cost Margin	26.6%	25.2%	23.4%	22.9%
Relative Employment	0.33%	0.25%	0.18%	0.17%
Relative Wage Rate	98.45%	99.94%	95.66%	97.79%
Relative Labour Productivity	128.15%	114.35%	102.48%	106.56%
Relative Price-Cost Margin	108.79%	106.22%	98.20%	95.99%

Source: Statistics Canada, Survey of Innovation 1999. Tabulated by author.

innovators and the non-innovating firms. There was little difference in relative wage rates across the different categories.

6.3 Innovation Activities

The survey asked respondents whether their firm had engaged in any of five innovation activities over the period 1997-99: R&D linked to new or significantly improved products or processes; acquisition of machinery, equipment or other technology linked to innovation; industrial engineering and industrial design for innovation; tooling up and production start-up linked to innovation; and training linked to innovation. Figure 1 shows that the percentage of firms reporting in-house R&D programs increased from around 20% for non-innovators to over 90% for world-first innovators. Clearly, R&D was an important function for the generation of world-first innovations. Table 8 shows that firms with R&D programs were larger and had higher labour productivity and price-cost margins.

Table 7: Performance and Innovation for Low-, Medium- and High-Performance Firms, 1997-99

Performance Variable	*World-First*	*Canada-First*	*Firm-First*	*Non-Innovator*
		(%)		
Market Share				
Low	27.8	27.4	38.3	43.5
Medium	31.1	36.2	36.2	33.8
High	41.1	36.5	25.5	22.7
Total	100.0	100.0	100.0	100.0
Relative Employment				
Low	25.4	28.9	39.0	47.4
Medium	35.7	34.9	35.1	30.9
High	38.9	36.3	25.9	21.7
Total	100.0	100.0	100.0	100.0
Relative Wage Rate				
Low	29.5	28.7	33.0	29.3
Medium	35.0	34.1	34.5	35.9
High	35.5	37.2	32.5	34.8
Total	100.0	100.0	100.0	100.0
Relative Labour Productivity				
Low	25.7	29.6	32.8	35.4
Medium	33.0	32.7	35.1	31.1
High	40.3	37.8	32.1	33.5
Total	100.0	100.0	100.0	100.0
Relative Price-Cost Margin				
Low	23.2	30.7	32.6	36.3
Medium	30.5	35.2	34.1	32.3
High	4[illegible].4	34.1	33.3	31.4
Total	100.0	100.0	100.0	100.0

Source: Statistics Canada, Survey of Innovation 1999. Tabulated by author.

Note: Totals may not sum due to rounding.

Figure 1: World-, Canada- and Firm-First Innovators with In-House R&D Programs, 1997-99

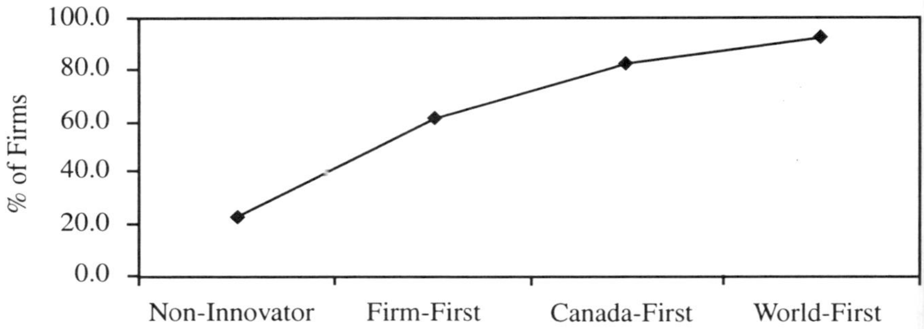

Source: Statistics Canada, Survey of Innovation 1999

Table 8: R&D Activity and Performance, 1997-99

Variable	*No R&D*	*With R&D*
	(Mean of Performance Variable)	
Market Share	0.152%	0.269%
Total Employment	120	200
Wage Rate	14.67	15.46
Labour Productivity	80,813	93,394
Price-Cost Margin	22.5%	24.8%
Relative Employment	0.144%	0.236%
Relative Wage Rate	96.5%	97.3%
Relative Labour Productivity	105.9%	108.1%
Relative Price-Cost Margin	95.2%	103.0%

Source: Statistics Canada, Survey of Innovation 1999. Tabulated by author.

Firms with a large market share and high relative employment were the most likely to have in-house R&D programs. They also engaged in the greatest number of innovation activities (3.6 per firm on average; see Appendix 4 for details). Table 9 shows that firms undertaking all five innovation activities had the highest market share. Table 10, on the other hand, indicates that world-first innovators tended to engage in more activities on average than the other innovators. For instance, 81.2% of firms in the world-first group reported performing four or five innovation activities compared to 78.8% of those in the Canada-first group.

Table 9: Number of Innovation Activities and Performance, 1997-99

Variable	*None*	*1*	*2*	*3*	*4*	*5*
			(Mean of Performance Variable)			
Market Share	0.138%	0.156%	0.179%	0.177%	0.243%	0.282%
Total Employment	105	131	128	123	181	218
Wage Rate	14.76	15.01	14.96	14.94	15.30	15.36
Labour Productivity	83,153	88,285	85,428	83,242	86,515	93,514
Price-Cost Margin	22.4%	23.7%	22.5%	23.0%	24.7%	24.7%
Relative Employment	0.134%	0.139%	0.173%	0.153%	0.217%	0.250%
Relative Wage Rate	96.8%	98.5%	97.1%	97.5%	96.7%	96.5%
Relative Labour Productivity	103.4%	111.1%	109.9%	106.3%	103.8%	108.9%
Relative Price-Cost Margin	91.8%	99.8%	94.1%	97.2%	104.1%	103.5%

Source: Statistics Canada, Survey of Innovation 1999. Tabulated by author.

Table 10: Number of Innovation Activities and Innovation Novelty, 1997-99

Number of Activities Related to New Products or Processes	*World-First*	*Canada-First*	*Firm-First*	*Non-Innovator*
		(%)		
None	0.7	0.6	3.2	63.5
1	3.2	2.0	7.8	11.9
2	5.5	7.1	11.9	9.1
3	9.4	11.5	16.8	6.7
4	15.3	17.1	20.7	4.6
5	65.9	61.7	39.6	4.1
Total	100.0	100.0	100.0	100.0
4 or 5 Activities	81.2	78.8	60.3	8.7

Source: Statistics Canada, Survey of Innovation 1999. Tabulated by author.
Note: Totals may not sum due to rounding.

6.4 Technology Sector and Innovation

Table 11 compares the absolute and relative performance of firms grouped in high-tech, medium-tech and low-tech industries.[6] The market share of firms in the high-tech and low-tech sectors was on average larger than that of firms in the medium-tech sector. Total employment in the high-tech sector was much higher than that in the medium- and low-tech sectors. The characteristic that most differentiated the high-tech sector from the other sectors was probably its higher price-cost margins in both absolute and relative terms.

Table 11: Technology Sector and Performance, 1997-99

Variable	*High-Tech*	*Medium-Tech*	*Low-Tech*
		(Mean of Performance Variable)	
Market Share	0.320%	0.198%	0.243%
Total Employment	369	158	163
Wage Rate	16.92	16.08	13.84
Labour Productivity	114,303	96,086	76,601
Price-Cost Margin	27.1%	25.4%	21.7%
Relative Employment	0.30%	0.17%	0.23%
Relative Wage Rate	93.1%	97.6%	96.4%
Relative Labour Productivity	91.6%	109.2%	106.0%
Relative Price-Cost Margin	105.5%	102.0%	96.7%

Source: Statistics Canada, Survey of Innovation 1999. Tabulated by author.

Table 12 shows that the vast majority of firms reported firm-first innovations. The table also shows that 16.5% of firms in the high-tech sector had world-first innovations; in comparison, 11% of firms in the medium-tech sector and only 4.9% of firms in the low-tech sector had world-first innovations.

Within the population of manufacturing firms, the high-tech sector included 3.4% of all firms, the low-tech sector had about 43% of firms and

Table 12: Percentage of High-, Medium- and Low-Tech Firms, by Innovation Novelty, 1997-99

Industry Group	*World-First*	*Canada-First*	*Firm-First*	*Non-Innovator*	*Total*
			(%)		
High-Tech	16.5	16.7	56.4	10.5	100
Medium-Tech	11.0	16.3	54.4	18.3	100
Low-Tech	4.9	12.3	61.6	21.2	100

Source: Statistics Canada, Survey of Innovation 1999. Tabulated by author.

Note: Totals may not sum due to rounding.

the medium-tech sector contained the bulk of firms — 53.5% (Table 13). Both the medium- and high-tech sectors produced world-first innovations that exceeded their proportion of the population of manufacturing firms. The low-tech sector did poorly in terms of world-first innovations, but accounted for nearly half of the firm-first innovations.

Table 13: Percentage of Innovators in High-, Medium- and Low-Tech Sectors, 1997-99

Industry Group	*Population*	*World-First*	*Canada-First*	*Firm-First*	*Non-Innovator*
			(%)		
High-Tech	3.4	6.5	3.9	3.3	1.8
Medium-Tech	53.5	68.8	59.8	50.6	50.7
Low-Tech	43.1	24.8	36.4	46.2	47.4
Total	100.0	100.0	100.0	100.0	100.0

Source: Statistics Canada, Survey of Innovation 1999. Tabulated by author.

Note: Totals may not sum due to rounding.

6.5 Firm Size and Performance

World-first and Canada-first innovators tended to be large firms employing more than 250 people (Table 14), while firm-first innovators and non-innovators dominated the category of small firms — those with between 20 and 49 employees. It is not surprising that, as firm size increased, market share increased (another measure of size within an industry), and price-cost margins were higher in absolute and relative terms too (Table 15). Thus, it is apparent that large firms on average outperformed small firms on all fronts, even after correction for industry differences. It was a different story, however, when firms were divided into three quantiles on the basis of each of the performance variables. Large firms had higher labour productivity on average than did medium-sized and small firms, yet on average they did not have higher price-cost margins (see Appendix 4).

Table 14: Novelty of Innovation, by Firm Size, 1997-99

Firm Size	*World-First*	*Canada-First*	*Firm-First*	*Non-Innovator*	*Total*
			(%)		
Small (20-49 Employees)	5.3	10.0	59.9	24.8	100.0
Medium (50-249 Employees)	9.2	14.9	57.9	18.0	100.0
Large (250+ Employees)	13.1	23.5	51.1	12.3	100.0
All Firms	8.6	14.6	57.6	19.3	100.0

Source: Statistics Canada, Survey of Innovation 1999. Tabulated by author.

Note: Totals may not sum due to rounding.

6.6 Time Lag and Innovation Novelty

Economists and policy-makers are often interested in the length of time it takes from the idea stage until a new product or process is sold or implemented within a firm. The Survey of Innovation asked respondents to report how long (in years and months) it took from an initial

Table 15: Firm Size and Performance, 1997-99

Variable	*Firm Size* *20-49 Employees*	 *50-249 Employees*	 *250+ Employees*
	(Mean of Performance Variable)		
Market Share	0.05%	0.14%	0.94%
Total Employment	34	106	712
Wage Rate	14.34	14.84	18.13
Labour Productivity	85,121	83,973	112,781
Price-Cost Margin	23.4%	23.9%	24.7%
Relative Employment	0.04%	0.13%	0.80%
Relative Wage Rate	93.5%	96.2%	107.7%
Relative Labour Productivity	110.2%	103.3%	116.2%
Relative Price-Cost Margin	96.3%	100.2%	106.4%

Source: Statistics Canada, Survey of Innovation 1999. Tabulated by author.

idea to the implementation of a full-fledged innovation. Of course, the question only pertained to firms that reported an innovation, so non-innovators were excluded. Figure 2 shows that world-first innovators reported longer time frames for developing and implementing their innovations. Of firms that reported times of more than two years for the

Figure 2: Time Lag and Innovation Novelty, 1997-99

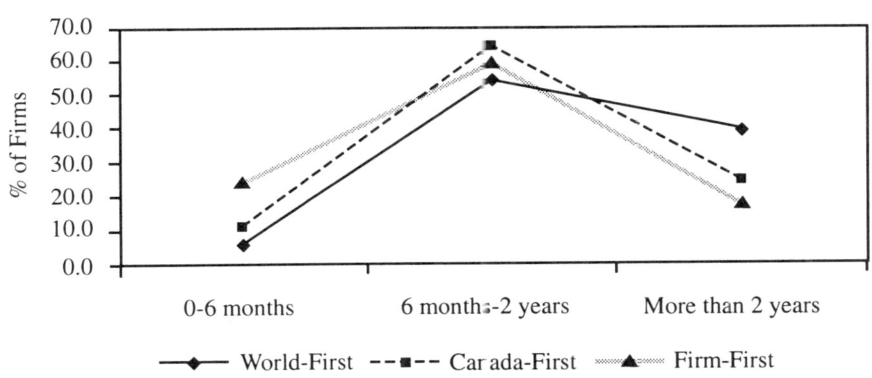

Source: Statistics Canada, Survey of Innovation 1999

development of an innovation, 40% were world-first innovators. This finding makes intrinsic sense, since world-first innovations would be expected to take longer to develop — they are more likely to be new or revolutionary, as opposed to firm-first innovations, which would normally involve the adoption of some existing technology.

An examination of the last column of Table 16 reveals that firms that took more than two years to generate, develop and implement their most significant innovation were larger in terms of total and relative employment, and also in terms of market share. Figure 2 and Table 16 provide evidence for a link between firm size and innovation novelty. Large firms can afford to wait longer because they know that they can spread the costs of innovation over a larger output. The argument is a variation on the cost-spreading hypothesis that explains the size–R&D correlation put forth by Cohen and Klepper (1996). However, it is apparent that firm size confers more than just the incentive to undertake R&D; the rewards from R&D are innovation. Because the innovation costs can be spread over a larger output, and because world-first innovations have a higher payoff for a firm than any other type of innovation, firms are willing to wait longer for the rewards.

Table 16: Time Lag for Innovation and Performance, 1997-99

Variable	*0-6 Months*	*6 Months-2 Years*	*More Than 2 Years*
		(Mean of Performance Variable)	
Market Share	0.157%	0.239%	0.323%
Total Employment	119	173	239
Wage Rate	14.16	15.18	16.40
Labour Productivity	78,524	89,354	101,619
Price-Cost Margin	24.2%	24.5%	24.6%
Relative Employment	0.156%	0.212%	0.252%
Relative Wage Rate	94.2%	96.5%	100.9%
Relative Labour Productivity	102.7%	108.6%	112.9%
Relative Price-Cost Margin	103.1%	102.5%	101.7%

Source: Statistics Canada, Survey of Innovation 1999. Tabulated by author.

6.7 Number of New Product Introductions

Respondents to the Survey of Innovation were asked to report the number of new products that they had introduced from 1997 to 1999 (Table 17). The firms that introduced the highest number of new products (over 50) were the largest (in terms of market share, and total and relative employment). Labour productivity in these firms was not necessarily the highest, but price-cost margins (both absolute and relative) were. This means, quite simply, that the introduction of new products (despite the failures) and high price-cost margins were associated. Causality cannot be inferred, however, since it may have been profitability that led to product innovations or vice versa.

Table 17: Number of New Product Introductions and Performance, 1997-99

	Number of New Product Introductions						
Variable	*1-2*	*3-5*	*6-10*	*11-20*	*21-50*	*50+*	*None*
			(Mean of Performance Variable)				
Market Share	0.249%	0.206%	0.231%	0.215%	0.290%	0.375%	0.180%
Total Employment	192	161	168	166	213	247	140
Wage Rate	15.66	15.52	14.72	14.30	13.81	13.59	15.42
Labour Productivity	91,182	90,330	85,598	85,357	86,253	82,867	88,328
Price-Cost Margin	24.0%	24.1%	22.[illegible]%	23.3%	22.8%	25.5%	24.0%
Relative Employment	0.204%	0.198%	0.192%	0.225%	0.257%	0.337%	0.163%
Relative Wage Rate	96.4%	98.7%	95.6%	93.7%	92.7%	94.1%	98.2%
Relative Labour Productivity	107.1%	108.8%	104.5%	101.9%	107.2%	104.5%	108.7%
Relative Price-Cost Margin	101.3%	100.4%	94.6%	97.9%	98.8%	104.7%	100.6%

Source: Statistics Canada, Survey of Innovation 1999. Tabulated by author.

6.8 Objectives, Problems and Impact of Innovation

Using a five-point Likert scale, respondents were asked to indicate the main reasons why their firm had offered new products (Appendix 5). One of the most important objectives for world-first innovators was to extend their product range — 81.7% of world-first innovators gave this objective a score of 4 or 5 on the Likert scale, indicating that it was very important. Non-innovators, in contrast, were more interested in increasing production capacity, reducing production time, improving product quality and increasing speed to market. It would seem that pursuing these objectives was not necessarily conducive to realizing an innovation. In terms of problems with innovating, the majority of innovating firms (world-, Canada- and firm-first) indicated that high cost was an issue, and that they were unable to devote staff to developing new products or processes. Forty-three per cent of world-first innovators indicated that a lack of skilled workers hindered their innovation efforts. Lack of financing and organizational rigidities were also problems for world-first innovators, but to a lesser degree (see Appendix 5 for details). Using a five-point Likert scale, CEOs were asked to respond to eight statements to indicate their opinion of the impact of innovations on their firm. Scores of 4 or 5 were considered to indicate strong agreement with the statements. For world-first innovators, one impact was pre-eminent: an increase in international market share due to innovation. That is a noteworthy result, consistent with the world-first innovators' desire to develop export markets (Table 5).

6.9 Use of Intellectual Property Rights and Innovation

World-first innovators used intellectual property rights to a far greater degree than any of the other innovators (Table 18). They relied heavily on confidentiality agreements, while patenting was also very important. Only 40.7% of Canada-first innovators and 21.9% of firm-first innovators reported the use of patents. The use of intellectual property rights was a monotonically decreasing function of world-, Canada- and firm-first innovators and non-innovating firms. Some non-innovators reported using intellectual property rights over the three-year period covered by the survey. Because the Oslo definition of innovation was used in the study reported here, it may be that some of the firms categorized as non-innovators were simply unsuccessful in their efforts from 1997 to 1999, but had had innovations prior to 1997.

Table 18: Use of Intellectual Property Rights, by Innovation Novelty, 1997-99

Intellectual Property Rights	*World-First*	*Canada-First*	*Firm-First*	*Non-Innovator*
		(%)		
Patents	62.8	40.7	21.9	10.4
Trademarks	57.9	49.1	34.8	21.4
Copyrights	26.7	18.7	10.5	5.2
Confidentiality Agreements	72.1	60.1	42.8	22.7
Trade Secrets	44.3	37.5	23.8	10.0
Other	2.9	2.6	2.7	2.1

Source: Statistics Canada, Survey of Innovation 1999. Tabulated by author.

6.10 Use of Government Support Programs and Innovation

Government support programs, in order of their importance for all firms, were: R&D tax credits; support for training; R&D grants; and information or Internet services. The three remaining programs (venture capital support, technology support and other) were relatively unimportant. Table 19 shows that world-first innovators were heavy users of government programs, since 77.1% of these firms used at least one government program. Since world-first innovators are very likely to engage in R&D, it is not surprising that 65.5% of them reported using R&D tax credits.

However, the percentage of firms that reported engaging in R&D was 92.4%, 81.8%, 61.1% and 23.1% for world-, Canada- and firm-first innovators and non-innovators, respectively. This is at odds with the percentage of firms in each category that reported using R&D tax credits. The discrepancy may be due to restrictive legislation that makes obtaining tax credits difficult, or to the possibility that the cost of putting claims into the Revenue Canada system exceeds the benefits to a firm. Whichever is the case, further exploration of the lack of use of government programs is warranted.

Table 19: Use of Government Programs, by Innovation Novelty, 1997-99

Government Program	*World-First*	*Canada-First*	*Firm-First*	*Non-Innovator*
		(%)		
R&D Tax Credits	65.5	51.6	33.4	13.8
R&D Grants	19.8	16.3	9.5	1.9
Venture Capital Support	4.3	3.5	2.9	1.6
Technology Support	14.9	12.0	8.0	3.0
Information or Internet Services	15.6	16.3	9.9	8.1
Support for Training	25.2	24.8	21.7	14.0
Other	2.6	1.8	2.6	1.8
Use of Any Government Program	77.1	69.3	53.5	31.6

Source: Statistics Canada, Survey of Innovation 1999. Tabulated by author.

7. ESTIMATION OF PERFORMANCE EQUATIONS[7]

All estimating equations included three dummy variables for innovation novelty: world-first, Canada-first and firm-first. The innovation categories were mutually exclusive. Thus, if the non-innovators were included in the regressions, the X matrix would not have full column rank (one of the four columns would be a linear combination of the other three; the matrix would be non-invertible). So, all regression results were compared to those for a non-innovating firm.

Demand and strategy variables were used in each equation as well. Principal components analysis was used to reduce the dimensions of questions 1 and 2. Consequently, the demand variable (or variables) would be one or more of the principal components from the competitive environment (question 1). The strategy variable (or variables) would be based on principal components analysis of question 2 (firm success factors). It was hypothesized that a high incidence of intellectual property rights usage in a firm would be an indicator of the intensity of innovation. Question 25 of the survey asked if, in the period 1997-99, a firm used patents, trademarks, copyrights, confidentiality agreements or trade secrets to protect its intellectual property. Thus, all regressions included dummy variables for the five methods of protecting a firm's knowledge. Another key variable in terms of innovation intensity was whether R&D was conducted by a firm; a dummy variable for this was also included in the equations. To address Schumpeter's hypothesis that large firms undertake more innovation than small firms, a dummy variable for medium size and another for large size were included. Finally, the regressions also included either NAICS industry dummy variables, or two other dummy variables that denoted whether a firm belonged to a high-tech or medium-tech sector.

7.1 Employment

The form of the employment demand equation was just a conditional input demand function for labour. The total number of workers employed would be a function of output (in this case, shipments), wage rate[8] and other variables:

$$(1)\ N_i = \alpha + \beta_1 W_i + \beta_2 S_i + \beta_3 R_i + \beta_4 I_{ij} + \beta_5 D_{ik} + \beta_6 F_{il} + \beta_7 P_{im} + \beta_8 Z_{in} + \beta_9 T_{io}$$

where:

N_I = total employment for firm i
W_i = wage rate ($/hr.) defined as production wages divided by production workers
S_i = total shipments
R_i = R&D conducted within the firm
I_{ij} = innovation type j composed of three types: world-first, Canada-first, firm-first
D_{ik} = demand conditioning variables
F_{il} = strategy conditioning variables
P_{im} = dummy variables for use of intellectual property rights
Z_{in} = dummy variables for size
T_{io} = dummy variable for technology sector: high-tech, medium-tech, low-tech

7.2 Labour Productivity

Rearranging the conditional input demand function for labour gave the labour productivity equation:

$$(2)\ L_i = \alpha + \beta_1 W_i + \beta_2 R_i + \beta_3 I_{ij} + \beta_4 D_{ik} + \beta_5 F_{il} + \beta_6 P_{im} + \beta_7 Z_{in} + \beta_8 T_{io}$$

where:

L_i = manufacturing value added divided by total employees ($/employee)

It was expected that the dichotomous innovation variables would be statistically different from zero (reject H_0), and it was also expected that the parameter estimates would be greater for world-first innovations than for firm-first innovations.

7.3 Market Share

The form of the market share equation was derived from the Lerner index. C was the price-cost margin of the firm:

$C_i = \frac{M_i}{\varepsilon}$, where M_i was firm i's market share, and ε was elasticity of demand

Rearranging this expression and taking logarithms produced:

$$\ln M_i = \ln C_i - \ln \varepsilon$$

Elasticity of demand for each of the m industries was unavailable; instead, the principal components of the firm-level demand conditioning variables were used. The equation for estimation was:

$$(3)\ M_i = \alpha + \beta_1 C_i + \beta_2 R_i + \beta_3 I_{ij} + \beta_4 D_{ik} + \beta_5 F_{il} + \beta_6 P_{im} + \beta_7 Z_{in} + \beta_8 T_{io}$$

7.4 Total Value Added

Following Griliches (1986) and Acs, Audretsch and Feldman (1994), a knowledge production function was employed to estimate innovation's contribution to output. Both studies used R&D expenditure as an input to the knowledge generation function. Output was measured as total value added in Griliches and by the number of patents per state in Acs, Audretsch and Feldman. The former method, using total value added as the dependent variable, was used in the study reported here as well.[9] Here, innovation novelty was used instead of R&D expenditure as an input to the knowledge generation function (innovation novelty served as a proxy for the introduction of a new product or process or a new combined product-process). The functional form was first specified as Cobb-Douglas and then a translog specification was tried. The Cobb-Douglas estimating equation in natural logarithms was:

$$(4)\ \ln V_i = \alpha_0 + \beta_K \ln K_i + \beta_N \ln N_i + \beta_R R_i + \beta_I I_{ij} + \beta_D D_{ik} + \beta_F F_{il} + \beta_P P_{im} + \beta_Z Z_{in} + \beta_T T_{io}$$

where:

V_i = total value added of firm i
K_i = fixed year-end net capital stock of firm i

The translog specification was:

$$(5)\ \ln V_i = \alpha_0 + \alpha_K \ln K_i + \alpha_N \ln N_i + \frac{1}{2}\alpha_{NL} \ln K_i \ln N_i + \frac{1}{2}\alpha_{KK}(\ln K_i)^2 + \frac{1}{2}\alpha_{NN}(\ln N_i)^2 + \beta_R R_i + \beta_I I_{ij} + \beta_D D_{ik} + \beta_F F_{il} + \beta_P P_{im} + \beta_Z Z_{in} + \beta_T T_{io}$$

8. VARIABLE DEFINITIONS

8.1 Innovation Novelty

Three binary variables represented the novelty of an innovation: world-first, Canada-first, firm-first. Non-innovators were the omitted category for estimation.

8.2 Firm Size

Firm size was measured by the number of production and salaried employees. Two binary variables for firm size were used for estimation: medium, representing 50-249 employees; and large, representing firms with 250 employees or more. Small firms with between 20 and 49 employees were the omitted category for estimation.

8.3 Technology Sector

As mentioned previously, three technology sectors have been identified by Hatzichronoglou (1997): high-tech, medium-tech and low-tech. For estimation purposes, binary variables for high-tech and medium-tech were used; low-tech was the omitted category for estimation.

8.4 Intellectual Property Rights

Baldwin, Hanel and Sabourin (2000) found that the use of intellectual property rights was important to firms in the manufacturing sector. In their probit regression analysis, equations were estimated for product-only innovation, product with process innovation and process-only innovation. It was found that the use of patents as an explanatory variable was significant and positive for all three equations, and that the use of other rights was also significant and positive. It would appear from the survey that the use, importance and, probably, effectiveness of intellectual property rights vary by industry. CEOs were asked whether their firm used any of five means of protecting their intellectual capital. Five binary variables were constructed to represent the use of patents, trademarks, copyrights, confidentiality agreements and trade secrets.

8.5 R&D Activity Within a Firm

One binary variable was constructed to represent whether a firm had an in-house R&D program or not.

8.6 Demand Conditioning Variables

Principal components analysis was done using the 11 non-forced Likert statements concerning a firm's competitive environment. The first four principal components were chosen for regression analysis because they had eigenvalues exceeding 1. The minimum factor (eigenvector) loading must exceed ±0.30, loadings of ±0.40 are considered more important, while those in excess of ±0.50 are very important in explaining the total variance (Hair, Anderson, Tatham, Black and Anderson 1995). The first principal component had eigenvectors exceeding 0.30 for statements 4, 5, 7, 8, 10 and 11. Statements 4 and 5 related to the entry of new firms and the possible substitution of existing products by new entrants; statements 7 and 8 related to a firm's inability to hire and retain qualified workers; and statements 10 and 11 asserted that production and office technologies change rapidly. The first principal component meant that the firms faced a tough competitive environment (easy substitution, strong labour market competition for good staff, rapid technological change). The second principal component had heavy loadings (greater than 0.60) on statements 1 and 3. This meant that consumer demand and competitors' actions were easy to predict. The third principal component had loadings on statements 5 (negative loading), 7 and 8, meaning that there was no threat from new products, but it was difficult to hire and retain qualified staff. The fourth principal component had heavy loadings (greater than 0.50) on statements 10 and 11, meaning that both production and office technologies changed rapidly or, in other words, the environment was technologically unstable.

8.7 Strategy Conditioning Variables

Principal components analysis was done using the 15 non-forced Likert statements concerning firm success factors. The first four principal components were chosen, since they had eigenvalues exceeding 1. The first principal component had loadings exceeding 0.30 on statements 13 and 15, meaning that the firms focused on performing R&D and on developing new products and processes; in other words, the firms were strongly oriented towards innovation. The second principal component had weightings exceeding 0.30 on statements 1, 2, 5 and 7 (negative). These firms thus sought to establish new markets, satisfy existing clients and promote their reputation, but did not hire new graduates from universities. The third principal component had loadings on statements 2, 9, 11, 13 (negative) and 15 (negative). These firms concentrated on satisfying existing clients, hiring experienced employees and training their employees, but did not perform R&D or develop new products and processes. In other words, these firms were not interested in innovation per se, but were more intent on getting the job done and maintaining the status quo. The fourth principal component had positive loadings on "developing export markets," "hiring new graduates from technical schools and colleges" and "recruiting skilled people from outside of Canada." There was a negative loading on "training employees," meaning that the firms looked to foreign markets for growth and wanted to hire well-trained domestic and foreign employees, but were not interested in doing the training themselves.

8.8 ASM Economic Variables

The construction of the ASM economic variables — price-cost margin, labour productivity, market share and employment — was discussed earlier. However, one serious shortcoming of the ASM data is the lack of two measures: one for the use of capital (which is an implicit rental rate for capital services) and another for the stock of capital owned by a firm. These data are not included in the ASM, but are collected in a separate survey by the Capital Stock Division within Statistics Canada. The capital stock survey and the ASM are not currently linked, and, according to Statistics Canada, there is no plan to link them in the near future.

To estimate long-run cost or production functions, implicit rental rates for capital are required (so that capital can be substituted for or by other factors of production). For short-run cost or production functions, a firm is assumed to face a fixed capital stock (net of depreciation). This is the approach that was used for the study reported here. For a pseudo-capital stock variable to be constructed, industry end-of-year net capital stock data (using geometric depreciation) were obtained from CANSIM. Then, the industry-level data were multiplied by a firm's market share (the firm's total value of shipments/industry value of shipments) to give the pseudo-capital stock for each firm. In other words, the assumption was that a firm's capital stock was directly proportional to its share of the total value of shipments within an industry (NAICS three-digit level). This assumption was clearly violated and biased against firms that relied on significant contracting out for parts, assembly or services. It was also violated for firms that underutilized their capital (due to slack economic periods, layoffs or loss of market share). However, it was the best approach for the job at hand. A summary of the variables, their means and standard deviations is provided in Table 20.

Table 20: Summary Statistics for Dependent and Independent Variables (Population Weighted)

Variable	*Label*	*N*	*Mean*	*Standard Deviation*
Innovation Novelty				
B_WORLD	World-first innovation, binary variable	5220	0.08546	0.3570
B_CDA	Canada-first innovation, binary variable	4740	0.15945	0.4691
B_FIRM	Firm-first innovation, binary variable	3951	0.74882	0.5581
Firm Size				
B_SIZELG	Large firm (250+ employees), binary variable	5220	0.13780	0.4401
B_SIZEMD	Medium-sized firm (50-249 employees), binary variable	5220	0.55673	0.6343
Technology Sector				
B_TECHH	High-tech sector, binary variable	5220	0.03367	0.2303
B_TECHM	Medium-tech sector, binary variable	5220	0.53492	0.6369
Intellectual Property Rights				
IP_PA	Patents, binary variable	5220	0.25928	0.5596
IP_TK	Trademarks, binary variable	5220	0.36250	0.6138
IP_CR	Copyrights, binary variable	5220	0.12023	0.4153
IP_CF	Confidentiality agreements, binary variable	5220	0.43926	0.6337
IP_TR	Trade secrets, binary variable	5220	0.24911	0.5523
ASM Economic Variables				
L_CAP	Natural log of capital	5220	1.37319	1.8716
L_CAP2	Natural log of capital squared	5220	4.03407	8.6305
L_E_C	Natural log of total employment times natural log of capital	5220	7.21955	11.4855
L_EMP	Natural log of total employment	5220	4.47184	1.2221
L_EMP2	Natural log of total employment squared	5220	20.91340	12.2262
L_SHIP	Natural log of total shipments	5220	9.56087	1.6375
L_TVA	Natural log of total value added	5220	8.71034	1.8014
L_WAGE	Natural log of wage rate	5212	2.66221	0.4211
LPROD	Labour productivity ($/employee)	5220	88293.39	112510.42
MSHR	Market share (multiply by 100 for %)	5220	0.00221	0.0085
PRFT	Price-cost margin (multiply by 100 for %)	5220	0.23860	0.2921
WAGERT	Wage rate ($/hr.)	5212	15.14199	6.7286
R&D Activity Within a Firm				
Rdact	In-house R&D program, binary variable	5220	0.59458	0.6269
Demand Conditioning Variables				
DEMAN_1	Demand variables, first principal component	5220	-0.00407	1.8935
DEMAN_2	Demand variables, second principal component	5220	-0.01352	1.5848
DEMAN_3	Demand variables, third principal component	5220	0.04467	1.4426
DEMAN_4	Demand variables, fourth principal component	5220	-0.02941	1.3769
Strategy Conditioning Variables				
STRAT_1	Strategy variables, first principal component	5220	-0.08157	2.7305
STRAT_2	Strategy variables, second principal component	5220	0.03892	1.5932
STRAT_3	Strategy variables, third principal component	5220	0.04842	1.4656
STRAT_4	Strategy variables, fourth principal component	5220	-0.01413	1.3341

Source: Statistics Canada, Survey of Innovation 1999. Tabulated by author.

9. STATISTICAL METHOD

Ordinary least-squares regressions were tried first and then the residuals were tested for heteroskedasticity. It was no surprise that the residuals did not follow a normal distribution, since most cross-sectional data are non-normal. Consequently, the generalized least-squares (GLS) procedure in SAS called SURVEYREG was used instead of ordinary least-squares regressions. The procedure is capable of using stratification information (that is, the number of sampled units within the population of each stratum) for estimation and calculation of standard errors; the Survey of Innovation had 31 strata. The procedure uses element-wise regression, and the variance-covariance matrix is estimated using a Taylor expansion for estimating the sampling errors of the estimates.

10. EMPIRICAL RESULTS[10]

It should be noted that causality and performance are difficult to disentangle. For instance, both of Schumpeter's hypotheses (that large firms conduct more R&D than small firms, and that greater industry concentration will stimulate R&D and innovation) predicted that firm size and concentration would lead to R&D and, hence, to positive performance. However, the opposite could also be true: that innovation causes firm growth and positive performance. With empirical regression estimates, all that can be said is that there is a statistically valid correlation between the dependent variable and the independent variables.

The conditional labour input demand equation was estimated as a Cobb-Douglas function (in natural logarithms). Total shipments had the expected positive sign, while wage rate had the expected negative sign, and both variables were significant at the 1% level (Table 21). The binary variable for world-first innovation was positive, which had a positive effect on employment. The magnitude of the world-first coefficient was greater than the coefficient for the firm-first binary variable. The Canada-first innovation measure was not statistically significant. The first and third principal components for the demand variables were significant, as were the third and fourth principal components for the strategy variables. The use of intellectual property rights

Table 21: Weighted GLS Equations

Parameter	*L_EMP Estimate*	*Standard Error*	*LPROD Estimate*	*Standard Error*	*MSHR Estimate*	*Standard Error*
Intercept	1.156*	0.0441	-10630.4*	3738.1	0.0007*	0.00008
L_SHIP	0.340*	0.0055				
L_WAGE	-0.219*	0.0133				
WAGERT			6173.9*	261.7		
PRFT					0.0004**	0.00018
B_WORLD	0.051*	0.0163	5129.0	5777.3	0.0008*	0.00030
B_CDA	-0.016	0.0136	-1118.0	3090.7	-0.0003***	0.00016
B_FIRM	0.023**	0.0104	-4588.1**	2333.1	-0.0002*	0.00009
DEMAN_1	0.006**	0.0026	-3609.4*	623.8	-0.0001*	0.00003
DEMAN_2	-0.004	0.0028	2365.5*	605.4	0.0001*	0.00004
DEMAN_3	0.009*	0.0030	-4661.7*	533.6	-0.0002*	0.00004
DEMAN_4	0.005	0.0032	-1675.2*	645.8	0.0000	0.00003
STRAT_1	0.003	0.0020	528.0	643.0	0.0001*	0.00002
STRAT_2	-0.005	0.0028	-687.6	796.0	-0.0002*	0.00003
STRAT_3	0.019*	0.0033	700.7	548.2	0.0001*	0.00003
STRAT_4	0.011*	0.0034	-502.4	904.5	0.0000	0.00003
IP_PA	-0.005	0.0088	9480.5*	1800.8	0.0006*	0.00013
IP_TK	0.003	0.0080	4149.9*	1420.3	0.0000	0.00011
IP_CR	0.024***	0.0123	8063.5**	3338.6	0.0007*	0.00019
IP_CF	0.011	0.0081	9609.0*	1630.9	-0.0002	0.00011
IP_TR	-0.023*	0.0087	10950.7*	2009.7	0.0006*	0.00012
Rdact	0.017**	0.0082	-576.1	1869.1	0.0000	0.00008
B_SIZEMD	0.713*	0.0089	-6135.9*	1707.5	0.0008*	0.00004
B_SIZELG	1.718*	0.0164	-5617.4**	2360.2	0.0084*	0.00027
B_TECHH	0.026***	0.0142	10097.8*	3238.7	-0.0011*	0.00030
B_TECHM	-0.036*	0.0076	309.1	1414.2	-0.0005*	0.00009
N	5212		5212		5220	
F-value	1559.4		55.1		60.2	
Prob>F	<0.0001		<0.0001		<0.0001	
R-square	0.874		0.189		0.203	

Source: Statistics Canada, Survey of Innovation 1999. Tabulated by author.
*1% level of significance. **5% level of significance. ***10% level of significance.

and the conduct of R&D within a firm had almost no effect on labour demand. Not surprisingly, size mattered, and large firms employed more workers than medium-sized and small firms. The medium-tech sector had fewer employees than the high-tech and low-tech sectors (this is confirmed by Table 11).

The labour productivity equation was estimated in levels. Wage rate had a positive and very significant effect on labour productivity. Its exact contribution to labour productivity was a firm's wage rate multiplied by 6,174. Both the world-first and Canada-first innovation variables were insignificant; however, firm-first innovators had lower productivity (in the order of $4,588) than non-innovators. All the demand variables were significant, while none of the strategy variables were. The use of intellectual property rights was very important in terms of explaining labour productivity, since all five types of intellectual property protection were significant at the 1% level. Labour productivity for large firms was $5,617 lower than for small firms, and it was $6,135 lower for medium-sized firms. Labour productivity for the high-tech sector was $10,098 more than for the low-tech sector.

The market share equation showed that profits (price-cost margins) contributed to market share at the 5% level of significance. A world-first innovation was correlated with firms that had 8% more market share than non-innovators. A firm-first innovation was correlated with firms that had 2% less market share than non-innovators. Three of the four demand variables were important (at the 1% level of significance), while three of the strategy variables were important. Patents, copyrights and trade secrets had a positive and significant effect on market share.

The value-added equations both had a suitable fit, although the simpler Cobb-Douglas function had a higher F-value than the translog specification (Table 22). In both cases, the economic variables possessed the expected signs. Returns to scale for the Cobb-Douglas equation were 1.02; the translog specification was very close to being linearly homogeneous (since the sum of L_EMP and L_CAP coefficients was very near 1, and the sum of L_EMP2, L_CAP2 and L_E_C was very near zero — which reduced the translog to a Cobb-Douglas). A world-first innovation would increase total value added (significant at the 5% level), while Canada-first and firm-first innovations would

Table 22: Weighted GLS Estimates for Total Value Added

Parameter	L_TVA (Cobb-Douglas)		L_TVA (Translog)	
	Estimate	Standard Error	Estimate	Standard Error
Intercept	5.262*	0.0656	5.125*	0.2264
L_EMP	0.594*	0.0138	0.738*	0.1118
L_EMP2			-0.029**	0.0145
L_CAP	0.428**	0.0097	0.263*	0.0698
L_CAP2			-0.049*	0.0082
L_E_C			0.074*	0.0199
B_WORLD	0.098**	0.0409	0.094**	0.0410
B_CDA	0.074**	0.0374	0.070***	0.0374
B_FIRM	0.056***	0.0287	0.055***	0.0288
DEMAN_1	-0.033*	0.0061	-0.032*	0.0061
DEMAN_2	-0.014***	0.0077	-0.012	0.0076
DEMAN_3	-0.015**	0.0069	-0.016**	0.0069
DEMAN_4	0.002	0.0072	0.002	0.0072
STRAT_1	0.018*	0.0042	0.016*	0.0041
STRAT_2	0.009	0.0074	0.009	0.0073
STRAT_3	0.051*	0.0094	0.052*	0.0095
STRAT_4	0.014	0.0107	0.014	0.0108
IP_PA	0.095*	0.0164	0.091*	0.0164
IP_TK	0.025	0.0179	0.026	0.0176
IP_CR	0.086*	0.0224	0.086*	0.0222
IP_CF	0.031	0.0195	0.035***	0.0192
IP_TR	-0.046**	0.0217	-0.038***	0.0214
Rdact	-0.009	0.0182	-0.013	0.0180
B_TECHH	0.325*	0.0197	0.306*	0.0198
B_TECHM	0.186*	0.0162	0.178*	0.0164
N	5220		5220	
F-value	516.6		457.7	
Prob>F	<0.0001		<0.0001	
R-square	0.6761		0.6789	

Source: Statistics Canada, Survey of Innovation 1999. Tabulated by author.
*1% level of significance. **5% level of significance. ***10% level of significance.

also increase value added but to a lesser extent (both were significant at the 10% level). The first and third demand variables were significant, as were the first and third strategy variables. The use of patents and copyrights was associated with higher value added. High-tech firms enjoyed higher value added than medium- and low-tech firms.

Table 23 summarizes the relationship between the demand and strategy variables and the four dependent variables for performance. There was a strong negative correlation between the first and third principal components for the demand variables and labour productivity, market share and value added. This means that firms whose demand environment comprised (1) threats from competitors, difficulty in hiring and retaining staff, and rapid production and office technology changes, or (2) no product threats, but difficulty in hiring and retaining employees had a lower performance level. For labour productivity and market share, the performance of firms that reported a demand environment in which demand and competitors' actions were easy to predict (represented by the DEMAN_2 variable) went up (i.e., labour productivity was enhanced and market share increased). With respect to the strategy variables, firms reporting strategies characterized by STRAT_1 were very concerned with R&D and innovation. These firms performed well in terms of

Table 23: Effects of Demand and Strategy Variables in Performance (GLS) Regressions

Variable	*Employment*	*Labour Productivity*	*Market Share*	*Total Value Added*
		(Effect on Performance Variable)		
DEMAN_1	(+)	(–)	(–)	(–)
DEMAN_2		(+)	(+)	
DEMAN_3	(+)	(–)	(–)	(–)
DEMAN_4		(–)		
STRAT_1			(+)	(+)
STRAT_2			(–)	
STRAT_3	(+)		(+)	(+)
STRAT_4				

Source: Statistics Canada, Survey of Innovation 1999. Tabulated by author.

market share and total value added. Firms with a strategy profile represented by STRAT_3 wanted to satisfy existing customers and to hire experienced employees, and were involved in training, but they were not very interested in R&D and innovation. This type of firm did well in terms of employment level, market share and total value added, despite not being strategically oriented towards innovation.

11. CONCLUSION

Evidence in support of Schumpeter's first hypcthesis can be found in Appendix 4 and in Figures 3 and 4. Two measures of size are market share and relative employment. When firms were separated into three quantile groups, it became evident that in-house R&D and the number of innovation activities performed by a firm increased as market share increased (Figure 3) and as relative employment increased (Figure 4). There was no apparent relationship between quantiles of labour productivity and in-house R&D or between quantiles of price-cost margins and in-house R&D. The number of innovation activities undertaken within a firm did not increase relative to quantiles of labour productivity or of price-cost margins. Cohen and Levin (1989) have argued that attention should be paid to the effects of innovation on firm growth. Unfortunately, the dataset did not allow for the testing of changes in

Figure 3: Market Share and Innovation Activities, 1997-99

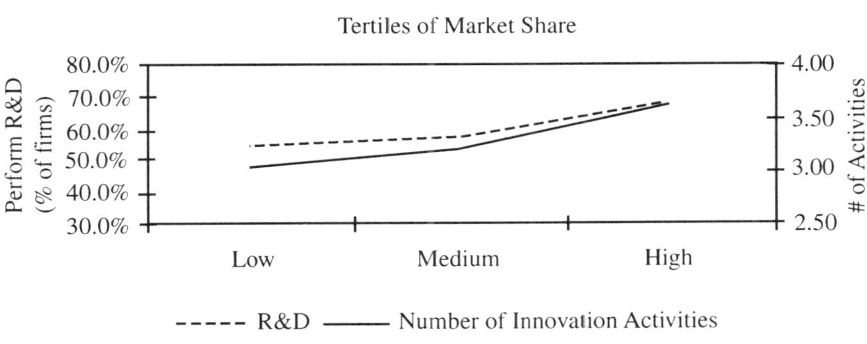

Source: Statistics Canada, Survey of Innovation 1999

Figure 4: Relative Employment and Innovation Activities, 1997-99

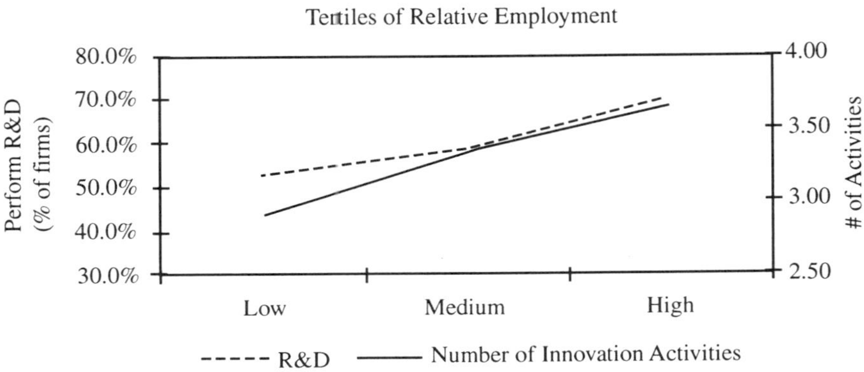

Source: Statistics Canada, Survey of Innovation 1999

growth of the performance variables against the existence of in-house R&D programs. An even more interesting test would be to see whether differential growth rates are related to innovation novelty. When the ASM data for 1999 are linked to the Survey of Innovation, it will be possible to calculate growth rates in the performance variables from 1997 to 1999, and these tests of the hypothesis can then be performed.

Schumpeter's second hypothesis was explored indirectly in the study presented here. Table 9 shows that the number of innovation activities increased with market share. Almost 66% of world-first innovators and almost 40% of firm-first innovators engaged in five innovation activities. Table 17 shows that the number of new product introductions increased monotonically with market share. It also increased with price-cost margin, both absolute and relative, but not monotonically.

Four main hypotheses were proposed earlier, relating firm performance to the novelty of an innovation. The first hypothesis was that innovative firms (world-first, Canada-first, firm-first) would have higher performance (in terms of specific performance measures) than non-innovative firms. Tables 6 and 7 show that market share, total employment, labour productivity and price-cost margins were all related to increased innovation novelty. Table 7 shows that, when firms were divided into three quantile groups on the basis of performance, monotonically increasing performance relative to innovation novelty persisted (the only exception being relative wages). In the multivariate analysis, world-first innovators did better than the rest on three of the four performance measures: employment, market share and total value added.

The second hypothesis was that the dichotomous innovation variables would be statistically different from zero (reject H_0) in the multivariate analysis. The results of the multivariate analysis were somewhat mixed, because the world-first innovator was more often than not very significant, but the Canada-first innovator was not significant beyond the 10% level in any of the performance (GLS) equations. The firm-first innovator was significant at the 10% level twice, at the 5% level twice and at the 1% level once.

The third hypothesis was that the estimated coefficients in the performance regressions would be greater for world-first innovations compared to firm-first innovations. In all cases, the ordering on coefficient size for the performance variables was world-, Canada- and firm-first (with world-first being the largest and firm-first being the smallest).

The fourth hypothesis was that firms in high-tech industries would perform better in terms of both the number of world-first innovations and overall performance than firms in medium- or low-tech industries. Table 11 shows that labour productivity and price-cost margin increased monotonically from the low- to the high-tech sector, but market share and total employment did not. When relative measures were considered, however, only price-cost margin increased monotonically from the low- to the high-tech sector. The regression analysis demonstrated that employment, labour productivity and total value added were greater in the high-tech sector than in the medium- and low-tech sectors. The exception was market share: high-tech firms had 0.11% lower market share than low-tech firms.

The contribution of the study reported here is that it used a novel measure for innovation that accounted for world-first innovations and examined them in relation to less noteworthy innovations. In general, world-first innovators are large firms that significantly outperform other firms within their own industry. They are very likely to undertake R&D, and are willing to spend over two years developing and bringing innovations to market (product innovations) or implementing them within the firm (process innovations). World-first innovators see intellectual property rights as very important — especially patents. They are also more apt to follow a strategy that involves aggressive pursuit of R&D and other innovation activities. World-first innovators perform better than other innovators and non-innovators with respect to market size, employment, labour productivity and profitability. They are, indeed, a distinct and important group of firms deserving of further exploration.

Appendix 1: Firm Performance Variables

Table A1.1: Means for Firm Performance Variables, by Industry

		Firm Performance Variable								
NAICS	*Industry*	*Market Share*	*Total Employees*	*Wage Rate* ($/hr.)*	*Labour Productivity* ($/employee)*	*Price-Cost Margin*	*Relative Employment*	*Relative Wage Rate*	*Relative Labour Productivity*	*Relative Price-Cost Margin*
311	Food	0.114%	198	14.00	94,000	20.9%	0.099%	97.3%	115.1%	106.6%
312	Beverage & Tobacco	0.965%	269	19.00	191,000	27.1%	0.927%	86.5%	88.9%	57.7%
313	Textile Mills	0.607%	144	13.00	67,000	22.6%	0.547%	99.9%	104.4%	99.5%
314	Textile Product	0.673%	112	10.00	48,000	14.1%	0.642%	92.8%	90.2%	66.6%
315	Apparel	0.166%	134	9.00	48,000	20.0%	0.160%	102.7%	114.7%	83.0%
316	Leather	1.269%	144	11.00	45,000	20.2%	1.238%	96.6%	111.7%	118.6%
321	Wood	0.127%	140	14.00	73,000	18.3%	0.117%	86.2%	90.1%	93.2%
322	Paper	0.333%	324	19.00	106,000	23.3%	0.319%	84.0%	84.0%	90.4%
323	Printing	0.169%	119	17.00	73,000	24.3%	0.151%	105.6%	117.8%	104.6%
324	Petroleum & Coal	2.064%	267	22.00	197,000	14.8%	1.901%	88.0%	114.5%	202.9%
325	Chemical	0.212%	169	18.00	178,000	28.3%	0.186%	100.1%	107.2%	84.8%
326	Plastics & Rubber	0.179%	163	14.00	85,000	27.2%	0.152%	94.3%	119.2%	111.6%
327	Non-metallic Mineral	0.272%	114	16.00	100,000	26.8%	0.246%	92.7%	103.2%	84.5%
331	Primary Metal	0.414%	370	19.00	113,000	25.1%	0.402%	84.9%	89.7%	107.8%
332	Fabricated Metal	0.072%	87	16.00	79,000	25.2%	0.059%	104.9%	116.3%	108.2%
333	Machinery	0.097%	115	17.00	91,000	26.9%	0.087%	98.7%	107.6%	102.0%
334	Computer & Electronics	0.286%	220	15.00	101,000	26.4%	0.241%	91.4%	92.4%	98.6%
335	Electrical Equipment	0.446%	174	16.00	89,000	14.8%	0.385%	99.3%	110.2%	62.4%
336	Transportation Equipment	0.229%	486	17.00	86,000	23.7%	0.215%	81.8%	65.7%	120.3%
337	Furniture	0.190%	120	12.00	62,000	26.3%	0.167%	102.1%	111.6%	105.7%
339	Miscellaneous Manufacturing	0.245%	89	12.00	81,000	25.8%	0.178%	105.4%	142.2%	94.3%

Source: Statistics Canada, Survey of Innovation 1999. Tabulated by author.

*These figures are rounded. They have been used to establish the ranking of the industries. Data from the Annual Survey of Manufactures can be found in CANSIM table 301-0003.

Table A1.2: Ranking of Means for Firm Performance Variables, by Industry

		Ranking of Firm Performance Variable								
NAICS	*Industry*	*Market Share*	*Total Employees*	*Wage Rate*	*Labour Productivity*	*Price-Cost Margin*	*Relative Employment*	*Relative Wage Rate*	*Relative Labour Productivity*	*Relative Price-Cost Margin*
311	Food	19	7	15	8	15	19	10	5	7
312	Beverage & Tobacco	3	4	3	2	3	3	17	19	21
313	Textile Mills	5	11	16	17	14	5	7	13	11
314	Textile Product	4	19	20	20	21	4	13	16	19
315	Apparel	17	14	21	19	17	15	4	6	18
316	Leather	2	12	19	21	16	2	11	8	3
321	Wood	18	13	13	16	18	18	18	17	14
322	Paper	8	3	4	5	13	8	20	20	15
323	Printing	16	16	6	15	11	17	1	3	9
324	Petroleum & Coal	1	5	1	1	19	1	16	7	1
325	Chemical	13	9	5	3	1	12	6	12	16
326	Plastics & Rubber	15	10	14	12	2	16	12	2	4
327	Non-metallic Mineral	10	18	9	7	5	9	14	14	17
331	Primary Metal	7	2	2	4	10	6	19	18	6
332	Fabricated Metal	21	21	10	14	9	21	3	4	5
333	Machinery	20	17	8	9	4	20	9	11	10
334	Computer & Electronics	9	6	12	6	6	10	15	15	12
335	Electrical Equipment	6	8	11	10	20	7	8	10	20
336	Transportation Equipment	12	1	7	11	12	11	21	21	2
337	Furniture	14	15	18	18	7	14	5	9	8
339	Miscellaneous Manufacturing	11	20	17	13	8	13	2	1	13

Source: Statistics Canada, Survey of Innovation 1999. Tabulated by author.

Appendix 2: Principal Components Analysis

Table A2.1: Eigenvalues for 11 Demand Categories – Question 1 (N = 5,220, Variables = 11)

Principal Component	*Eigenvalue*	*Difference*	*Proportion*	*Cumulative*
1	2.190	0.706	0.199	0.199
2	1.484	0.220	0.135	0.334
3	1.264	0.120	0.115	0.449
4	1.143	0.199	0.104	0.553
5	0.945	0.036	0.086	0.639
6	0.909	0.088	0.083	0.721
7	0.821	0.191	0.075	0.796
8	0.630	0.032	0.057	0.853
9	0.598	0.035	0.054	0.908
10	0.564	0.110	0.051	0.959
11	0.453		0.041	1.000

Source: Statistics Canada, Survey of Innovation 1999. Tabulated by author.

Table A2.2: Eigenvalues for 16 Success Factors – Question 2 (N = 5,220, Variables = 16)

Principal Component	*Eigenvalue*	*Difference*	*Proportion*	*Cumulative*
1	4.507	2.956	0.282	0.282
2	1.551	0.287	0.097	0.379
3	1.264	0.171	0.079	0.458
4	1.093	0.164	0.068	0.526
5	0.929	0.101	0.058	0.584
6	0.828	0.075	0.052	0.636
7	0.753	0.021	0.047	0.683
8	0.732	0.031	0.046	0.729
9	0.701	0.058	0.044	0.772
10	0.644	0.020	0.040	0.813
11	0.624	0.047	0.039	0.852
12	0.577	0.035	0.036	0.888
13	0.542	0.026	0.034	0.922
14	0.516	0.119	0.032	0.954
15	0.397	0.053	0.025	0.979
16	0.344		0.022	1.000

Source: Statistics Canada, Survey of Innovation 1999. Tabulated by author.

Appendix 3: Tests for Normal Distribution and Wilcoxon Rank Score Test for Difference in Means

Table A3.1: Tests for Normal Distribution of Questions 1-2

Variable	*Kolmogorov-Smirnov*	*Pr > D*	*Cramer-von Mises*	*Pr > W^2*	*Anderson-Darling*	*Pr > A^2*	*H_0: ~N*
MSHR	0.363	0.010	225.72	0.005	1105.88	0.005	Reject All 3
R_EMP	0.348	0.010	204.63	0.005	1010.57	0.005	Reject All 3
R_LPROD	0.195	0.010	83.61	0.005	469.02	0.005	Reject All 3
R_PRFT	0.120	0.010	28.09	0.005	170.37	0.005	Reject All 3
DEMAN_1	0.039	0.010	2.22	0.005	15.62	0.005	Reject All 3
DEMAN_2	0.024	0.010	0.82	0.005	6.01	0.005	Reject All 3
DEMAN_3	0.010	>0.150	0.08	0.211	0.60	0.120	Cannot Reject All 3
DEMAN_4	0.028	0.010	1.20	0.005	7.10	0.005	Reject All 3
STRAT_1	0.054	0.010	4.40	0.005	27.76	0.005	Reject All 3
STRAT_2	0.040	0.010	2.68	0.005	18.50	0.005	Reject All 3
STRAT_3	0.030	0.010	1.58	0.005	9.78	0.005	Reject All 3
STRAT_4	0.0235	0.010	0.981	0.005	6.561	0.005	Reject All 3

Source: Statistics Canada, Survey of Innovation 1999. Tabulated by author.

Table A3.2: Wilcoxon Rank Score Test for Difference in Means

Comparison of Five Innovation Categories	*Kruskal-Wallis Statistic (Chi-Square)*	*Pr > Chi-Square*	H_0: $\mu_1 = \mu_2$
World-First vs. Canada-First			
MSHR	1.698	0.193	Cannot Reject
R_EMP	2.536	0.111	Cannot Reject
R_LPROD	0.715	0.398	Cannot Reject
R_PRFT	1.689	0.194	Cannot Reject
DEMAN_1	0.193	0.660	Cannot Reject
DEMAN_2	0.542	0.462	Cannot Reject
DEMAN_3	2.577	0.108	Cannot Reject
DEMAN_4	0.392	0.531	Cannot Reject
STRAT_1	20.928	<0.0001	Reject
STRAT_2	3.169	0.075	Reject
STRAT_3	8.008	0.005	Reject
STRAT_4	2.708	0.100	Cannot Reject
World-First vs. Firm-First			
MSHR	47.028	<0.0001	Reject
R_EMP	46.116	<0.0001	Reject
R_LPROD	16.552	<0.0001	Reject
R_PRFT	7.481	0.006	Reject
DEMAN_1	2.978	0.084	Reject
DEMAN_2	0.277	0.599	Cannot Reject
DEMAN_3	4.057	0.044	Reject
DEMAN_4	7.424	0.006	Reject
STRAT_1	134.142	<0.0001	Reject
STRAT_2	42.509	<0.0001	Reject
STRAT_3	45.793	<0.0001	Reject
STRAT_4	1.707	0.191	Cannot Reject
World-First vs. "Unknown"			
MSHR	43.975	<0.0001	Reject
R_EMP	42.005	<0.0001	Reject
R_LPROD	19.078	<0.0001	Reject
R_PRFT	7.508	0.006	Reject
DEMAN_1	0.468	0.494	Cannot Reject
DEMAN_2	0.271	0.603	Cannot Reject
DEMAN_3	2.734	0.098	Reject
DEMAN_4	3.392	0.066	Reject
STRAT_1	129.940	<0.0001	Reject
STRAT_2	27.555	<0.0001	Reject
STRAT_3	47.402	<0.0001	Reject
STRAT_4	1.006	0.316	Cannot Reject

Source: Statistics Canada, Survey of Innovation 1999. Tabulated by author.

Table A3.2: Wilcoxon Rank Score Test for Difference in Means – Continued

Comparison of Five Innovation Categories	*Kruskal-Wallis Statistic (Chi-Square)*	*Pr > Chi-Square*	$H_0: \mu_1 = \mu_2$
World-First vs. Non-Innovator			
MSHR	64.7473	<0.0001	Reject
R_EMP	84.939	<0.0001	Reject
R_LPROD	13.2006	0.0003	Reject
R_PRFT	12.764	0.0004	Reject
DEMAN_1	30.6131	<0.0001	Reject
DEMAN_2	1.661	0.1975	Cannot Reject
DEMAN_3	11.1413	0.0008	Reject
DEMAN_4	39.7261	<0.0001	Reject
STRAT_1	450.7302	<0.0001	Reject
STRAT_2	22.7135	<0.0001	Reject
STRAT_3	106.0893	<0.0001	Reject
STRAT_4	2.278	0.131	Cannot Reject
Canada-First vs. Firm-First			
MSHR	44.7166	<0.0001	Reject
R_EMP	36.8253	<0.0001	Reject
R_LPROD	13.9764	0.0002	Reject
R_PRFT	2.2578	0.1329	Cannot Reject
DEMAN_1	2.4271	0.1193	Cannot Reject
DEMAN_2	2.5344	0.1114	Cannot Reject
DEMAN_3	0.0398	0.8418	Cannot Reject
DEMAN_4	5.5448	0.0185	Reject
STRAT_1	72.9843	<0.0001	Reject
STRAT_2	31.3443	<0.0001	Reject
STRAT_3	23.6612	<0.0001	Reject
STRAT_4	0.280	0.597	Cannot Reject
Canada-First vs. "Unknown"			
MSHR	41.0413	<0.0001	Reject
R_EMP	32.8254	<0.0001	Reject
R_LPROD	16.5252	<0.0001	Reject
R_PRFT	2.8113	0.0936	Reject
DEMAN_1	0.0699	0.7915	Cannot Reject
DEMAN_2	0.0932	0.7601	Cannot Reject
DEMAN_3	0.0003	0.9873	Cannot Reject
DEMAN_4	2.0285	0.1544	Cannot Reject
STRAT_1	65.5584	<0.0001	Reject
STRAT_2	17.3536	<0.0001	Reject
STRAT_3	25.2537	<0.0001	Reject
STRAT_4	0.623	0.430	Cannot Reject

Source: Statistics Canada, Survey of Innovation 1999. Tabulated by author.

Table A3.2: Wilcoxon Rank Score Test for Difference in Means – Continued

Comparison of Five Innovation Categories	*Kruskal-Wallis Statistic (Chi-Square)*	*Pr > Chi-Square*	H_0: $\mu_1 = \mu_2$
Canada-First vs. Non-Innovator			
MSHR	65.3258	<0.0001	Reject
R_EMP	81.028	<0.0001	Reject
R_LPROD	10.4462	0.0012	Reject
R_PRFT	7.2277	0.0072	Reject
DEMAN_1	52.094	<0.0001	Reject
DEMAN_2	5.0313	<0.0249	Reject
DEMAN_3	4.2928	0.0383	Reject
DEMAN_4	44.3696	<0.0001	Reject
STRAT_1	487.9384	<0.0001	Reject
STRAT_2	14.4121	0.0001	Reject
STRAT_3	95.2053	<0.0001	Reject
STRAT_4	12.3301	0.0004	Reject
Firm-First vs. "Unknown"			
MSHR	0.288	0.591	Cannot Reject
R_EMP	0.152	0.697	Cannot Reject
R_LPROD	1.090	0.297	Cannot Reject
R_PRFT	0.115	0.734	Cannot Reject
DEMAN_1	1.936	0.164	Cannot Reject
DEMAN_2	1.947	0.163	Cannot Reject
DEMAN_3	0.044	0.833	Cannot Reject
DEMAN_4	0.745	0.388	Cannot Reject
STRAT_1	0.025	0.874	Cannot Reject
STRAT_2	0.808	0.369	Cannot Reject
STRAT_3	0.349	0.555	Cannot Reject
STRAT_4	0.0798	0.7776	Cannot Reject
Firm-First vs. Non-Innovator			
MSHR	7.427	0.006	Reject
R_EMP	22.452	<0.0001	Reject
R_LPROD	0.022	0.883	Cannot Reject
R_PRFT	2.820	0.093	Reject
DEMAN_1	105.1451	<0.001	Reject
DEMAN_2	1.0917	0.2961	Cannot Reject
DEMAN_3	5.1572	0.0232	Reject
DEMAN_4	32.2743	<0.0001	Reject
STRAT_1	366.0939	<0.0001	Reject
STRAT_2	0.4723	0.4919	Cannot Reject
STRAT_3	49.8039	<0.001	Reject
STRAT_4	12.3848	0.0004	Reject

Source: Statistics Canada, Survey of Innovation 1999. Tabulated by author.

Table A3.2: Wilcoxon Rank Score Test for Difference in Means – Concluded

Comparison of Five Innovation Categories	*Kruskal-Wallis Statistic (Chi-Square)*	*Pr > Chi-Square*	H_0: $\mu_1 = \mu_2$
"Unknown" vs. Non-Innovator			
MSHR	3.793	0.052	Reject
R_EMP	15.271	<0.0001	Reject
R_LPROD	0.619	0.431	Cannot Reject
R_PRFT	1.514	0.219	Cannot Reject
DEMAN_1	70.7808	<0.0001	Reject
DEMAN_2	4.4326	0.0353	Reject
DEMAN_3	4.7532	0.0292	Reject
DEMAN_4	32.943	<0.0001	Reject
STRAT_1	325.4051	<0.0001	Reject
STRAT_2	0.0029	0.9567	Cannot Reject
STRAT_3	34.5524	<0.0001	Reject
STRAT_4	8.9281	0.0028	Reject

Source: Statistics Canada, Survey of Innovation 1999. Tabulated by author.

Appendix 4: Quantiles of Performance Variables, Innovation Activities and Firm Size

Table A4.1: Three Quantiles of Performance Variables

Market Share	*Mean Market Share*	*R&D In-House*	*Number of Innovation Activities*
Low	0.02%	54.8%	3.03
Medium	0.08%	57.3%	3.21
High	0.66%	68.3%	3.62
Relative Employment	*Mean Relative Employment*	*R&D In-House*	*Number of Innovation Activities*
Low	0.03%	52.2%	2.91
Medium	0.09%	59.0%	3.33
High	0.56%	70.1%	3.64
Relative Wage Rate	*Mean Relative Wage Rate*	*R&D In-House*	*Number of Innovation Activities*
Low	68.3%	59.1%	3.32
Medium	92.3%	60.2%	3.22
High	128.2%	59.2%	3.23
Relative Labour Productivity	*Mean Relative Labour Productivity*	*R&D In-House*	*Number of Innovation Activities*
Low	45.9%	59.0%	3.20
Medium	89.4%	58.1%	3.28
High	183.3%	61.3%	3.29
Relative Price-Cost Margin	*Mean Relative Price-Cost Margin*	*R&D In-House*	*Number of Innovation Activities*
Low	16.5%	56.3%	3.08
Medium	100.4%	60.3%	3.36
High	180.1%	61.7%	3.32

Source: Statistics Canada, Survey of Innovation 1999. Tabulated by author.

Table A4.2: Three Quantiles of Performance Variables and Firm Size (Number of Employees)

Performance Variables	*20-49*	*50-249*	*250+*
		(%)	
Market Share			
Low	66.6	29.1	1.5
Medium	28.6	44.6	12.7
High	4.7	26.3	85.8
Total	100.0	100.0	100.0
Relative Employment			
Low	81.0	23.8	0.0
Medium	15.6	51.9	4.8
High	3.4	24.3	95.2
Total	100.0	100.0	100.0
Relative Wage Rate			
Low	35.8	31.9	19.2
Medium	36.4	35.8	27.0
High	27.9	32.2	53.9
Total	100.0	100.0	100.0
Relative Labour Productivity			
Low	34.9	32.8	24.8
Medium	30.6	35.8	33.0
High	34.6	31.5	42.2
Total	100.0	100.0	100.0
Relative Price-Cost Margin			
Low	36.1	31.4	30.2
Medium	31.0	34.7	34.5
High	32.9	33.8	35.2
Total	100.0	100.0	100.0

Source: Statistics Canada, Survey of Innovation 1999. Tabulated by author.

Appendix 5: Objectives, Problems and Impact of Innovation

Table 5A.1: Objectives of Innovation, by Novelty of Innovation, 1997-99

Objectives of Innovation	*World-First*	*Canada-First*	*Firm-First*	*Non-Innovator*
	(% of Firms in Strong Agreement)			
Reduce labour costs	56.8	57.2	59.9	18.9
Increase production capacity	67.5	74.0	77.0	23.5
Reduce production time	67.2	63.3	68.2	20.3
Improve production flexibility	61.0	63.5	67.1	18.9
Extend product range	81.7	78.1	68.9	14.7
Improve product quality	83.1	84.1	82.2	23.7
Increase speed to market	63.5	61.6	64.1	20.5
Replace products being phased out	42.2	42.1	34.9	6.3
Reduce materials consumption	34.8	39.9	34.6	10.7
Reduce environmental damage	27.6	28.9	23.0	9.3
Reduce energy consumption	23.8	25.8	22.0	7.7
Deal with new government regulations	18.2	19.8	17.3	6.4

Source: Statistics Canada, Survey of Innovation 1999. Tabulated by author.

Table 5A.2: Problems with Innovation, by Novelty of Innovation, 1997-99

Problems with Innovation	*World-First*	*Canada-First*	*Firm-First*	*Non-Innovator*
	(% of Firms in Strong Agreement)			
High cost	60.0	60.7	50.5	15.7
Inability to devote staff to projects	60.2	59.2	54.7	15.5
Unable to qualify for government assistance programs	16.7	14.8	15.5	4.8
Lack of skilled personnel	43.2	37.7	37.2	11.7
Lack of financing	29.3	27.4	25.0	8.2
Lack of marketing capability	18.8	18.5	17.1	5.6
Lack of information on technology	16.4	16.0	13.6	4.5
Lack of external technical support services	13.3	13.9	11.7	4.1
Lack of access to expertise in universities	7.4	6.4	4.4	1.8
Lack of access to expertise in government laboratories	5.9	5.6	3.6	0.9
Lack of cooperation with other firms	6.5	8.1	5.8	1.5
Lack of customer responsiveness to new products	22.1	22.4	17.9	4.8
Organizational rigidities	20.7	16.5	20.1	7.5
Government regulations	11.5	11.9	9.9	2.1
Other	14.2	13.4	10.2	2.7

Source: Statistics Canada, Survey of Innovation 1999. Tabulated by author.

Table 5A.3: Impact of Innovation, by Novelty of Innovation, 1997-99

Impact of Innovation	*World-First*	*Canada-First*	*Firm-First*	*Non-Innovator*
	(% of Firms in Strong Agreement)			
Increase productivity	49.5	56.5	56.9	15.1
Increase profitability	60.7	61.7	55.2	14.6
Increase delivery speed	38.7	40.0	45.4	12.0
Increase flexibility	65.2	66.1	62.6	12.4
Increase domestic market share	48.1	50.1	43.6	8.4
Increase international market share	56.0	47.0	33.6	6.4
Maintain profit margins	64.0	67.4	56.0	11.7
Keep up with competitors	77.2	81.4	75.6	16.5

Source: Statistics Canada, Survey of Innovation 1999. Tabulated by author.

Notes

The author would like to thank Can D. Le of the Marketplace Innovation Division within Industry Canada for the invitation to participate in the analysis of the Survey of Innovation. Without his commitment and support, this project would not have been completed. Thanks to Pierre Therrien (also from Industry Canada) and to Sean Cahill (from Agriculture and Agri-Food Canada) for providing helpful comments and suggestions.

Thanks as well to the staff of the Science, Innovation and Electronic Information Division within Statistics Canada for their generous assistance: Frances Anderson, Fred Gault, Brian Nemes, Guy Sabourin and Susan Schaan.

The author alone is responsible for the opinions expressed in this chapter and for any errors and omissions that may remain.

[1]According to Cohen and Levin (1989), the notable feature of empirical work testing Schumpeter's first hypothesis is its inconclusiveness. They cite the use of non-random samples for testing, and the fact that there may be sample selection bias present — for instance, some studies concentrate on 500-1,000 of the largest manufacturing firms. Scott (1984) has pointed out that firm and industry effects may be collinear. One way around the collinearity problem is to use separate regressions or correlates for different industries (Scherer 1984). It is also possible to use fixed effects specifications for different industries (Bound, Cummins, Griliches, Hall and Jaffe 1984; Scott 1984; Cohen, Levin and Mowery 1987). Cohen and Levin (1989) have pointed out that size and innovation studies have a common problem: they overlook the effect of innovation on firm growth, and the effect that growth has on a firm's size.

[2]The output of R&D expenditure is (hopefully) innovation (new knowledge). Innovations can be embodied in new products or new processes that enhance profits. However, the transformation from R&D expenditure to innovation is a stochastic, idiosyncratic process. In other words, the R&D–innovation transformation is not nearly as deterministic as is the production of machinery in a factory.

[3]CEOs from 5,944 firms were asked to complete questionnaires and 5,455 responded (some firms were out of scope, and some no longer existed). However, 50 firms that were sampled had fewer than 20 employees or a gross income of less than $250,000; they were removed. Firms in Logging (NAICS 1133), Mining (NAICS 212) and Utilities (NAICS 221) were also removed at the request of Statistics Canada. The final number of firms included for analysis was 5,220.

[4]Question 5 of the survey asked if, in the period 1997-99, the firm had "any unsuccessful or not yet completed projects to develop or introduce new or significantly improved products (goods or services) or production/manufacturing processes." In accordance with the Oslo definition of innovation (OECD/Eurostat 1997), firms that indicated that they had not had any innovations other than unsuccessful or uncompleted projects were considered to be non-innovators.

[5]The Kruskal-Wallis statistic was checked for H_0: $\mu_1 = \mu_2$. The test statistic follows a chi-square distribution.

[6]The classification of NAICS industries into high-, medium- and low-tech sectors follows the OECD (Hatzichronoglou 1997) and is based on R&D intensity. Industries in the high-tech sector include: Pharmaceutical and Medicine (3254);

Computer and Electronics (334); and Aerospace Product and Parts (3364). The low-tech sector includes: Food (311); Beverage and Tobacco (312); Textile Mills (313); Textile Product (314); Apparel (315); Leather (316); Wood (321); Paper (322); Printing (323); and Furniture (337). The medium-tech sector consists of: Petroleum and Coal (324); Chemical (except Pharmaceutical and Medicine) (325); Plastics and Rubber (326); Non-metallic Mineral (327); Primary Metal (331); Fabricated Metal (332); Machinery (333); Navigational, Measuring, Medical and Control Instruments (3345); Magnetic and Optical Media (3346); Electrical Equipment (335); Transportation Equipment (except Aerospace Product and Parts) (336); and Miscellaneous Manufacturing (339).

[7]It should be noted that it was unnecessary to use the relative performance measures described in the previous section as dependent variables as long as industry dummy variables were employed in the regressions.

[8]The production worker wage rate was used here as a proxy for the wage rate of all workers. A wage rate for salaried employees could not be constructed, since their hours worked were not reported.

[9]Total value added was defined as total shipments minus the costs of fuel, electricity and materials. Therefore, instead of a K (capital), L (labour), E (energy), M (materials) function being estimated, only a K, L function was estimated.

[10]Regressions were performed at the three-digit NAICS level, using 21 industry dummies. They are not included here for the sake of brevity. Instead, they can be obtained from the author upon request.

Objectives, Outcomes and Firm Characteristics

Section Overview

Fred Gault

The issues in this section are aspects of why innovation happens and the outcomes that result.

Tang looks at the influence of business objectives of firms and of government support programmes on innovation-related activities. The two broad innovation-related activities are technology invention and technology adoption. Technology invention consists of the doing of research and development, the use of patents, trademarks, or copyright, and engaging in industrial engineering and design. Technology adoption also consists of the doing of research and development as well as engaging in industrial engineering and design and in technology acquisition. The government support programmes are R&D tax credits, R&D grants, and technology support and assistance programmes. The business objectives are two: to develop products; or, to improve production processes. In the course of the analysis, firms are also classified as large or small, and as high-tech or low-tech.

Tang finds that the use of government support programmes, such as R&D tax credits and grants, and technology support and assistance programmes have a positive effect on both the innovation-related activities of technology adoption and of technology invention. The effect is pervasive across size and industrial categories. As for the business objectives, developing products is linked to technology invention and improving production processes is linked to technology adoption.

Le and Tang look at the different outcomes of innovation-related activities in the firm. They find that R&D is more related to product innovation, while technology acquisition is important for process innovation. Collaboration, or co-operation, is significantly related to innovation, especially for larger firms, but not for sales, in the case of product innovations. All of these results vary with industry.

Le and Tang also find that small firms are as effective as large firms at converting innovation-related activities, such as R&D, into innovations, but small firms have a lower propensity to engage in these innovation-related activities. The result, seen elsewhere in this book, is that small firms are less likely to be innovative than large firms.

7

Business Objectives and Innovation-Related Activities: Evidence from Canadian Manufacturing Firms

Jianmin Tang

1. INTRODUCTION

Innovation has become an important focus of virtually every country's policy agenda. Indeed, the Canadian Speech from the Throne in 2001 reiterated the importance of innovation and set up an ambitious innovation agenda for Canadians for the following 10 years: "An innovation economy is essential to creating economic opportunity for Canadians. ... Our objective should be no less than to be recognized as one of the most innovative countries in the world. ... We must strive for Canada to become one of the top five countries for research and development performance by 2010." Canada is currently lagging behind many of its key competitors in innovation according to measures such as patents (Trajtenberg 2000), R&D intensity (Le and Tang 2001) and technology adoption (Baldwin and Sabourin 1996). In addition, Canada is a small open economy with a high degree of foreign ownership and an important resource sector. It is a challenge for Canada to strive to become a more innovative country, and business communities as well as governments have a role to play. To encourage firms to become more innovative, it is first necessary to understand why firms innovate.

A firm's business operation consists of a series of business objectives. Innovation is a vehicle to realize those objectives. There have, however, been few studies that directly link business objectives to innovation-related activities. In the study presented in this chapter, some light was shed on the relationship between business objectives and innovation-related activities through the testing of three hypotheses:

(i) Innovation-related activities, business objectives and the use of government support programs vary across industries and among firms of different sizes.
(ii) Different innovation-related activities are linked to different business objectives.
(iii) Government support programs have a significant impact on firms' innovation, but the impact is different for different types of innovation-related activities.

Innovation has many dimensions. In the study presented here, only two very broad innovation-related activities were considered: technology invention and technology adoption.[1] These two innovation-related activities differ substantially in terms of their origin and commercialization. A firm's technology invention is its own initiative and is proprietary. It may lead to unique products or processes for the firm. However, it may take a fairly long time for the firm to reap benefits from the invention, if there are any. In contrast, technology adoption means that a firm buys the technology from another firm and integrates the bought technology into its overall business. Often, such technology is a commercialized invention and its impact is immediate. Despite the differences, technology invention and technology adoption share some similarities: for example, both need R&D and industrial engineering and industrial design as innovation inputs. In the case of technology adoption, a firm needs to engage in R&D and industrial engineering and industrial design in order to absorb and integrate an acquired technology into its overall business operation.

A firm may strive to achieve several business objectives over a given period. These objectives could be related to products or production processes, but could also be associated with human resources and managerial practices. In the study presented here, attention was restricted to business objectives related to products and production processes. For simplicity's sake, two types of business objectives were identified: "to develop products" and "to improve production processes." The first is associated with extending product range or replacing products being phased out; the second is related to increasing production capacity, reducing production time, improving production flexibility and product quality, and reducing labour costs, materials consumption and energy consumption. Clearly, the business objective not only aims to increase profit margins, but also is driven by such non-cost-related initiatives as increasing production capacity and reducing production time.

In addition to business objectives, innovation is motivated and facilitated by various innovation-related government support programs. In the study presented here, government support programs (hereafter, government support) were restricted to R&D tax credits, R&D grants, and government technology support and assistance programs.

To examine the impact of business objectives and government support on innovation-related activities, the relationship between innovation-related activities, business objectives and government support needed to be determined. To this end, innovation-related activities, business objectives and government support were modelled as latent variables. A latent variable is unobservable, but is indicated by multiple observed variables. Business objectives and government support were linked to innovation-related activities through a structural equation model.

The data used for the study presented in this chapter were obtained from the Statistics Canada Survey of Innovation 1999, which has been linked to the 1997 Annual Survey of Manufacturers (ASM). The focus of the Survey of Innovation was the Canadian manufacturing sector and selected natural resources industries for which the sample unit was the provincial enterprise. A provincial enterprise (hereafter, firm) included all establishments in the same province and industry at the four-digit NAICS (North American Industry Classification System) level. To reduce the response burden, the Survey of Innovation surveyed firms with a gross business income of at least $250,000 and more than 19 employees in 1999.[2] All information in the survey concerns firms' innovation environment and related activities during the period 1997-99.[3] After linkage to the 1997 ASM, the Survey of Innovation offers additional information on firms' operational activities, such as employment, in 1997. The linked Survey of Innovation database was used for the study described here.

The linked database contains data on 5,455 manufacturing firms. Each firm is assigned a weight, which allows that firm to represent other firms in the population with similar characteristics. Thus, for example, if the weight given to firm X is 5, firm X represents five firms in the sample. To represent these five firms, the data obtained from the sampled firm are multiplied by its weight, in this case 5. The total population of manufacturing firms is 8,921, a number that is equal to the sum of the population weights of the in-sample firms. For the purposes of the study presented here, however, 654 in-sample firms with incomplete information were excluded from the total sample.[4] Thus, the final sample contained data on 4,801 in-sample firms, which represented a subpopulation of 7,800 manufacturing firms.

2. INNOVATION-RELATED ACTIVITIES, BUSINESS OBJECTIVES AND GOVERNMENT SUPPORT

In this section, variables in the linked Survey of Innovation that are relevant to innovation-related activities, business objectives and government support are identified and described.

2.1 Innovation-Related Activities

As mentioned earlier, technology invention and technology adoption were the two innovation-related activities that were considered. From the linked Survey of Innovation, six main innovation-related activity indicators associated with these two innovation-related activities were identified. Denoted as y-variables, these indicators are listed below, with the Survey of Innovation question number in parentheses ({ }):

y_1 patents {25a}

y_2 trademarks {25b}

y_3 copyrights {25c}

y_4 R&D {24a}

y_5 industrial engineering and industrial design {6c}

y_6 acquisition of machinery, equipment or other technologies {6b}.

All of these six indicators were binary ordinal variables. For instance, if a firm used patents to protect its intellectual property during the period 1997-99, the variable for patents equalled 1; otherwise, it equalled 0. Similarly, if a firm engaged in R&D activities during the period 1997-99, the R&D variable equalled 1; otherwise, it equalled 0. The same applied to other innovation-related activity indicators.

However, none of them was a perfect measure of technology invention or technology adoption. Each of them was subject to a measurement error. That is why they are called indicators. For instance, a patent is a good indicator for technology invention, but it is not perfect. Not all inventions are patentable, and not all patentable inventions are patented. Some firms may find it more desirable to use trademarks, copyrights and other methods to protect their intellectual property. Similarly, acquisition of technologies is an excellent indicator for technology adoption, but it is not perfect since acquisition of technologies is only part of technology adoption. There are other aspects of technology adoption, such as industrial engineering and industrial design. Finally, some of those indicators can be good indicators for both technology invention and technology adoption. Those are R&D and industrial engineering and industrial design, since they are innovation inputs for both technology invention and technology adoption.

The correlation coefficients between the innovation indicators are reported in Table 1. First, patents, trademarks and copyrights are highly correlated with both R&D and engineering and design. This is not surprising, given that R&D and industrial engineering and industrial design are innovation inputs for technology invention. Second, acquisition of technologies is highly correlated with engineering and design, but less correlated with R&D. This suggests that R&D may be less important than engineering and design for technology adoption. However, further analysis is required to reach a conclusion.

Because these indicators were associated with measurement errors and multi-collinearity problems, technology invention was measured as an unobservable latent variable (η_1). The latent variable was indicated by the following observed variables: patents (y_1), trademarks (y_2), copyrights (y_3), R&D (y_4), and industrial engineering and industrial design (y_5). Similarly, technology adoption was measured as an unobservable latent variable (η_2), indicated by the following observed variables: R&D (y_4), industrial engineering and industrial design (y_5), and acquisition of technologies (y_6). The latent variables and their indicators are presented in Table 2.

Table 1: Correlation Coefficients Between Innovation Indicators

Indicators	y_1	y_2	y_3	y_4	y_5	y_6
y_1: Patents	1.00					
y_2: Trademarks	0.56	1.00				
y_3: Copyrights	0.52	0.64	1.00			
y_4: R&D	0.47	0.30	0.28	1.00		
y_5: Engineering & Design	0.29	0.12	0.13	0.42	1.00	
y_6: Acquisition of Technologies	0.06	0.05	0.05	0.11	0.42	1.00

Source: Statistics Canada, Survey of Innovation 1999, the linked database. Tabulated by author.

Table 2: Latent Variables and Their Indicators

Activities	*Latent Variable*	*Indicators*
Innovation-Related Activities		
Technology Invention	η_1	y_1, y_2, y_3, y_4, y_5
Technology Adoption	η_2	y_4, y_5, y_6
Business Objectives		
To Develop Products	ξ_1	x_1, x_2
To Improve Production Processes	ξ_2	$x_3, x_4, x_5, x_6, x_7, x_8, x_9$
Government Support	ξ_3	x_{10}, x_{11}, x_{12}

2.2 Business Objectives

As discussed earlier, attention was restricted to two types of business objectives: "to develop products" and "to improve production processes." From the linked Survey of Innovation, nine indicators associated with at least one of the two business objectives were identified. Denoted as x-variables, these indicators are listed below, with the Survey of Innovation question number in parentheses ({ }):

x_1 to extend product range {9e}

x_2 to replace products being phased out {9h}

x_3 to increase production capacity {9b}

x_4 to reduce production time {9c}

x_5 to improve production flexibility {9d}

x_6 to improve product quality {9f}

x_7 to reduce labour costs {9a}

x_8 to reduce materials consumption {9i}

x_9 to reduce energy consumption {9k}.

These indicators were the main reasons given by firms for innovation during the period 1997-99. Firms were asked to indicate the degree of importance attached to each indicator by using numbers from 0 (irrelevant) to 5 (highly important). Thus, all these indicators were six-point ordinal variables.

However, each of these indicators was only a partial measure of its business objective. For instance, "to extend product range" (x_1) and "to replace products being phased out" (x_2) both indicate product development, but they are dealing with two different product initiatives. In addition, for each business objective there are two or more indicators that are highly correlated, as shown in Table 3. For instance, "to increase production capacity" (x_3) is highly correlated with "to reduce production time" (x_4). Thus, each business objective was measured as a latent variable, as was done for innovation-related activities. The unobservable latent variables for the objectives "to develop products" and "to improve production processes" were denoted as ξ_1 and ξ_2, respectively. They were indicated by two or more observed variables. As shown in Table 2, the latent variable for the objective "to develop products" (ξ_1) is indicated by the initiatives "to extend product range" (x_1) and "to replace products being phased out" (x_2). Similarly, the latent variable for the objective "to improve production processes" (ξ_2) is indicated by the initiatives "to increase production capacity" (x_3), "to reduce production time" (x_4), "to improve production flexibility" (x_5), "to improve product quality" (x_6), "to reduce labour costs" (x_7), "to reduce materials consumption" (x_8) and "to reduce energy consumption" (x_9).

2.3 Government Support

Government, at both the federal and provincial levels, runs various programs to support innovation-related activities. From the linked Survey of Innovation, three main indicators for government support were identified. They are listed below, with the Survey of Innovation question number in parentheses ({ }):

x_{10} R&D tax credits {29a}

x_{11} R&D grants {29b}

x_{12} government technology support and assistance programs {29d}.

All of these three indicators were binary ordinal variables. For instance, if a firm indicated that it used R&D tax credit programs from either the federal or provincial government during the period 1997-99, the R&D tax credits indicator equalled 1; otherwise, it equalled 0. The same applied to the indicators for R&D grants, and government technology support and assistance programs.

Each of these indicators was only a partial measure of government support. For instance, R&D tax credits are a partial measure of government support since not all eligible firms will apply for them because of the complex paperwork involved. In addition, these indicators are highly correlated, as shown in Table 3. For instance, the correlation coefficient between R&D tax credits and R&D grants is 0.63.

It is important to note that the indicators for business objectives are insignificantly correlated with government support programs (Table 3),

Table 3: Correlation Coefficients Between Indicators for Business Objectives and Government Support

Variable	x_1	x_2	x_3	x_4	x_5	x_6	x_7	x_8	x_9	x_{10}	x_{11}	x_{12}
x_1: To extend product range	1.00											
x_2: To replace products being phased out	0.39	1.00										
x_3: To increase production capacity	0.09	0.03	1.00									
x_4: To reduce production time	0.03	0.10	0.67	1.00								
x_5: To improve production flexibility	0.17	0.15	0.61	0.65	1.00							
x_6: To improve product quality	0.33	0.23	0.34	0.36	0.37	1.00						
x_7: To reduce labour costs	0.02	0.11	0.58	0.65	0.50	0.32	1.00					
x_8: To reduce materials consumption	0.08	0.21	0.33	0.35	0.31	0.27	0.37	1.00				
x_9: To reduce energy consumption	0.07	0.21	0.28	0.28	0.29	0.27	0.29	0.58	1.00			
x_{10}: R&D tax credits	0.11	0.12	-0.06	0.00	-0.02	0.04	-0.03	0.04	0.06	1.00		
x_{11}: R&D grants	0.11	0.09	0.01	0.04	0.05	0.04	0.01	0.09	0.12	0.63	1.00	
x_{12}: Government technology support and assistance programs	0.04	0.03	0.01	0.00	0.04	0.02	-0.02	0.00	0.05	0.33	0.48	1.00

Source: Statistics Canada, Survey of Innovation 1999, the linked database. Tabulated by author.

implying that business objectives are initiated independently of those government programs.

As with innovation-related activities and business objectives, government support was measured as an unobservable latent variable (ξ_3). As shown in Table 2, the latent variable is indicated by R&D tax credits (x_{10}), R&D grants (x_{11}), and government technology support and assistance programs (x_{12}).

3. STRUCTURAL EQUATION MODEL

To determine the relationship between innovation-related activities, as dependent variables, and business objectives and government support, as independent variables, a regression model was needed. However, unlike conventional regression estimation, both the dependent and independent variables in question were latent variables. Because of this, a structural equation model was employed. A general structural model is discussed in Jöreskog and Sörbom (1996).

For the study described here, the structural equation model consisted of three equation systems: two measurement error equations and one structural equation. The measurement error equation for *y*-variables was:

(1) $y = M\eta + \varepsilon$.

The measurement error equation for *x*-variables was:

(2) $x = \Lambda\xi + \delta$.

The structural equation was:

(3) $\eta = \Gamma\xi + \varsigma$.

In this model, the variables were defined as follows.

y is a $p \times 1$ ($p = 6$) vector of observed variables for innovation-related activities.

x is a $q \times 1$ ($q = 12$) vector of observed variables for business objectives or government support.

η is an $m \times 1$ ($m = 2$) random vector of latent dependent variables for innovation-related activities.

ξ is an $n \times 1$ ($n = 3$) random vector of latent independent variables for business objectives or government support.

M is a $p \times m$ matrix of coefficients of the regression of y on η.

Λ is a $q \times n$ matrix of coefficients of the regression of x on ξ.

Γ is an $m \times n$ matrix of coefficients of the regression of η on ξ.

ε is a $p \times 1$ vector of measurement errors in y.

δ is an $n \times 1$ vector of measurement errors in x.

ς is an $m \times 1$ vector of equation errors in the structural relationship between η and ξ.

As in traditional ordinary least-squares regression models, the errors for the measurement error equations and structural equation were assumed to satisfy the following assumptions: ε is uncorrelated with η, δ is uncorrelated with ξ, and ς is uncorrelated with ξ; however, the correlation between and among ς, ε, and δ is permissible.

In the above model, the parameter matrices were defined as follows:

$$M = \begin{pmatrix} \mu_{1,1} & 0 \\ \mu_{2,1} & 0 \\ \mu_{3,1} & 0 \\ \mu_{4,1} & \mu_{4,2} \\ \mu_{5,1} & \mu_{5,2} \\ 0 & \mu_{6,2} \end{pmatrix}, \Lambda = \begin{pmatrix} \lambda_{1,1} & 0 & 0 \\ \lambda_{2,1} & 0 & 0 \\ 0 & \lambda_{3,2} & 0 \\ 0 & \lambda_{4,2} & 0 \\ 0 & \lambda_{5,2} & 0 \\ 0 & \lambda_{6,2} & 0 \\ 0 & \lambda_{7,2} & 0 \\ 0 & \lambda_{8,2} & 0 \\ 0 & \lambda_{9,2} & 0 \\ 0 & 0 & \lambda_{10,3} \\ 0 & 0 & \lambda_{11,3} \\ 0 & 0 & \lambda_{12,3} \end{pmatrix}, \text{ and } \Gamma = \begin{pmatrix} \gamma_{1,1} & \gamma_{1,2} & \gamma_{1,3} \\ \gamma_{2,1} & \gamma_{2,2} & \gamma_{2,3} \end{pmatrix}$$

Not all elements of the parameter matrices M and Λ are free. If an indicator is related to a latent variable, the parameter between the indicator and the latent variable is free; otherwise, it is fixed to be 0. The free elements were identified on the basis of information from Table 2. The path diagram of this model is presented in Figure 1, which provides a better visual picture of the relationship between innovation-related activities, business objectives and government support.

The parameters of the model were estimated by minimizing the difference between the sample covariances of all indicators (innovation-related activities, business objectives and government support) and the covariances predicted by the model. A generally weighted least-squares method was used for the estimation since all observed variables were ordinal.[5] The weight matrix was the inverse of the estimated asymptotic covariance matrix of the polychoric correlations of the observed variables.

Figure 1: The Path Diagram of the Structural Equation Model

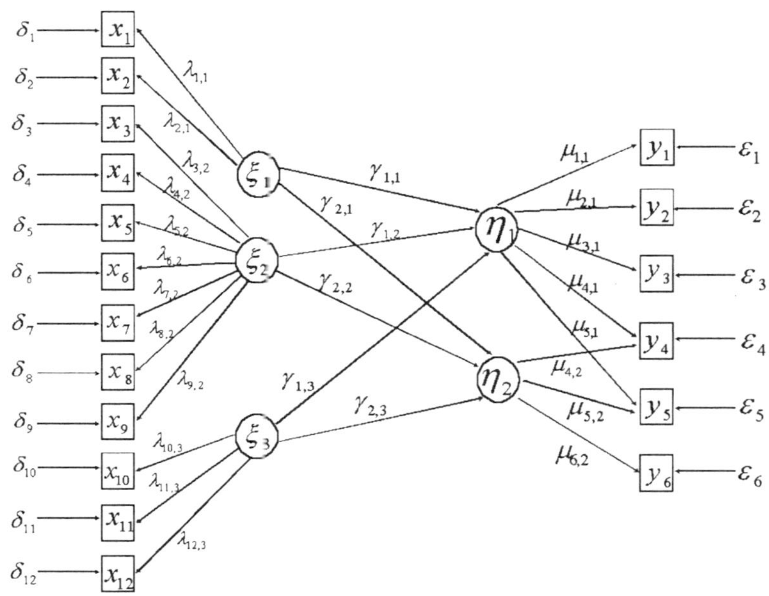

Note: The correlations between error terms are omitted from the diagram.

4. EMPIRICAL ANALYSIS

In this section, the estimation results are discussed, using the linked Survey of Innovation data.

4.1 Sample Profile

As mentioned earlier, the data for the study reported here were from the linked Survey of Innovation. The final sample for analysis contained data on 4,801 in-sample manufacturing firms. The firms in the sample were divided into four subsamples: small low-tech, large low-tech, small high-tech and large high-tech.[6] The aim was to capture variations in innovation across industry and firm size. There is heterogeneity in innovation among industries. Firms in information and communications technology or pharmaceutical industries are highly innovative; in contrast, firms in paper and other resource-based industries are less innovative. By the nature of their businesses, firms in chemical industries tend to rely more on technology invention than on technology adoption, as shown by Gu and Tang (2001). The sample for the study described here was also divided into two groups: low-tech firms, which were firms with NAICS codes 311-323 and 337, and medium- and high-tech firms (hereafter, high-tech), which were firms with NAICS codes 324-336 and 339. The industry classification by level of technology is consistent with the OECD definition (Le 2001).

There is also heterogeneity in innovation among different-sized firms. More large firms than small firms engage in R&D due to high fixed costs and economies of scale (Baldwin 1997). In addition, large firms are more likely than small firms to adopt advanced technology (Baldwin, Rama and Sabourin 1999). The sample for the study reported here was also divided into two groups by size: small firms (those with no more than 100 employees) and large firms (those with more than 100 employees).

4.2 Population Characteristics

The most common innovation-related activity for the manufacturing sector as a whole was acquisition of technologies, followed by R&D

and engineering and design (Table 4). Small firms in both low- and high-tech industries lagged significantly behind large ones on all fronts, from using patents to protect intellectual property to acquisition of technologies. At the industry level, the picture was not uniform. Low-tech firms on average used patents, copyrights, R&D and engineering and design less than high-tech firms, but outperformed high-tech firms in the use of trademarks and in acquisition of technologies. However, when the results were also weighted by revenue from production, high-tech firms outperformed low-tech firms in the use of trademarks as well. The percentage of firms engaging in innovation-related activities is reported at a more detailed industry level (three-digit NAICS codes) in Tables A1.1 and A1.2 in Appendix 1.

Table 4: Percentage of Firms Engaging in Innovation-Related Activities, 1997-99

Part 1: Weighted by Population Weight									
Observed Variables	*Low-Tech*		*High-Tech*				*Low-*	*High-*	*All*
	Small	*Large*	*Small*	*Large*	*Small*	*Large*	*Tech*	*Tech*	*Firms*
y_1: Patents	12.3	21.6	29.3	50.3	22.3	37.5	15.9	36.8	27.9
y_2: Trademarks	33.6	48.5	33.8	45.5	33.7	46.8	39.5	37.9	38.6
y_3: Copyrights	7.7	14.9	12.2	19.6	10.4	17.5	10.6	14.8	13.0
y_4: R&D	49.5	67.7	66.6	80.3	59.6	74.7	56.6	71.5	65.2
y_5: Engineering & Design	47.7	66.1	63.5	78.5	57.0	72.8	54.9	68.7	62.9
y_6: Acquisition of Technologies	85.7	90.2	78.9	88.1	81.7	89.0	87.5	82.1	84.4
Part 2: Weighted by Population Weight and Revenue from Production									
Observed Variables	*Low-Tech*		*High-Tech*				*Low-*	*High-*	*All*
	Small	*Large*	*Small*	*Large*	*Small*	*Large*	*Tech*	*Tech*	*Firms*
y_1: Patents	15.6	26.1	37.0	72.9	28.9	59.4	24.6	69.4	56.0
y_2: Trademarks	38.3	51.0	39.3	62.8	38.9	59.4	49.2	60.5	57.1
y_3: Copyrights	10.5	16.7	16.6	37.7	14.3	31.6	15.8	35.6	29.7
y_4: R&D	57.1	75.7	66.1	86.7	62.7	83.5	73.0	84.7	81.2
y_5: Engineering & Design	51.8	73.7	65.6	74.9	60.3	74.5	70.6	73.9	72.9
y_6: Acquisition of Technologies	85.9	89.3	78.7	80.0	81.5	82.7	88.8	79.9	82.5

Source: Statistics Canada, Survey of Innovation 1999, the linked database. Tabulated by author.

For business objectives, "to improve product quality" was considered to be of the highest importance for innovation (with a score of 4 or 5) by most manufacturing firms, followed by "to increase production capacity" and "to extend product range" (Table 5). More large firms than small firms indicated that all the objectives were highly important for innovation, the most significant being "to reduce labour costs" and "to reduce materials consumption." But at the industry level, the picture was not uniform. More high-tech firms than low-tech firms considered "to replace products being phased out," "to reduce production time" and "to reduce energy consumption" to be highly important; on the other hand, more low-tech firms than high-tech firms considered "to improve production flexibility" and "to reduce materials consumption" to be highly important. However, when the results were also weighted by revenue from production, more large high-tech firms than small high-tech firms considered "to improve production flexibility" to be highly important. The picture at a more detailed industry level is presented in Tables A2.1 and A2.2 in Appendix 2.

Finally, the most popular form of government support was R&D tax credits, followed by R&D grants (Table 6). On average, large firms used government support programs more often than small firms. Similarly, high-tech firms used government support programs more often than low-tech firms. At the more detailed industry level, Computer and Electronics led in the use of R&D tax credits, followed by Machinery and Chemical (Tables A3.1 and A3.2 in Appendix 3). As for R&D grants, Computer and Electronics led, followed by Food and Plastics and Rubber. In the use of government support programs, Computer and Electronics ranked first, followed by Electrical Equipment and Food. Clearly, the greatest beneficiary of government support programs was Computer and Electronics.

When the results were also weighted by revenue from production, Petroleum and Coal led in the use of R&D tax credits, followed by Transportation Equipment and Chemical. Transportation Equipment led in getting R&D grants, followed by Food and Chemical. Transportation Equipment was also the leader in using government support programs, followed by Electrical Equipment and Paper.

Table 5: Percentage of Firms Indicating Business Objectives as Highly Important* for Innovation, 1997-99

Part 1: Weighted by Population Weight									
Observed Variables	*Low-Tech*		*High-Tech*						
	Small	*Large*	*Small*	*Large*	*Small*	*Large*	*Low-Tech*	*High-Tech*	*All Firms*
x_1: To extend product range	65.4	71.4	68.8	71.9	67.4	71.7	67.7	69.9	69.0
x_2: To replace products being phased out	31.0	35.9	35.0	40.3	33.4	38.3	32.9	36.9	35.2
x_3: To increase production capacity	75.8	75.8	72.0	75.8	73.6	75.8	75.8	73.4	74.4
x_4: To reduce production time	63.6	62.2	66.8	70.1	65.5	66.6	63.1	68.0	65.9
x_5: To improve production flexibility	66.2	67.3	60.1	67.1	62.6	67.2	66.6	62.5	64.3
x_6: To improve product quality	80.7	81.8	80.9	81.1	80.8	81.4	81.1	81.0	81.0
x_7: To reduce labour costs	55.4	60.7	56.4	62.6	56.0	61.7	57.5	58.6	58.1
x_8: To reduce materials consumption	36.0	37.2	30.3	39.2	32.6	38.3	36.5	33.4	34.7
x_9: To reduce energy consumption	18.5	23.6	23.0	26.6	21.2	25.3	20.5	24.3	22.7
Part 2: Weighted by Population Weight and Revenue from Production									
Observed Variables	*Low-Tech*		*High-Tech*						
	Small	*Large*	*Small*	*Large*	*Small*	*Large*	*Low-Tech*	*High-Tech*	*All Firms*
x_1: To extend product range	64.5	72.7	68.7	66.8	67.1	68.5	71.5	67.0	68.4
x_2: To replace products being phased out	29.1	31.2	36.7	49.4	33.8	44.2	30.9	48.2	43.0
x_3: To increase production capacity	75.5	71.3	73.3	82.0	74.1	78.9	71.9	81.1	78.4
x_4: To reduce production time	60.0	54.9	65.4	73.1	63.3	67.9	55.6	72.4	67.4
x_5: To improve production flexibility	65.7	63.3	63.5	78.4	64.3	74.0	63.6	76.9	72.9
x_6: To improve product quality	82.7	81.7	78.6	90.5	80.2	87.9	81.8	89.3	87.1
x_7: To reduce labour costs	54.0	60.7	57.6	70.0	56.2	67.3	59.7	68.8	66.1
x_8: To reduce materials consumption	36.1	42.5	30.3	41.5	32.5	41.8	41.6	40.4	40.7
x_9: To reduce energy consumption	22.5	28.8	23.0	41.7	22.8	38.0	27.9	39.9	36.3

Source: Statistics Canada, Survey of Innovation 1999, the linked database. Tabulated by author.

*A business objective was considered highly important if a firm scored the objective as 4 or 5.

Table 6: Percentage of Firms Using One or More Government Support Programs, 1997-99

Part 1: Weighted by Population Weight									
Observed Variables	*Low-Tech*		*High-Tech*						
	Small	*Large*	*Small*	*Large*	*Small*	*Large*	*Low-Tech*	*High-Tech*	*All Firms*
x_{10}: R&D tax credits	23.2	36.1	41.2	55.1	33.8	46.6	28.2	46.1	38.5
x_{11}: R&D grants	7.0	10.6	12.9	13.6	10.5	12.2	8.4	13.1	11.1
x_{12}: Government technology support and assistance programs	7.5	9.5	8.6	11.2	8.1	10.5	8.3	9.5	9.0
Part 2: Weighted by Population Weight and Revenue from Production									
Observed Variables	*Low-Tech*		*High-Tech*						
	Small	*Large*	*Small*	*Large*	*Small*	*Large*	*Low-Tech*	*High-Tech*	*All Firms*
x_{10}: R&D tax credits	28.3	45.3	39.9	74.3	35.5	66.0	42.9	71.0	62.6
x_{11}: R&D grants	10.1	13.7	11.4	24.1	10.9	21.1	13.2	22.9	20.0
x_{12}: Government technology support and assistance programs	6.9	12.1	7.2	21.7	7.1	18.9	11.3	20.3	17.6

Source: Statistics Canada, Survey of Innovation 1999, the linked database. Tabulated by author.

4.3 Empirical Results

The structural equation model presented in section 3 was estimated using data from the linked Survey of Innovation. To improve efficiency, the structural equation model was estimated for all four groups simultaneously: small low-tech, large low-tech, small high-tech and large high-tech. In addition, the relationship parameter between the latent variable and the indicator was fixed, so that each latent variable was scaled to one of its indicators. For the estimation, $\mu_{1,1}$, $\mu_{6,2}$, $\lambda_{1,1}$, $\lambda_{3,2}$ and $\lambda_{10,3}$ were set to unity.[7] This was done subjectively, but a latent variable could be scaled in other ways and it would not change the results.

The estimation was done gradually. First, it was assumed that all coefficients across the four groups were the same and that the covariances between and among ζ, ε, and δ were equal to 0. Then, the constraints that were statistically most significant were gradually relaxed until the remaining constraints were not significant and the estimation achieved an excellent goodness of fit.[8] The estimation results are reported in Table 7.

Several interesting results emerged. First, patents, trademarks, copyrights and R&D, as well as engineering and design, were highly significant indicators for technology invention for all groups. The significance of R&D and engineering and design implies that they are important innovation inputs for technology invention.

Second, engineering and design was a very significant indicator for technology adoption for all groups, implying that it is an important innovation input for technology adoption. R&D was highly significant for technology adoption for high-tech firms and insignificant for low-tech firms. This implies that, in terms of technology adoption, R&D is more important for high-tech firms than for low-tech firms.

Third, all x-indicators were significantly correlated with their latent variables for business objectives and government support, implying that they are significant indicators of their latent variables.

Fourth, it is also clear from Table 7 that the business objective "to develop products" was the driving force for technology invention in all groups. However, it impeded technology adoption in small low-tech firms, although it had no significant impact on technology adoption in large low-tech and large and small high-tech firms. This result may

Table 7: Estimation Results of the Structural Equation Model Using Multiple-Sample Analysis

Parameters	*Small Low-Tech*	*Large Low-Tech*	*Small High-Tech*	*Large High-Tech*
$\mu_{1,1}$: Technology invention → patents	0.54 (8.15)*	1.00 (—)	1.00 (—)	1.00 (—)
$\mu_{2,1}$: Technology invention → trademarks	0.76 (12.17)*	0.76 (12.17)*	0.76 (12.17)*	0.76 (12.17)*
$\mu_{3,1}$: Technology invention → copyrights	0.50 (9.01)*	0.50 (9.01)*	0.50 (9.01)*	0.50 (9.01)*
$\mu_{4,1}$: Technology invention → R&D	0.55 (8.63)*	0.90 (4.90)*	1.38 (9.86)*	0.41 (6.84)*
$\mu_{5,1}$: Technology invention → engineering & design	0.16 (3.79)*	0.47 (2.85)*	0.69 (7.73)*	0.16 (3.79)*
$\mu_{4,2}$: Technology adoption → R&D	0.09 (1.49)	0.09 (1.49)	0.17 (3.06)*	0.17 (3.06)*
$\mu_{5,2}$: Technology adoption → engineering & design	0.77 (7.44)*	1.00 (5.33)*	0.65 (6.86)*	0.75 (5.80)*
$\mu_{6,2}$: Technology adoption → acquisition of technologies	1.00 (—)	1.00 (—)	1.00 (—)	1.00 (—)
$\lambda_{1,1}$: To develop products → to extend product range	1.00 (—)	1.00 (—)	1.00 (—)	1.00 (—)
$\lambda_{2,1}$: To develop products → to replace products being phased out	2.18 (5.22)*	0.85 (5.36)*	1.68 (6.70)*	2.13 (4.39)*
$\lambda_{3,2}$: To improve production processes → to increase production capacity	1.00 (—)	1.00 (—)	1.00 (—)	1.00 (—)
$\lambda_{4,2}$: To improve production processes → to reduce production time	0.99 (24.03)*	1.00 (21.56)*	1.08 (21.24)*	1.26 (12.92)*
$\lambda_{5,2}$: To improve production processes → to improve production flexibility	0.90 (24.80)*	0.86 (18.82)*	0.97 (29.10)*	1.46 (10.79)*
$\lambda_{6,2}$: To improve production processes → to improve product quality	0.60 (13.95)*	0.59 (13.06)*	0.61 (16.84)*	0.83 (10.26)*
$\lambda_{7,2}$: To improve production processes → to reduce labour costs	0.78 (18.62)*	0.90 (20.97)*	1.02 (19.61)*	1.07 (12.43)*
$\lambda_{8,2}$: To improve production processes → to reduce materials consumption	0.52 (13.02)*	0.64 (13.59)*	0.57 (16.53)*	0.96 (8.99)*
$\lambda_{9,2}$: To improve production processes → to reduce energy consumption	0.45 (11.57)*	0.65 (13.85)*	0.47 (13.46)*	0.80 (8.57)*

Source: Statistics Canada, Survey of Innovation 1999, the linked database. Tabulated by author.

All four groups were estimated simultaneously using multiple-sample analysis. Variance and covariance matrices for error terms, as well as goodness-of-fit statistics, are not reported.

*Significant at the 5% level.

Table 7: Estimation Results of the Structural Equation Model Using Multiple-Sample Analysis – Concluded

Parameters	*Small Low-Tech*	*Large Low-Tech*	*Small High-Tech*	*Large High-Tech*
$\lambda_{10,3}$: Government support → R&D tax credits	1.00 (—)	1.00 (—)	1.00 (—)	1.00 (—)
$\lambda_{11,3}$: Government support → R&D grants	0.78 (6.40)*	0.97 (4.39)*	0.84 (9.11)*	1.29 (7.32)*
$\lambda_{12,3}$: Government support → government technology support and assistance programs	0.34 (4.63)*	0.74 (6.63)*	0.64 (6.56)*	0.86 (7.61)*
$\gamma_{1,1}$: To develop products → technology invention	0.89 (5.85)*	0.55 (4.95)*	0.63 (6.14)*	1.19 (5.12)*
$\gamma_{1,2}$: To improve production processes → technology invention	-0.14 (-5.08)*	-0.18 (-2.72)*	-0.14 (-5.08)*	-0.14 (-5.08)*
$\gamma_{1,3}$: Government support → technology invention	0.37 (4.09)*	0.20 (2.89)*	0.43 (8.58)*	0.26 (3.57)*
$\gamma_{2,1}$: To develop products → technology adoption	-0.54 (-3.82)*	0.01 (0.17)	0.01 (0.17)	0.01 (0.17)
$\gamma_{2,2}$: To improve production processes → technology adoption	0.49 (7.75)*	0.29 (4.39)*	0.45 (9.49)*	0.49 (6.34)*
$\gamma_{2,3}$: Government support → technology adoption	0.54 (4.47)*	0.29 (3.42)*	0.13 (2.50)*	0.23 (3.04)*

Source: Statistics Canada, Survey of Innovation 1999, the linked database. Tabulated by author.

All four groups were estimated simultaneously using multiple-sample analysis. Variance and covariance matrices for error terms, as well as goodness-of-fit statistics, are not reported.
*Significant at the 5% level.

indicate that, when a firm is at the stage of developing products, it is less concerned with, or may even reduce (in the case of small low-tech firms) its activities related to, technology adoption because of limited resources.[9]

Fifth, the business objective "to improve production processes" was the driving force for technology adoption, but it impeded technology invention in all groups. This result may indicate that, once firms start to deal with their production processes, they have the products in hand. They reduce activities associated with technology invention and put more resources into improving production processes through technology adoption.

Finally, government support programs had a significant impact on both technology invention and technology adoption.

4.4 Summary of Results

Several interesting results emerged from the study reported here.

(i) Large firms outperformed small ones on all fronts. More large firms than small firms evaluated each business objective as highly important for innovation. Large firms used government support programs more often than small firms, and they were more likely to undertake innovation-related activities.

(ii) At the industry level, the picture was not uniform. More high-tech firms than low-tech firms considered some business objectives, such as "to replace products being phased out," to be highly important for innovation; conversely, more low-tech firms than high-tech firms considered other objectives, such as "to improve production flexibility," to be highly important. High-tech firms used patents and copyrights more often to protect their intellectual property, undertook more R&D and engineering and design, and used government support programs more than low-tech firms, but they used trademarks and acquired technologies less often than low-tech firms.

(iii) R&D was a very important innovation input for technology invention for all firms and for technology adoption for high-tech firms. It was less important for technology adoption for low-tech firms.

(iv) Engineering and design was a very important innovation input for both technology invention and technology adoption for all firms.

(v) The government support program that was used most was R&D tax credits, followed by R&D grants.

(vi) The business objective "to develop products" was the driving force for technology invention. However, it impeded technology adoption in small low-tech firms, although it had no significant impact on technology adoption in large and small high-tech and large low-tech firms.

(vii) The business objective "to improve production processes" was the driving force for technology adoption. However, it impeded technology invention in all groups.

(viii) Government support programs had a significant impact on both technology invention and technology adoption.

5. CONCLUSION

The study presented in this chapter has three main implications. First, Canadian firms are often criticized for competing on the basis of replication rather than on the basis of innovation. To compete on the basis of innovation, Canadian firms have to develop new products. This can be achieved only through technology invention, facilitated by government support programs. Second, innovation has many dimensions and takes different forms. Neither R&D nor the use of patents, the most common indicators for innovation performance, is a perfect measure of such performance. Evaluation of innovation performance should take into account industrial structures and should be based on a combination of many indicators, including acquisition of technologies, which is the most popular innovation-related activity. Finally, to improve Canada's innovation performance, the innovation performance of small firms must first be enhanced, since small firms lag behind large firms in innovation-related activities.

Appendix1: Innovation-Related Activities Profile at the Three-Digit NAICS Industry Level

Table A1.1: Percentage of Firms Engaging in Innovation-Related Activities, 1997-99 (Weighted by Population Weight)

	NAICS	*Industry*	y_1	y_2	y_3	y_4	y_5	y_6
Low-Tech	311	Food	15.9	**56.2**	**13.4**	**72.6**	61.8	**86.8**
	312	Bever. & Tobacco	**29.4**	**73.8**	**28.5**	**68.5**	**65.3**	83.1
	313	Textile Mills	17.4	34.9	**15.4**	**82.4**	**66.4**	**88.1**
	314	Textile Product	27.5	**57.5**	**13.1**	63.8	45.1	**86.7**
	315	Apparel	8.8	**56.1**	6.8	39.2	40.2	82.6
	316	Leather	21.3	**54.3**	7.3	**74.3**	52.2	**84.9**
	321	Wood	7.9	18.3	2.6	52.6	61.8	**87.8**
	322	Paper	22.8	27.4	7.3	**65.9**	**66.0**	**86.8**
	323	Printing	12.8	22.9	**16.7**	37.5	35.9	**93.3**
	337	Furniture	26.1	38.6	9.8	53.4	61.1	**87.4**
	All Low-Tech Industries		15.9	**39.5**	10.6	56.6	54.9	**87.5**
High-Tech	324	Petroleum & Coal	27.7	**48.9**	12.8	**74.5**	**66.0**	78.7
	325	Chemical	**43.1**	**58.5**	**18.1**	**80.6**	61.8	78.4
	326	Plastics & Rubber	**41.1**	**41.6**	11.6	**73.0**	**74.3**	**86.6**
	327	Non-metal. Mineral	**31.1**	34.0	11.4	**67.8**	53.0	74.1
	331	Primary Metal	23.0	19.9	9.7	**70.8**	**79.9**	**90.3**
	332	Fabricated Metal	24.7	27.5	8.7	57.9	**63.4**	82.7
	333	Machinery	**46.9**	**39.8**	**15.2**	**82.7**	**75.8**	80.6
	334	Comp. & Electro.	**45.9**	**48.0**	**30.6**	**85.0**	**71.5**	81.8
	335	Electrical Equip.	**51.6**	**42.8**	**17.4**	**79.5**	**77.4**	79.7
	336	Transport. Equip.	**34.6**	27.8	**16.2**	**71.1**	**76.5**	**85.1**
	339	Miscel. Manufact.	**35.9**	**45.6**	**22.7**	61.0	61.1	83.1
	All High-Tech Industries		**36.8**	37.9	**14.8**	**71.5**	**68.7**	82.1
		Total	27.9	38.6	13.0	65.2	62.9	84.4

Source: Statistics Canada, Survey of Innovation 1999, the linked database. Tabulated by author.
*Numbers larger than the total average are in bold.

Table A1.2: Percentage of Firms Engaging in Innovation-Related Activities, 1997-99 (Weighted by Population Weight and Revenue from Production)

	NAICS	*Industry*	y_1	y_2	y_3	y_4	y_5	y_6
Low-Tech	311	Food	23.4	**65.4**	22.5	**82.0**	70.9	**83.5**
	312	Bever. & Tobacco	**63.8**	**86.5**	**38.2**	**86.7**	**89.1**	**97.1**
	313	Textile Mills	26.2	42.8	21.9	**87.5**	69.2	**91.4**
	314	Textile Product	30.4	54.4	8.2	74.5	60.5	**92.3**
	315	Apparel	9.1	**60.2**	9.1	44.3	52.7	**83.0**
	316	Leather	36.7	**63.1**	14.6	**83.7**	60.4	73.8
	321	Wood	8.5	19.9	1.8	64.1	71.6	**93.3**
	322	Paper	21.1	30.8	6.6	75.1	**74.2**	**88.0**
	323	Printing	14.8	23.6	16.4	38.5	44.3	**96.8**
	337	Furniture	47.5	**62.4**	16.5	65.3	71.4	**94.2**
	All Low-Tech Industries		24.6	49.2	15.8	73.0	70.6	**88.8**
High- Tech	324	Petroleum & Coal	49.0	53.1	**30.7**	71.5	72.7	**92.2**
	325	Chemical	**63.2**	**63.6**	23.7	81.1	**77.8**	**83.1**
	326	Plastics & Rubber	**70.8**	**63.5**	**44.7**	69.3	**85.7**	**93.5**
	327	Non-metal. Mineral	43.2	53.4	24.3	74.8	**73.5**	**89.1**
	331	Primary Metal	43.1	39.6	29.2	**88.7**	**87.3**	**95.3**
	332	Fabricated Metal	47.4	37.7	11.2	70.8	72.6	**89.5**
	333	Machinery	**67.3**	53.8	26.9	**89.1**	**79.2**	**85.4**
	334	Comp. & Electro.	**71.9**	41.0	**43.8**	**87.6**	**82.0**	**93.7**
	335	Electrical Equip.	**74.9**	**61.8**	29.4	**87.4**	**81.2**	**87.4**
	336	Transport. Equip.	**83.3**	**71.7**	**43.9**	**90.3**	66.7	67.1
	339	Miscel. Manufact.	55.0	52.7	24.0	75.0	**75.9**	**87.8**
	All High-Tech Industries		**69.4**	**60.5**	**35.6**	**84.7**	**73.9**	79.9
		Total	56.0	57.1	29.7	81.2	72.9	82.5

Source: Statistics Canada, Survey of Innovation 1999, the linked database. Tabulated by author.
*Numbers larger than the total average are in bold.

Appendix 2: Business Objectives Profile at the Three-Digit NAICS Industry Level

Table A2.1: Percentage of Firms Indicating Business Objectives as Highly Important* for Innovation, 1997-99 (Weighted by Population Weight)

	NAICS	*Industry*	x_1	x_2	x_3	x_4	x_5	x_6	x_7	x_8	x_9
Low-Tech	311	Food	**72.4**	30.5	70.6	53.7	58.4	**81.9**	57.6	20.1	20.1
	312	Bever. & Tobacco	**77.8**	31.2	73.4	53.6	**76.5**	72.5	**64.5**	**42.5**	**35.7**
	313	Textile Mills	**75.1**	**50.7**	73.6	**67.8**	**71.7**	**85.0**	55.5	**36.4**	**27.3**
	314	Textile Product	**70.8**	**44.8**	69.5	59.6	49.6	68.8	**72.7**	23.9	14.1
	315	Apparel	66.7	**43.7**	67.4	60.4	64.2	77.0	55.8	34.4	19.9
	316	Leather	**73.0**	**57.4**	57.7	**68.1**	59.2	**81.8**	51.7	**41.8**	21.3
	321	Wood	61.8	17.4	**77.0**	56.4	62.6	76.3	56.0	**48.7**	**24.2**
	322	Paper	65.5	25.6	74.0	55.7	**66.6**	**84.6**	50.3	**41.9**	**24.0**
	323	Printing	66.9	31.4	**88.7**	**84.0**	**83.2**	**86.0**	55.6	**41.8**	13.8
	337	Furniture	65.0	**41.1**	**82.3**	**70.6**	**71.3**	**86.0**	**63.6**	**41.7**	19.3
	All Low-Tech Industries		67.7	32.9	**75.8**	63.1	**66.6**	**81.1**	57.5	**36.5**	20.5
High- Tech	324	Petroleum & Coal	63.8	**38.3**	70.2	38.3	53.2	72.3	36.2	**38.3**	**42.6**
	325	Chemical	**75.4**	**45.3**	70.1	53.2	55.8	79.2	49.4	27.7	**23.7**
	326	Plastics & Rubber	**79.3**	**41.3**	**78.8**	**70.5**	**67.5**	**82.2**	**59.2**	**46.3**	**31.9**
	327	Non-metal. Mineral	**70.7**	26.4	70.6	59.3	52.0	**83.2**	55.4	34.2	**36.4**
	331	Primary Metal	52.1	21.6	**79.5**	**74.6**	64.3	77.6	**72.5**	**35.6**	**35.5**
	332	Fabricated Metal	61.4	29.8	**80.7**	**75.4**	**64.5**	80.1	**61.7**	30.2	19.6
	333	Machinery	**73.8**	35.0	67.8	64.8	64.1	**82.0**	53.7	26.9	22.6
	334	Comp. & Electro.	**79.4**	**54.9**	59.5	65.2	56.7	76.4	57.1	26.0	12.0
	335	Electrical Equip.	**77.5**	**48.2**	60.8	**70.3**	**68.5**	77.9	**61.0**	**49.2**	**27.4**
	336	Transport. Equip.	62.0	**39.0**	**78.4**	**74.8**	**68.0**	**83.6**	**67.5**	**37.2**	**28.8**
	339	Miscel. Manufact.	**72.4**	**38.5**	72.8	**68.7**	59.1	**85.0**	**58.7**	**36.5**	16.6
	All High-Tech Industries		**69.9**	**36.9**	73.4	**68.0**	62.5	81.0	**58.6**	33.4	**24.3**
		Total	69.0	35.2	74.4	65.9	64.3	81.0	58.1	34.7	22.7

Source: Statistics Canada, Survey of Innovation 1999, the linked database. Tabulated by author.

*A business objective was considered highly important if a firm scored the objective as 4 or 5. Numbers larger than the total average are in bold.

Table A2.2: Percentage of Firms Indicating Business Objectives as Highly Important* for Innovation, 1997-99 (Weighted by Population Weight and Revenue from Production)

	NAICS	*Industry*	x_1	x_2	x_3	x_4	x_5	x_6	x_7	x_8	x_9
Low-Tech	311	Food	**73.6**	32.0	71.4	52.5	55.9	79.2	66.0	26.0	20.4
	312	Bever. & Tobacco	**91.9**	38.3	71.6	**70.3**	**87.6**	**87.5**	**79.5**	**61.5**	**47.5**
	313	Textile Mills	**72.4**	**61.0**	**82.7**	**70.6**	**74.9**	81.0	54.1	37.7	35.3
	314	Textile Product	64.1	**43.7**	**78.6**	**70.9**	**73.0**	83.3	**77.7**	32.9	30.5
	315	Apparel	67.8	**53.9**	60.9	**71.4**	62.4	73.1	**67.2**	34.4	19.9
	316	Leather	**77.7**	**65.3**	41.5	57.8	50.4	64.5	58.3	29.2	19.2
	321	Wood	**75.0**	16.5	71.0	47.6	63.5	78.5	55.9	**59.5**	32.9
	322	Paper	61.4	25.1	65.2	39.6	54.7	84.1	35.1	**42.7**	26.2
	323	Printing	54.9	33.0	**87.7**	**84.8**	**84.6**	**90.7**	**69.1**	**47.6**	28.6
	337	Furniture	**79.0**	32.7	**90.0**	**78.7**	**78.7**	**87.9**	**77.3**	**58.8**	**39.3**
	All Low-Tech Industries		**71.5**	30.9	71.9	55.6	63.6	81.8	59.7	**41.6**	27.9
High-Tech	324	Petroleum & Coal	**79.0**	40.3	**81.2**	20.5	56.5	84.7	35.6	33.4	**48.9**
	325	Chemical	**80.1**	**45.4**	74.2	44.6	63.9	85.3	47.8	27.5	27.2
	326	Plastics & Rubber	**85.5**	**46.4**	**80.7**	**68.3**	70.5	**91.9**	**66.9**	39.7	**47.2**
	327	Non-metal. Mineral	**71.7**	30.3	69.6	45.9	53.7	86.6	54.3	36.9	**39.7**
	331	Primary Metal	66.7	34.1	69.4	62.8	56.3	78.3	53.9	**43.3**	**41.4**
	332	Fabricated Metal	67.3	36.1	**82.6**	**70.2**	69.4	82.8	56.0	**44.2**	15.3
	333	Machinery	**78.1**	**46.9**	74.1	**77.5**	70.4	84.8	59.7	29.6	25.9
	334	Comp. & Electro.	**78.2**	**63.9**	78.1	**83.5**	72.9	**88.2**	65.5	32.5	9.2
	335	Electrical Equip.	**75.1**	**53.1**	70.9	**74.5**	**82.6**	71.4	62.1	**59.9**	27.5
	336	Transport. Equip.	55.1	**53.1**	**87.8**	**88.6**	**91.6**	**95.7**	**87.0**	**45.5**	**51.8**
	339	Miscel. Manufact.	**79.0**	**47.8**	**78.5**	**75.3**	**73.7**	**88.3**	64.8	36.0	14.1
	All High-Tech Industries		67.0	**48.2**	**81.1**	**72.4**	**76.9**	**89.3**	**68.8**	40.4	**39.9**
		Total	68.4	43.0	78.4	67.4	72.9	87.1	66.1	40.7	36.3

Source: Statistics Canada, Survey of Innovation 1999, the linked database. Tabulated by author.

*A business objective was considered highly important if a firm scored the objective as 4 or 5. Numbers larger than the total average are in bold.

Appendix 3: Government Support Programs Profile at the Three-Digit NAICS Industry Level

Table A3.1: Percentage of Firms Using One or More Government Support Programs, 1997-99 (Weighted by Population Weight)

	NAICS	*Industry*	x_{10}	x_{11}	x_{12}
Low-Tech	311	Food	**38.8**	**15.6**	**13.0**
	312	Bever. & Tobacco	26.4	2.8	3.0
	313	Textile Mills	**55.0**	7.8	6.4
	314	Textile Product	**45.3**	6.7	**9.7**
	315	Apparel	15.1	4.5	8.2
	316	Leather	35.1	1.6	6.2
	321	Wood	26.8	11.1	**10.5**
	322	Paper	**38.6**	8.3	8.6
	323	Printing	10.4	1.7	1.7
	337	Furniture	25.2	7.3	6.5
	All Low-Tech Industries		28.2	8.4	8.3
High-Tech	324	Petroleum & Coal	**57.4**	8.5	8.5
	325	Chemical	**58.3**	**15.1**	**9.4**
	326	Plastics & Rubber	**43.8**	**15.4**	**9.8**
	327	Non-metal. Mineral	28.2	8.8	2.7
	331	Primary Metal	**42.5**	9.0	8.9
	332	Fabricated Metal	32.3	9.8	7.4
	333	Machinery	**60.0**	**14.0**	**11.7**
	334	Comp. & Electro.	**68.3**	**25.8**	**16.8**
	335	Electrical Equip.	**56.9**	**14.7**	**13.6**
	336	Transport. Equip.	**50.7**	**11.4**	**12.8**
	339	Miscel. Manufact.	30.3	**12.0**	5.0
	All High-Tech Industries		**46.1**	**13.1**	**9.5**
	Total		38.5	11.1	9.0

Source: Statistics Canada, Survey of Innovation 1999, the linked database. Tabulated by author.

*Numbers larger than the total average are in bold.

Table A3.2: Percentage of Firms Using One or More Government Support Programs, 1997-99 (Weighted by Population Weight and Revenue from Production)

	NAICS	*Industry*	x_{10}	x_{11}	x_{12}
Low-Tech	311	Food	48.5	**20.4**	9.4
	312	Bever. & Tobacco	29.9	0.5	14.0
	313	Textile Mills	60.3	9.4	9.8
	314	Textile Product	49.9	10.5	10.8
	315	Apparel	24.1	5.4	4.7
	316	Leather	36.8	1.1	4.4
	321	Wood	35.6	16.0	14.2
	322	Paper	53.8	11.8	16.9
	323	Printing	22.8	5.2	4.4
	337	Furniture	34.0	5.0	7.9
	All Low-Tech Industries		42.9	13.2	11.3
High- Tech	324	Petroleum & Coal	**85.3**	7.4	7.9
	325	Chemical	**75.2**	17.7	12.1
	326	Plastics & Rubber	43.6	16.0	7.0
	327	Non-metal. Mineral	27.9	4.1	5.0
	331	Primary Metal	**70.1**	8.5	11.8
	332	Fabricated Metal	37.2	8.5	4.5
	333	Machinery	57.5	8.7	10.5
	334	Comp. & Electro.	**72.8**	13.0	11.5
	335	Electrical Equip.	**69.0**	9.4	**23.7**
	336	Transport. Equip.	**81.0**	**37.8**	**33.5**
	339	Miscel. Manufact.	49.3	10.0	3.2
	All High-Tech Industries		**71.0**	**22.9**	**20.3**
	Total		62.6	20.0	17.6

Source: Statistics Canada, Survey of Innovation 1999, the linked database. Tabulated by author.
*Numbers larger than the total average are in bold.

Notes

The author wishes to thank Frances Anderson, Brian Nemes, Guy Sabourin and Susan Schaan for their excellent support in providing access to the linked Survey of Innovation 1999 database. He is also grateful to Frances Anderson, Fred Gault, Can Le, Brian Nemes, Someshwar Rao and Pierre Therrien for helpful comments and suggestions. The author is responsible for any errors and omissions that may remain. All opinions expressed in this chapter are entirely those of the author, and they should in no way be attributed to Industry Canada.

[1]Technology invention is often called fundamental innovation, and technology adoption is called applied innovation (Rao, Ahmad, Horsman and Kaptein-Russell 2001).

[2]However, after linkage to the 1997 ASM, some firms with fewer than 20 employees in 1997 were also in the linked database.

[3]For methodological issues and an overall description of the survey, see Schaan and Nemes (2002).

[4]Firms that did not engage in any innovation-related activities, such as R&D or acquisition of technologies, during the period 1997-99 skipped many survey questions by design, including some variables analyzed here.

[5]If other methods such as maximum likelihood or generalized least squares were used, parameter estimates might be distorted and the chi-square goodness-of-fit measure and standard errors might be unreliable (Jöreskog and Sörbom 1996).

[6]This study could not divide the sample further since the estimation methodology, the weighted least squares, required a fairly large sample size for estimating the asymptotic covariance matrix of the polychoric correlations of the observed variables.

[7]When multiple samples are estimated simultaneously, the fixed parameter may be significantly different from unity for some samples. In this case, such fixed parameters had to be relaxed for those samples.

[8]To evaluate the estimation, a number of goodness-of-fit statistics were used, including the chi-square measure, the root mean square error of approximation (RMSEA) and the expected cross-validation index (ECVI) (Jöreskog and Sörbom 1999). These statistics are not reported, but can be supplied on request.

[9]Technology adoption has more to do with the introduction or improvement of production processes than with the introduction or improvement of products. This is consistent with the observation that one firm's product is often adopted by another firm to improve a production process. For instance, the NEBULA Performance Monitor is a product of LINMOR Technologies Inc., an Ottawa-based high-tech company, and it has been adopted by AT&T to monitor and improve its services production.

8

Innovation Inputs and Innovation Outputs: A Firm-Level Analysis

Can D. Le and Jianmin Tang

1. INTRODUCTION

Innovation has long been recognized as one of the driving forces in economic performance. Recent innovation surveys in Europe have shown that over 30% of manufacturing sales are derived from new or improved products (UK Department of Trade and Industry 1999). The role of innovation in economic growth has therefore become increasingly important, especially in the information technology sector.

The relationship between innovation and a firm's other activities has been analyzed extensively in the last two decades. Acs and Audretsch (1988) used data collected by the US Small Business Administration in 1982 to show that the total number of innovations in an industry is positively related to R&D expenditures (especially after the exclusion of federal R&D), skilled labour and the proportion of large firms in the industry, and negatively related to industry concentration and unionization. Later on, Audretsch (1995) used the same database to show that the availability of funds, as indicated by firm profitability, is more conducive to subsequent innovative input in firms in an environment of high technological opportunity than in those in an environment of low technological opportunity. The same database was subsequently used by Acs and Gifford (1996), in conjunction with data obtained from the *Business Week* survey of company-financed R&D

expenditures, to show that for firms which conduct product improvement, and thus face low obsolescence, innovation decreases with firm size, while for firms facing high obsolescence innovation increases with firm size.

On the European scene, Brouwer and Kleinknecht (1996) used data collected in 1992 in the Netherlands' National Innovation Survey of 8,000 firms with 10 or more employees to show that large firms are more likely to innovate. However, among innovative firms, smaller firms tend to be more innovative than large firms. In other words, the share of sales of innovative products in a firm's total sales tends to be higher in smaller firms. Leiponen (2000b) used panel data obtained through a 1991 survey of 209 Finnish manufacturing firms to show that complementary capabilities are necessary for firms to profit from innovation. She also showed that factors influencing the profitability of innovating firms differ from those influencing the profitability of non-innovators. Sandven (2000) used data from the Norwegian Innovation Survey of 1992, together with accounting data for the period 1991-97, to investigate the relationship between innovation and economic performance at the firm level. Among other results, this author found a clear positive relationship between innovation and sales growth.

More recently, Cozzarin (2001) used data from the Survey of Innovation 1999 conducted by Statistics Canada to examine the relationship between innovation quality and firm performance. Defining "innovation quality" on the basis of whether an innovation is "world-first" (high quality), "Canada-first" (medium quality) or "firm-first" (low quality), he showed that there is a strong and positive relationship between innovation quality and a firm's performance measures, such as price-cost margin, employment, labour productivity and wages. He also found that the price-cost margins of innovators are higher than those of non-innovators. In addition, he found that firms in high-tech industries[1] perform better than those in medium- or low-tech industries.

2. OBJECTIVE OF THE STUDY

The objective of the study presented in this chapter was to assess the effectiveness of firms' innovation inputs by linking their innovation inputs to innovation outputs, using the Survey of Innovation 1999 conducted by Statistics Canada. In particular, the following three questions were addressed:

(i) Do firms' innovation inputs have a significant impact on their innovation outputs and, if so, which type of innovation input is most effective?

(ii) Do different industries rely on different types of innovation inputs?

(iii) Are small firms as effective as large firms in converting innovation inputs into innovation outputs?

3. HYPOTHESIS

The hypothesis was that firms engage in many innovation inputs and that their effectiveness, in terms of generating innovation outputs, differs across those inputs. Two innovation outputs were examined: product innovation and process innovation. Product or process innovation for a firm was defined as the introduction of a new or significantly improved product or process during a given period.[2] Five innovation inputs were identified: R&D; acquisition of technologies; industrial engineering and industrial design; tooling up; and training. A co-operation variable, representing a firm's involvement in collaboration and co-operation with other firms, was also identified.

The work presented in this chapter is similar to that of Therrien (2002), who, using the same survey data, investigated why some firms are innovators and others are not. He defined innovators as firms that introduced a new or significantly improved product or process during the survey period. The potential explanatory variables he explored included a firm's success factors, government support programs, and variables indicating the competitive environment.

The work presented here differs from Therrien's in two respects. First, a distinction is made here between the two innovation outputs to be examined: product innovation offers consumers new choices and services; process innovation aims to improve production efficiency, and is concerned with the techniques and methods used for making goods and services for the market. This distinction is important since product and process innovation may require different innovation inputs. For instance, R&D may be more important than the acquisition of technologies for product innovation, while the opposite may be true for process innovation. Furthermore, while innovation is important, the impact on the market is the ultimate goal.[3] Thus, the impact of innovation inputs on sales from product innovation is also analyzed, taking into account the significance of each product innovation.

Second, the focus here is on innovation inputs as explanatory variables for innovation outputs. Variables such as competitive environment and government innovation support, which were included in the regression in Therrien's work, are excluded on the basis that their inclusion would cause a bias against innovation inputs, since those variables themselves are part of the reason why firms engage in innovation inputs in the first place.

4. DATA

The data used for the study presented in this chapter were obtained from the Statistics Canada's Survey of Innovation 1999, which has been linked to the 1997 Annual Survey of Manufacturers (ASM). The focus of the Survey of Innovation was the Canadian manufacturing sector and selected natural resources industries for which the sample unit was the provincial enterprise. A provincial enterprise (hereafter, firm) included all establishments in the same province and industry at the four-digit NAICS (North American Industry Classification System) level. To reduce the response burden, the Survey of Innovation 1999 surveyed firms with a gross business income of at least $250,000 and more than 19 employees in 1999.[4] All data in the survey concern firms' innovation environment and related activities during the period 1997-99.[5] After linkage to the 1997 ASM, the Survey of Innovation offers additional information on firms' operational activities, such as employment, in 1997. The linked Survey of Innovation database was used for the study described here.

The linked database contains data on 5,455 manufacturing firms. Each firm is assigned a weight, which allows that firm to represent other firms in the population with similar characteristics. Thus, for example, if the weight given to firm X is 5, firm X represents five firms in the sample. To represent these five firms, the data obtained from the sampled firm are multiplied by its weight, in this case 5. The total population of manufacturing firms is 8,921, a number that is equal to the sum of the population weights of the in-sample firms. For the purposes of the study presented here, however, four in-sample firms without employment information were excluded.[6] Thus, the final sample contained data on 5,451 in-sample firms, representing a subpopulation of 8,916 manufacturing firms.

As mentioned earlier, in this chapter "innovation outputs" refers to product innovation and process innovation. The two variables corresponding to product innovation and process innovation in the Survey of Innovation are listed below, with the Survey of Innovation question number in parentheses ({ }):

- y_1 introduced a new or significantly improved product {3a}
  - y_{1a} sales from new products {12b}
  - y_{1b} sales from significantly improved products {12c}
- y_2 introduced a new or significantly improved process {4}.

For product innovation (y_1), firms were asked: "During the last three years, 1997 to 1999, did your firm offer new or significantly improved products to your clients?"[7] The variable was binary: 1 for "yes" and 0 for "no." Similarly, for process innovation firms were asked: "During the last three years, 1997 to 1999, did your firm introduce new or significantly improved production/manufacturing processes?"[8] Similarly, the variable was binary: 1 for "yes" and 0 for "no."

Clearly, these variables treated all product innovations equally and performed no quality control. To take into account the significance of each product innovation, variable y_{1a} was introduced to represent sales from new products and y_{1b} to represent sales from significantly improved products.[9] For sales from new or significantly improved products, a firm was first asked if it had introduced any new or significantly improved products between 1997 and 1999. If the answer was "yes," the firm was then asked to estimate the percentage of its sales in 1999 that resulted from new or significantly improved products that had been introduced during the period 1997-99. For simplicity, the two variables were designed to be binary: 1 for more than 15% of sales in 1999 that were due to new or significantly improved products introduced during the period 1997-99; and, otherwise, 0.[10]

For the study described here, five innovation inputs and a co-operation variable, which represented a firm's involvement in collaboration and co-operation with other firms, were identified as potential explanatory variables for the innovation outputs. These variables are listed below, with the Survey of Innovation question number in parentheses ({ }):

x_1 R&D {6a}

x_2 acquisition of machinery, equipment or other technologies {6b}

x_3 industrial engineering and industrial design {6c}

x_4 tooling up and production start-up {6d}

x_5 innovation-related training {6e}

x_6 involvement in collaboration and co-operation with other firms {2n}.

The variable for tooling up and production start-up represented the innovation input associated with the improvement and modification of existing facilities to accommodate new products or processes. For all innovation inputs, a firm was asked: "During the last three years, 1997 to 1999, did your firm engage in the above activities which are linked to offering new or significantly improved products or to introducing new or significantly improved production/manufacturing processes?" All five variables were binary: 1 for "yes" and 0 for "no." For the co-operation variable, a firm was asked to indicate its opinion about the factor's importance for the success of the firm, using numbers from 0 (irrelevant) to 5 (highly important).[11]

5. EMPIRICAL ANALYSIS

Cross-section data on Canadian manufacturing firms were used to shed light on the relationship between innovation inputs and innovation outputs. First, a regression model was set up.

5.1 Regression Model

To estimate the impact of innovation inputs on the likely generation of an innovation output, a logit model was set up as follows:

$$(1)\quad \text{Prob}[y_{i,j}=1]=\left[1+\exp\left(-\mathrm{X}_j\beta\right)\right]^{-1}, \text{ and}$$

$$(2)\quad \mathrm{X}_j\beta=\beta_0+\beta_1 RD_j+\beta_2 AT_j+\beta_3 ED_j+\beta_4 TS_j+\beta_5 TN_j+\beta_6 CP_j + \beta_6 SM_j+\beta_7 SL_j+\sum_{n=1}^{19}\alpha_n I_{nj}+\varepsilon_j,$$

where $y_{i,j}$ denotes innovation output i, β is a vector of coefficients to be estimated, and X_j is the set of explanatory variables influencing the probability of generating the innovation output, defined as follows.

RD_j is R&D for firm *j*.

AT_j is acquisition of machinery, equipment or other technologies for firm *j*.

ED_j is industrial engineering and industrial design for firm *j*.

TS_j is tooling up and production start-up for firm *j*.

TN_j is innovation-related training for firm *j*.

CP_j is involvement in collaboration and co-operation with other firms for firm *j*.

SM_j is a size dummy for medium firms, 1 for firm *j* being a medium firm and 0 otherwise.

SL_j is a size dummy for large firms, 1 for firm *j* being a large firm and 0 otherwise.

I_{nj} is a binary industry dummy, 1 for firm *j* belonging to industry *n* and 0 otherwise.

ε_j is the error term for firm *j*.

In the regression model, firm size and industry dummies were also introduced to capture size-related and industry-related residuals that were not captured by other variables. To capture the size effect, firms were divided into three groups based on size, similar to the classification used in the European Union's Community Innovation Survey 3. The reference group in the regression model consisted of small firms, defined as those with between 20 and 49 employees; medium firms were defined as those with between 50 and 249 employees; and large firms were defined as those with 250 employees or more. Industries were grouped at the three-digit NAICS level. There was a total of 20 manufacturing industries.[12] The reference industry in the regression model was Miscellaneous Manufacturing (NAICS 339).

5.2 Sample Profile

As mentioned earlier, the data used for the study reported in this chapter were from the linked Survey of Innovation. The final sample for analysis contained data on 5,451 in-sample manufacturing firms, including 1,623 small firms, 2,949 medium firms and 879 large firms.

Table 1 shows the percentage of firms that were product or process innovators.[13] During the period 1997-99, about 68% of firms introduced product innovation, 66% introduced process innovation and 54% introduced both product and process innovation. On average, fewer small firms than large firms were product or process innovators or both. Among firms that were product or process innovators, Computer and Electronics ranked first in product innovation, followed by Electrical Equipment and Textile Product. In process innovation, Printing ranked first, followed by Electrical Equipment and Textile Mills.

In terms of sales from product innovation, about 22% of firms had more than 15% of sales from new or significantly improved products (Table 2). Again, fewer small firms than large firms had such sales. Computer and Electronics led in sales from new products, followed by Electrical Equipment. This was also true for sales from significantly improved products.

With regard to innovation inputs, the most common innovation input in the manufacturing sector as a whole was acquisition of technologies, followed by training and R&D (Table 3). On average, fewer small firms than large firms engaged in those innovation inputs. However, the picture was not uniform across industries. Computer and Electronics, Electrical Equipment and Plastics and Rubber led in R&D;

Table 1: Percentage of Firms Engaged in Product or Process Innovation, 1997-99

NAICS	Industry	y_1 Product Innovation				y_2 Process Innovation			
		Small	*Medium*	*Large*	*All Firms*	*Small*	*Medium*	*Large*	*All Firms*
311	Food	73.1	71.5	78.9	73.5	63.2	65.4	72.7	66.3
312	Bever. & Tobacco	45.2	59.4	87.7	63.1	48.2	63.5	78.8	63.3
313	Textile Mills	73.2	71.7	96.3	76.1	67.4	71.2	88.7	73.0
314	Textile Product	74.3	87.5	85.7	82.7	43.6	67.1	85.7	61.0
315&316	Apparel & Leather	47.4	60.7	65.6	56.9	46.8	57.4	67.5	54.9
321	Wood	47.6	56.3	66.0	54.5	57.4	64.8	75.7	63.6
322	Paper	45.8	66.2	58.4	60.0	57.3	68.9	64.9	65.5
323	Printing	64.3	65.0	70.2	65.0	70.4	83.2	84.5	77.7
324	Petroleum & Coal	60.0	55.6	81.3	64.8	60.0	44.4	87.5	63.0
325	Chemical	80.7	72.8	90.6	78.4	55.1	65.8	68.7	62.5
326	Plastics & Rubber	74.1	77.3	88.2	77.8	67.1	73.7	80.7	72.8
327	Non-metal. Mineral	64.3	58.9	61.6	61.9	47.4	57.0	60.9	52.3
331	Primary Metal	46.9	58.0	63.8	57.8	43.8	74.0	69.6	67.2
332	Fabricated Metal	49.5	63.0	68.5	57.9	50.1	71.6	83.2	63.6
333	Machinery	75.0	74.1	93.8	76.0	59.2	72.6	73.0	68.2
334	Comp. & Electro.	80.0	92.3	80.2	86.4	44.1	71.9	79.3	65.7
335	Electrical Equip.	79.0	87.2	85.2	84.5	66.2	81.0	71.6	75.0
336	Transport. Equip.	58.8	68.1	78.0	68.8	41.8	67.8	80.2	65.4
337	Furniture	67.2	65.3	75.4	66.6	60.8	72.3	82.3	69.7
339	Miscel. Manufact.	75.1	78.2	86.6	77.2	57.1	74.9	75.3	66.6
	All Manufacturing	63.5	68.6	76.7	68.0	56.1	69.7	75.0	65.8

Source: Statistics Canada, Survey of Innovation 1999. Tabulated by authors.

Table 1: Percentage of Firms Engaged in Product or Process Innovation, 1997-99 – Concluded

NAICS	Industry	y_1 or y_2 Product or Process Innovation				y_1 and y_2 Product and Process Innovation			
		Small	*Medium*	*Large*	*All Firms*	*Small*	*Medium*	*Large*	*All Firms*
311	Food	80.5	77.9	86.9	80.5	55.8	59.1	64.7	59.2
312	Bever. & Tobacco	59.0	78.9	91.7	76.8	34.5	44.0	74.7	49.7
313	Textile Mills	85.4	83.0	96.3	85.8	55.2	59.8	88.7	63.3
314	Textile Product	74.3	87.5	92.8	83.5	43.6	67.1	78.5	60.2
315&316	Apparel & Leather	62.6	72.5	80.1	70.0	31.6	45.6	53.1	41.8
321	Wood	68.8	74.1	84.8	73.6	36.2	46.9	56.9	44.4
322	Paper	65.6	82.1	78.4	77.9	37.5	53.0	44.9	47.6
323	Printing	79.3	90.2	84.5	85.1	55.4	57.9	70.2	57.6
324	Petroleum & Coal	65.0	72.2	93.8	75.9	55.0	27.8	75.0	51.9
325	Chemical	87.1	85.6	94.7	87.5	48.7	53.1	64.6	53.4
326	Plastics & Rubber	82.7	85.6	92.6	85.7	58.4	65.5	76.3	64.9
327	Non-metal. Mineral	72.1	71.7	73.5	72.0	39.6	44.2	49.1	42.2
331	Primary Metal	58.5	78.7	80.3	75.5	32.2	53.4	53.1	49.4
332	Fabricated Metal	63.9	80.9	83.2	74.3	35.6	53.7	68.5	47.2
333	Machinery	82.8	88.4	96.8	87.2	51.5	58.3	70.0	57.0
334	Comp. & Electro.	89.8	95.0	90.7	92.7	34.3	69.2	68.8	59.5
335	Electrical Equip.	85.0	92.2	90.1	89.8	60.2	76.0	66.6	69.7
336	Transport. Equip.	63.1	79.3	90.4	78.8	37.5	56.6	67.9	55.4
337	Furniture	80.6	81.2	92.6	81.9	47.4	56.4	65.2	54.4
339	Miscel. Manufact.	79.0	83.9	94.9	82.2	53.2	69.3	67.0	61.7
	All Manufacturing	74.5	82.0	87.7	80.2	53.5	56.4	64.0	53.5

Source: Statistics Canada, Survey of Innovation 1999. Tabulated by authors.

Table 2: Percentage of Firms with More Than 15% of Sales Derived from Product Innovation, 1997-99

NAICS	*Industry*	y_{1a} *More Than 15% of Sales from New Products*				y_{1b} *More Than 15% of Sales from Significantly Improved Products*			
		Small	*Medium*	*Large*	*All Firms*	*Small*	*Medium*	*Large*	*All Firms*
311	Food	23.0	16.5	15.2	18.3	15.7	18.4	14.5	16.7
312	Bever. & Tobacco	16.2	23.8	0.0	15.2	6.3	15.7	0.0	8.7
313	Textile Mills	28.3	25.3	58.5	31.5	21.2	27.8	42.8	28.6
314	Textile Product	26.4	25.8	42.9	27.9	31.6	27.0	42.5	30.3
315&316	Apparel & Leather	18.4	23.0	26.2	21.8	19.8	24.8	25.3	23.2
321	Wood	10.7	17.8	11.8	14.6	9.5	17.6	15.0	14.4
322	Paper	13.9	16.4	11.0	14.3	22.8	20.1	26.7	22.7
323	Printing	18.5	20.2	22.8	19.6	21.8	21.2	22.8	21.6
324	Petroleum & Coal	10.0	11.1	6.3	9.3	10.0	11.1	25.0	14.8
325	Chemical	16.2	17.2	28.3	18.6	20.9	16.4	29.2	20.0
326	Plastics & Rubber	17.5	23.4	33.4	23.0	16.7	27.4	32.6	25.1
327	Non-metal. Mineral	19.5	13.2	14.8	16.6	21.0	20.2	24.9	21.0
331	Primary Metal	10.6	17.8	17.1	16.3	8.6	13.1	19.5	14.3
332	Fabricated Metal	11.4	18.5	19.0	15.7	9.8	18.9	20.9	15.4
333	Machinery	27.6	27.9	41.2	28.9	29.0	28.1	40.1	29.4
334	Comp. & Electro.	28.1	46.7	60.9	44.5	38.0	45.5	49.3	44.2
335	Electrical Equip.	32.7	40.9	29.6	36.4	42.7	38.5	26.0	37.3
336	Transport. Equip.	21.5	31.1	31.1	28.9	20.4	28.5	29.1	26.8
337	Furniture	15.9	31.5	29.1	26.8	22.3	26.7	26.0	25.3
339	Miscel. Manufact.	22.5	33.9	34.4	28.6	24.4	29.0	34.6	27.2
	All Manufacturing	18.6	23.5	25.3	22.1	19.3	23.6	25.9	22.5

Source: Statistics Canada, Survey of Innovation 1999. Tabulated by authors.

Table 2: Percentage of Firms with More Than 15% of Sales Derived from Product Innovation, 1997-99 – Concluded

NAICS	*Industry*	y_{1a} *or* y_{1b} *More Than 15% of Sales from New Products or More Than 15% of Sales from Significantly Improved Products*				y_{1a} *and* y_{1b} *More Than 15% of Sales from New Products and More Than 15% of Sales from Significantly Improved Products*			
		Small	*Medium*	*Large*	*All Firms*	*Small*	*Medium*	*Large*	*All Firms*
311	Food	28.0	25.0	23.1	25.5	10.7	9.9	6.5	9.5
312	Bever. & Tobacco	16.2	23.8	0.0	15.2	6.3	15.7	0.0	8.7
313	Textile Mills	36.8	35.1	73.6	41.8	12.7	18.0	27.7	18.2
314	Textile Product	44.1	44.3	71.4	47.2	13.9	8.5	14.0	11.0
315&316	Apparel & Leather	26.4	34.1	35.2	31.7	11.7	13.7	16.3	13.3
321	Wood	15.4	27.8	22.6	22.8	4.9	7.6	4.2	6.2
322	Paper	24.7	26.5	29.0	27.0	12.1	10.0	8.7	10.1
323	Printing	28.6	32.0	34.2	30.7	11.7	9.5	11.5	10.5
324	Petroleum & Coal	20.0	22.2	25.0	22.2	0.0	0.0	6.3	1.9
325	Chemical	26.1	25.7	50.0	29.6	11.0	7.9	7.4	8.9
326	Plastics & Rubber	25.7	36.8	55.6	36.1	8.5	14.0	10.4	12.0
327	Non-metal. Mineral	29.5	24.6	24.9	27.1	10.9	8.9	14.8	10.4
331	Primary Metal	16.7	20.1	25.5	21.2	2.6	10.8	11.1	9.4
332	Fabricated Metal	16.0	26.2	34.5	22.6	5.2	11.2	5.3	8.5
333	Machinery	41.7	41.5	61.5	43.2	14.9	14.5	19.8	15.1
334	Comp. & Electro.	48.9	59.5	71.3	59.0	17.2	32.7	39.0	29.7
335	Electrical Equip.	59.4	55.1	44.5	54.3	16.0	24.2	11.2	19.4
336	Transport. Equip.	35.4	41.0	44.8	40.8	6.6	18.6	15.4	15.0
337	Furniture	27.0	41.7	42.1	37.5	11.2	16.5	13.0	14.7
339	Miscel. Manufact.	30.4	41.4	44.3	36.4	16.5	21.5	24.7	19.3
	All Manufacturing	27.7	33.8	38.8	32.4	10.2	13.4	12.4	12.2

Source: Statistics Canada, Survey of Innovation 1999. Tabulated by authors.

Table 3: Percentage of Firms Engaging in Innovation Inputs or Indicating Collaboration and Co-operation with Other Firms to Be Highly Important, 1997-99

NAICS	*Industry*	x_1 *R&D*				x_2 *Acquisition of Technologies*				x_3 *Engineering & Design*			
		Small	*Medium*	*Large*	*All Firms*	*Small*	*Medium*	*Large*	*All Firms*	*Small*	*Medium*	*Large*	*All Firms*
311	Food	66.8	68.4	73.1	68.9	74.3	75.6	78.6	75.8	44.3	54.4	67.4	54.0
312	Bever. & Tobacco	51.7	65.5	75.4	64.3	40.6	79.1	82.9	69.2	31.8	54.0	78.7	54.4
313	Textile Mills	71.9	73.8	96.3	77.0	72.6	76.3	88.7	77.4	51.7	55.9	77.3	58.3
314	Textile Product	51.8	78.1	64.3	67.4	80.3	75.3	100.0	79.7	23.0	47.4	71.3	41.5
315&316	Apparel & Leather	41.6	55.8	62.6	51.9	55.7	71.7	69.4	66.3	17.0	37.9	62.1	33.5
321	Wood	39.1	55.7	66.4	51.2	62.7	78.8	82.4	73.6	37.4	58.0	66.0	51.8
322	Paper	57.8	60.0	72.0	63.3	65.9	76.3	86.8	77.6	45.0	60.2	65.5	59.0
323	Printing	45.6	54.4	73.0	51.8	78.7	88.1	90.2	84.1	30.4	31.0	57.9	32.4
324	Petroleum & Coal	60.0	72.2	62.5	64.8	65.0	55.6	87.5	68.5	55.0	50.0	68.8	57.4
325	Chemical	76.4	74.9	87.8	77.4	65.0	75.2	80.0	72.4	48.9	57.7	73.2	57.0
326	Plastics & Rubber	76.9	77.7	88.1	78.8	74.1	81.4	83.6	79.7	64.5	66.5	85.1	68.3
327	Non-metal. Mineral	52.1	57.1	73.8	55.9	52.6	59.3	83.5	57.8	34.5	45.1	65.9	41.3
331	Primary Metal	49.7	72.8	75.9	69.5	59.5	82.5	84.3	78.9	49.8	69.6	81.8	69.8
332	Fabricated Metal	50.8	59.1	86.3	57.0	58.2	74.9	92.2	69.0	39.9	59.7	83.2	52.9
333	Machinery	69.2	80.5	96.9	78.1	63.6	79.1	88.2	74.7	62.3	73.4	80.4	70.3
334	Comp. & Electro.	75.0	86.5	89.3	83.9	69.8	79.2	91.4	79.0	64.4	69.4	74.6	69.1
335	Electrical Equip.	65.7	83.7	90.2	79.8	53.5	80.0	85.2	73.4	58.8	76.7	75.3	71.3
336	Transport. Equip.	54.0	68.8	79.3	68.4	52.2	75.6	87.0	73.5	39.3	68.0	84.4	66.1
337	Furniture	57.5	61.7	75.2	61.5	67.7	78.6	89.8	76.3	39.6	57.4	72.3	53.4
339	Miscel. Manufact.	51.7	73.3	81.7	63.7	63.1	81.7	89.9	73.5	39.2	65.4	81.7	54.0
	All Manufacturing	56.6	67.0	78.6	65.0	64.1	77.2	84.5	73.8	42.4	58.3	73.7	55.0

Source: Statistics Canada, Survey of Innovation 1999. Tabulated by authors.

Table 3: Percentage of Firms Engaging in Innovation Inputs or Indicating Collaboration and Co-operation with Other Firms to Be Highly Important, 1997-99 – Concluded

NAICS	*Industry*	x_4 *Tooling Up*				x_5 *Training*				x_6 *Collaboration and Co-operation*			
		Small	*Medium*	*Large*	*All Firms*	*Small*	*Medium*	*Large*	*All Firms*	*Small*	*Medium*	*Large*	*All Firms*
311	Food	52.7	60.4	68.6	59.7	62.0	68.3	74.7	67.7	30.6	30.0	30.6	30.3
312	Bever. & Tobacco	36.8	66.9	83.3	62.9	38.2	71.4	87.1	66.3	3.7	29.6	28.8	22.1
313	Textile Mills	50.3	64.3	73.6	62.3	77.5	69.9	81.2	73.7	28.9	33.1	38.9	33.0
314	Textile Product	38.3	39.2	78.5	43.2	53.0	57.3	92.8	59.7	27.7	9.9	21.6	17.3
315&316	Apparel & Leather	27.8	38.5	56.7	36.8	45.3	57.2	57.4	53.4	14.5	21.0	26.8	19.5
321	Wood	46.3	61.7	68.0	57.1	53.7	70.6	75.1	65.3	20.6	21.6	47.2	24.6
322	Paper	49.7	53.2	64.5	56.1	62.9	69.3	79.0	71.1	29.6	24.4	26.1	25.9
323	Printing	46.1	49.2	67.4	49.0	70.4	75.6	90.2	74.3	27.5	30.2	45.8	30.0
324	Petroleum & Coal	60.0	55.6	62.5	59.3	65.0	72.2	87.5	74.1	25.0	55.6	43.8	40.7
325	Chemical	51.2	62.0	75.5	60.3	66.6	73.1	84.3	72.6	30.5	40.4	51.1	38.6
326	Plastics & Rubber	65.9	78.4	83.7	75.6	67.7	74.9	85.1	74.2	32.7	32.7	46.0	34.4
327	Non-metal. Mineral	47.9	54.4	62.8	51.7	48.4	57.0	76.0	54.1	28.8	20.3	20.6	24.7
331	Primary Metal	59.4	72.7	69.0	69.1	52.2	77.8	89.5	76.7	24.4	19.8	34.0	25.0
332	Fabricated Metal	48.0	66.4	86.3	60.0	51.3	72.0	89.1	64.5	23.3	24.9	33.8	24.6
333	Machinery	51.2	64.2	86.9	61.7	61.6	73.5	90.1	70.9	39.0	33.2	61.6	37.4
334	Comp. & Electro.	42.5	73.7	83.6	67.1	80.6	83.3	88.9	83.7	50.5	49.0	55.7	50.8
335	Electrical Equip.	53.0	73.4	84.0	69.6	58.6	82.6	88.9	77.0	34.6	45.2	29.6	39.2
336	Transport. Equip.	50.4	73.6	88.9	72.6	58.1	74.1	94.8	76.3	24.9	34.7	27.5	30.5
337	Furniture	48.0	63.6	76.4	60.0	63.3	69.8	79.2	68.6	25.3	20.8	48.8	24.2
339	Miscel. Manufact.	43.5	66.0	85.0	56.5	58.1	78.6	81.7	69.2	36.0	33.6	21.4	34.1
	All Manufacturing	48.0	62.3	75.2	59.2	58.7	71.3	82.8	68.6	28.2	28.9	37.4	29.8

Source: Statistics Canada, Survey of Innovation 1999. Tabulated by authors.

Printing, Textile Product and Plastics and Rubber in acquisition of technologies; Electrical Equipment, Machinery and Primary Metal in engineering and design; Plastics and Rubber, Transportation Equipment and Electrical Equipment in tooling up; and Computer and Electronics, Electrical Equipment and Primary Metal in training.

Finally, about 30% of firms, and more large firms than small, indicated that collaboration and co-operation with other firms were highly important for their success (Table 3).[14] At the industry level, Computer and Electronics was the leader, followed by Petroleum and Coal and Electrical Equipment.

5.3 Empirical Results

To link innovation inputs to innovation outputs, a logit model of innovators versus non-innovators was estimated, with non-innovators as the reference group. Innovators were those firms that had either product innovation or process innovation, and non-innovators were those that had no product or process innovation. The regression results show that engaging in any of the innovation inputs (R&D, acquisition of technologies, engineering and design, tooling up and training) or being involved in collaboration and co-operation with other firms significantly increased the likelihood of being an innovator (Table 4). In addition, after controlling for other factors, it was found that small firms were as likely as large firms to become innovators. This is a significant result since it implies that, for any given innovation input, small firms were as effective as large firms in converting innovation inputs into innovation outputs. The problem is that small firms were less likely than large firms to undertake activities that are inputs to innovation. Thus, fewer small firms than large firms were innovators. This is consistent with the results obtained by Brouwer and Kleinknecht (1996). In Table 4, the regression results for firms that were both product and process innovators versus non-innovators are also reported; firms that were product innovators only or process innovators only were excluded. The same results were obtained as in the first regression.

To find out whether product and process innovations rely on different innovation inputs, three regressions were run in which product innovators were separated from process innovators. The first regression

Table 4: Estimated Impacts of Innovation Inputs on the Likelihood of Being Product or Process Innovators

Variables	*Product or Process Innovators vs. Non-Innovators*		*Product and Process Innovators vs. Non-Innovators*	
	Coef.	*t-value*	*Coef.*	*t-value*
Intercept	-1.320	-4.78*	-2.736	-7.64*
R&D	1.707	15.74*	1.853	13.60*
Acquisition of Technologies	1.292	11.59*	1.678	11.09*
Engineering & Design	0.261	2.05*	0.255	1.63*
Tooling Up	0.981	7.05*	1.336	8.35*
Training	1.308	10.69*	1.635	11.03*
Co-operation	0.094	2.66*	0.107	2.28*
Medium Firms	-0.075	-0.71	-0.029	-0.20
Large Firms	-0.174	-1.02	-0.120	-0.55
Food	0.004	0.03	0.137	0.36
Bever. & Tobacco	-0.221	-0.48	-0.628	-1.03
Textile Mills	0.499	1.11	0.553	0.92
Textile Product	0.247	0.53	-0.007	0.01
Apparel & Leather	-0.119	-0.39	-0.345	-0.89
Wood	-0.394	-1.32	-0.774	-2.05*
Paper	-0.500	-1.48	-0.792	-1.84**
Printing	0.421	1.21	0.630	1.43
Petroleum & Coal	-0.554	-1.03	-0.790	-1.18
Chemical	0.248	0.75	-0.190	-0.44
Plastics & Rubber	-0.167	-0.50	-0.361	-0.86
Non-metal. Mineral	-0.069	-0.21	-0.378	-0.87
Primary Metal	-1.063	-2.85*	-1.557	-3.44*
Fabricated Metal	-0.416	-1.39	-0.749	-1.96**
Machinery	0.375	1.07	0.199	0.49
Comp. & Electro.	0.430	1.06	0.059	0.12
Electrical Equip.	0.596	1.37	0.593	1.07
Transport. Equip.	-0.559	-1.64	-0.921	-2.16*
Furniture	0.199	0.58	0.176	0.40
Number of In-Sample Firms	5,451		3,994	

Source: Statistics Canada, Survey of Innovation 1999. Tabulated by authors. The reference group is Miscellaneous Manufacturing.

*Significant at the 5% level.

**Significant at the 10% level.

concerned product innovators only versus non-innovators as a reference group, and it was based on a sample of 1,851 firms because process innovators were excluded. As shown in Table 5, all previous results remained the same, except that acquisition of technologies and engineering and design were not significant factors for product innovation. The second regression concerned process innovators only versus non-innovators, and it was based on a sample of 1,688 firms because product innovators were excluded. Again, the results remained the same, except that engineering and design was not a significant factor for process innovation. The third regression concerned product innovators only versus process innovators only, and it was based on a sample of 1,457 firms because all other firms were excluded. The results clearly show that R&D was more significant for product innovation, and all other innovation inputs, especially acquisition of technologies, were more important for process innovation (Table 5).

In the above analyses, there was no control for the quality of innovation. To see if innovation quality made any significant difference, "sales from new or significantly improved product innovation" was used as a dependent variable. Although this variable still pertained to product innovation in general, it also reflected the significance of each product innovation. A regression similar to the one for product innovators only versus non-innovators presented in Table 5 was run, except that in this regression firms with less than 15% of sales from product innovation were treated the same as non-innovators. The estimation results show that the general picture remained unchanged, although there were some exceptions (Table 6). The variables "tooling up" and "co-operation" became insignificant. The result implies that, although "tooling up" and "co-operation" had a significant impact on firms becoming product innovators, they were not significant in terms of the commercialization of product innovation.

Complementarities among the innovation inputs were also analyzed. The exercise was an attempt to find out whether firms undertaking several innovation inputs simultaneously were more likely than other firms to be product or process innovators. To capture the impact of complementarities systematically, each of three additional variables was added (one at a time) into the regression model:[15]

Table 5: Estimated Impacts of Innovation Inputs on the Likelihood of Doing Pure Product Innovation or Pure Process Innovation

Variables	*Product Innovators Only vs. Non-Innovators*		*Process Innovators Only vs. Non-Innovators*		*Product Innovators Only vs. Process Innovators Only*	
	Coef.	*t-value*	*Coef.*	*t-value*	*Coef.*	*t-value*
Intercept	-1.551	-4.73*	-3.932	-7.69*	-2.474	-5.45*
R&D	1.913	14.78*	0.379	2.43*	1.192	8.24*
Acquisition of Technologies	0.081	0.57	2.216	12.33*	-1.982	-11.51*
Engineering & Design	0.036	0.23	0.124	0.72	-0.364	-2.47*
Tooling Up	0.401	2.28*	0.878	4.94*	-0.527	-3.37*
Training	0.929	6.11*	1.042	6.10*	-0.320	-1.98**
Co-operation	0.108	2.51*	0.078	1.55	0.049	1.05
Medium Firms	-0.164	-1.28	0.240	1.52	-0.312	-2.11*
Large Firms	-0.190	-0.93	0.139	0.58	-0.195	-0.94
Food	-0.093	-0.25	0.475	0.89	-0.237	-0.51
Bever. & Tobacco	-0.480	-0.84	0.984	1.36	-1.253	-1.99**
Textile Mills	0.632	1.24	1.294	1.68**	-0.683	-1.14
Textile Product	0.717	1.37	-1.126	-0.93	1.995	1.76**
Apparel & Leather	-0.134	-0.37	0.682	1.32	-0.781	-1.69**
Wood	-0.728	-1.97**	0.886	1.77**	-1.591	-3.51*
Paper	-0.772	-1.83**	0.635	1.17	-1.558	-3.12*
Printing	-0.205	-0.44	1.765	3.18*	-1.898	-3.69*
Petroleum & Coal	-0.452	-0.69	0.078	0.09	-1.572	-2.03*
Chemical	0.496	1.33	0.798	1.40	-0.276	-0.59
Plastics & Rubber	-0.124	-0.31	0.668	1.16	-0.333	-0.68
Non-metal. Mineral	0.019	0.05	0.397	0.70	-0.628	-1.27
Primary Metal	-1.165	-2.47*	0.378	0.67	-1.478	-2.79*
Fabricated Metal	-0.629	-1.74**	0.672	1.32	-1.094	-2.41*
Machinery	0.388	1.05	1.125	2.07*	-0.406	-0.88
Comp. & Electro.	0.544	1.23	1.124	1.65**	0.238	0.46
Electrical Equip.	0.417	0.83	0.384	0.51	-0.447	-0.75
Transport. Equip.	-0.421	-1.05	0.357	0.64	-0.561	-1.15
Furniture	0.026	0.06	1.322	2.35*	-1.007	-2.05*
Number of In-Sample Firms	1,851		1,688		1,457	

Source: Statistics Canada, Survey of Innovation 1999. Tabulated by authors. The reference group is Miscellaneous Manufacturing.

*Significant at the 5% level.

**Significant at the 10% level.

Table 6: Estimated Linkage Between Innovation Inputs and Sales of More Than 15% from New or Significantly Improved Products

Variables	*Innovators with More Than 15% of Sales from New or Significantly Improved Products vs. Other Product Innovators or Non-Innovators*	
	Coef.	*t-value*
Intercept	-2.323	-6.44*
R&D	1.207	7.36*
Acquisition of Technologies	0.098	0.63
Engineering & Design	0.220	1.32
Tooling Up	0.074	0.41
Training	0.738	4.42*
Co-operation	0.017	0.33
Medium Firms	-0.048	-0.32
Large Firms	-0.054	-0.24
Food	-1.158	-2.68*
Bever. & Tobacco	-0.350	-0.57
Textile Mills	0.421	0.79
Textile Product	0.919	1.77
Apparel & Leather	-0.120	-0.31
Wood	-0.979	-2.29*
Paper	-0.737	-1.56
Printing	-0.335	-0.62
Petroleum & Coal	-1.732	-1.58
Chemical	-0.479	-1.21
Plastics & Rubber	-0.676	-1.55
Non-metal. Mineral	-0.193	-0.47
Primary Metal	-1.316	-2.25*
Fabricated Metal	-1.371	-3.18*
Machinery	-0.027	-0.07
Comp. & Electro.	0.537	1.33
Electrical Equip.	0.467	0.96
Transport. Equip.	-0.152	-0.36
Furniture	0.343	0.79
Number of In-Sample Firms	1.851	

Source: Statistics Canada, Survey of Innovation 1999. Tabulated by authors. The reference group is Miscellaneous Manufacturing.

*Significant at the 5% level.

**Significant at the 10% level.

X1X2 = 1 for firms undertaking both R&D and acquisition of technologies

= 0 for other firms;

X1X3 = 1 for firms undertaking R&D and engineering and design

= 0 for other firms;

X2X3 = 1 for firms undertaking acquisition of technologies and engineering and design

= 0 for other firms.

All these variables were negatively significant, implying non-complementarity among those innovation inputs (Table 7).

To estimate the complementarities among innovation inputs, it was important to include all innovation input variables, together with the multi-innovation input variable. Otherwise, the multi-innovation input variable would capture the effect caused by the absence of those innovation input variables.

To illustrate this point, Table 8 repeats the regression from Table 7, but without the following variables: acquisition of technologies, engineering and design, tooling up and training. For the two regressions of product or process innovators versus non-innovators and process innovators only versus non-innovators, all multi-innovation input variables were positive and significant. However, this does not imply complementarities among innovation inputs. Instead, these multi-innovation input variables captured the effect caused by the excluded innovation input variables. For the regression of product innovators only versus non-innovators, none of the multi-innovation input variables was positive or significant. This is understandable, given that acquisition of technologies and engineering and design were not significant factors for product innovation (see Table 5).

6. CONCLUSION

Engaging in innovation inputs is very important, but it is even more important to transform them into product and process innovation, and ultimately into sales. Following are the key findings of the analysis:

Table 7: Innovation Input Complementarities and Their Impacts on Innovation Outputs (with All Innovation Input Variables and One Multi-Innovation Input Variable)

Variables	*Group 1: Product or Process Innovators vs Non-Innovators*				*Group 2: Product Innovators Only vs. Non-Innovators*			
	Case 1	*Case 2*	*Case 3*	*Case 4*	*Case 1*	*Case 2*	*Case 3*	*Case 4*
R&D	1.707	2.524	1.982	1.704	1.913	2.611	2.238	1.892
	(15.74)*	(15.30)*	(14.26)*	(15.61)*	(14.78)*	(14.55)*	(14.20)*	(14.42)*
Acquisition of Technologies	1.292	1.827	1.287	1.615	0.081	0.832	0.041	0.563
	(11.59)	(13.36)	(11.45)	(12.48)	(0.57)	(4.55)*	(0.29)	(3.42)
Engineering & Design	0.261	0.277	0.627	1.063	0.036	0.013	0.689*	0.911
	(2.05)	(2.13)	(3.70)	(5.21)	(0.23)	(0.08)	(3.04)*	(4.04)*
Tooling Up	0.981	1.095	1.006	1.103	0.401	0.490	0.420	0.522
	(7.05)*	(7.72)*	(7.20)*	(7.69)*	(2.28)*	(2.83)*	(2.41)*	(2.91)
Training	1.308	1.263	1.293	1.275	0.929	0.834	0.898	0.872
	(10.69)*	(10.09)*	(10.49)*	(10.26)*	(6.11)*	(5.47)*	(5.89)*	(5.68)*
Co-operation	0.094	0.096	0.094	0.090	0.108	0.105	0.105	0.097
	(2.66)*	(2.65)*	(2.63)*	(2.52)*	(2.51)*	(2.41)*	(2.42)*	(2.23)*
Medium Firms	-0.075	-0.077	-0.076	-0.077	-0.164	-0.192	-0.160	-0.172
	(-0.71)	(-0.69)	(-0.71)	(-0.71)	(-1.28)	(-1.46)	(-1.23)	(-1.32)
Large Firms	-0.174	-0.146	-0.176	-0.127	-0.190	-0.142	-0.188	-0.141
	(1.02)	(-0.84)	(-1.03)	(-0.74)	(-0.93)	(-0.69)	(-0.92)	(-0.68)
X1X2: R&D and Acquisition of Technologies		-1.547				-1.559		
		(-7.20)*				(-6.17)*		
X1X3: R&D and Engineering & Design			-0.781				-1.078	
			(-3.40)*				(-3.89)*	
X2X3: Acquisition of Technologies and Engineering & Design				-1.316				-1.640
				(-5.32)*				(-5.71)*
Number of In-Sample Firms	5,451				1,851			

Source: Statistics Canada, Survey of Innovation 1999. Tabulated by authors.

Table 7: Innovation Input Complementarities and Their Impacts on Innovation Outputs (with All Innovation Input Variables and One Multi-Innovation Input Variable) – Concluded

Variables	*Group 3: Process Innovators Only vs. Non-Innovators*			
	Case 1	*Case 2*	*Case 3*	*Case 4*
R&D	0.379	2.022	0.773	0.322
	(2.43)*	(6.49)*	(3.63)*	(2.08)*
Acquisition of Technologies	2.216	2.947	2.174	2.781
	(12.33)*	(12.40)*	(12.04)*	(12.48)*
Engineering & Design	0.124	0.086	0.489	1.698
	(0.72)	(0.51)	(2.24)*	(5.05)*
Tooling Up	0.878	0.928	0.895	1.004
	(4.94)*	(5.25)*	(5.01)*	(5.64)*
Training	1.042	0.968	1.036	0.924
	(6.10)*	(5.69)*	(6.05)*	(5.41)*
Co-operation	0.078	0.083	0.079	0.073
	(1.55)	(1.63)	(1.56)	(1.43)
Medium Firms	0.240	0.228	0.237	0.240
	(1.52)	(1.43)	(1.49)	(1.51)
Large Firms	0.139	0.167	0.137	0.200
	(0.58)	(0.70)	(0.57)	(0.84)
X1X2: R&D and Acquisition of Technologies		-2.066		
		(-5.96)*		
X1X3: R&D and Engineering & Design			-0.841	
			(-2.73)*	
X2X3: Acquisition of Technologies and Engineering & Design				-1.940
				(-5.26)*
Number of In-Sample Firms	1,688			

Source: Statistics Canada, Survey of Innovation 1999. Tabulated by authors.

Coefficients for industry dummies are not reported, and t-values are in parentheses. Case 1, reported in Table 4, is included for completeness.
*Significant at the 5% level.
**Significant at the 10% level.

Table 8: Innovation Input Complementarities and Their Impacts on Innovation Outputs (with Only the R&D Variable and One Multi-Innovation Input Variable)

Variables	*Group 4: Product or Process Innovators vs. Non-Innovators*			*Group 5: Product Innovators Only vs. Non-Innovators*			*Group 6: Process Innovators Only vs. Non-Innovators*		
	Case 2	*Case 3*	*Case 4*	*Case 2*	*Case 3*	*Case 4*	*Case 2*	*Case 3*	*Case 4*
R&D	1.674	2.093	2.216	2.502	2.309	2.295	0.165	1.109	1.035
	(11.77)*	(17.22)*	(23.06)*	(15.61)*	(15.88)*	(18.53)*	(0.71)	(6.79)*	(7.94)
Co-operation	0.175	0.171	0.154	0.142	0.141	0.140	0.133	0.128	0.100
	(5.91)*	(5.78)*	(5.01)*	(3.38)*	(3.36)*	(3.34)*	(3.33)*	(3.24)*	(2.38)
Medium Firms	0.217	0.226	0.127	-0.105	-0.112	-0.116	0.470	0.480	0.469
	(2.42)*	(2.53)*	(1.38)	(-0.84)	(-0.89)	(-0.29)	(3.74)*	(3.87)*	(3.57)
Large Firms	0.344	0.340	0.102	-0.008	-0.038	-0.051	0.516	0.535	0.328
	(2.41)*	(2.38)*	(0.68)	(-0.14)	(-0.20)	(-0.26)	(2.67)*	(2.82)*	(1.62)
X1X2: R&D and Acquisition of Technologies	1.342			-0.316			1.725		
	(8.21)*			(-1.76)**			(6.92)*		
X1X3: R&D and Engineering & Design		0.988			0.019			0.717	
		(6.38)*			(0.11)			(3.67)*	
X2X3: Acquisition of Technologies and Engineering & Design			1.704			0.098			1.765
			(15.12)*			(0.61)			(12.76)*
Number of In-Sample Firms		5,451			1,851			1,688	

Source: Statistics Canada, Survey of Innovation 1999. Tabulated by authors.
Coefficients for industry dummies are not reported, and t-values are in parentheses.
*Significant at the 5% level.
**Significant at the 10% level.

(i) All innovation inputs are significant, but each innovation input is used to produce different innovation outputs. R&D is more important for product innovation than for process innovation, while other innovation inputs (acquisition of technologies, engineering and design, tooling up and training) are more important for process innovation than for product innovation.

(ii) This finding implies that, for those firms that compete on the basis of technological efficiency (process innovation), government support programs focusing on the acquisition of technologies, industrial engineering and industrial design, and training would be more beneficial than those focusing on R&D. On the other hand, firms that compete on the basis of product innovation would benefit more from government support programs related to R&D.

(iii) Collaboration and co-operation with other firms significantly contribute to all innovation outputs, but do not significantly contribute to sales from product innovation.

(iv) Among innovators, small firms are as likely as large firms to perform well. This implies that small firms are as effective as large firms in converting innovation inputs into innovation outputs. However, small firms are less likely than large firms to undertake innovation initiatives. This suggests that small firms need more help than large firms to become innovative.

(v) There is no evidence to support the hypothesis that there are complementarities among innovation inputs.

Notes

The authors wish to thank Frances Anderson, Brian Nemes, Guy Sabourin and Susan Schaan for their excellent support in providing access to the linked Survey of Innovation 1999 database. They are also grateful to Frances Anderson, Fred Gault, Brian Nemes and Someshwar Rao for helpful comments and suggestions. The authors are responsible for any errors and omissions that may remain. All opinions expressed in this chapter are those of the authors, and they should in no way be attributed to Industry Canada.

[1]See Le (2001) for definitions.

[2]In the *Oslo Manual* (OECD/Eurostat 1997), product or process innovation is considered to be the activity of innovation. Thus, the definition of innovation output

in this chapter is equivalent to that of innovation activity in the *Oslo Manual*, which gives guidelines for collecting and interpreting innovation data. According to the *Oslo Manual*, a product innovation is an innovation activity that has been introduced to the market, and a process innovation is an innovation activity that has been used in the production process. It should be noted that the definition of innovation output or activity is different from the definition of innovation outcome in the *Oslo Manual*. The latter is associated with improved economic performance such as greater market share or increased sales due to product innovation or, internally, changes in skill levels.

[3]There is evidence that Canadian firms lag in their ability to capitalize on the benefits from their innovations, compared to firms in some European countries (Mohnen and Therrien 2001).

[4]However, after linkage to the 1997 ASM, some firms with fewer than 20 employees in 1997 were also in the linked database.

[5]For methodological issues and an overall description of the survey, see Schaan and Anderson (2001).

[6]Employment information was used to divide firms into three groups of different sizes.

[7]As defined in the survey, a new product is a product that is new to a firm and whose characteristics or intended uses differ significantly from those of the firm's previously produced products. A significantly improved product is an existing product whose performance has been significantly enhanced or upgraded. A complex product that consists of a number of components or integrated subsystems may be improved by partial changes to one of the components or subsystems. Changes to a firm's existing products that are purely aesthetic or that involve only minor modifications are not to be included.

[8]As defined in the survey, a new production/manufacturing process is a process that is new to a firm. It involves the introduction into the firm of new production/manufacturing methods, procedures, systems, machinery or equipment which differ significantly from the firm's previous production/manufacturing processes. A significantly improved production/manufacturing process involves significant changes to a firm's existing processes that may be intended to produce new or significantly improved products or production/manufacturing processes. Minor or routine changes to processes are not to be included.

[9]No information on sales from process innovation was available.

[10]In the survey, the two variables were six-point ordinal variables, covering six sales ranges from 1%-5% to 76%-100%. However, the majority of firms were in the first two categories.

[11]This variable represented a firm's perception of the importance of co-operation and may not have reflected a firm's actual co-operation status. However, when this variable was replaced by an alternative variable representing a firm's actual co-operation status (the variable corresponding to question 14 in the survey), similar results were obtained. The alternative variable was not chosen here since it offered no information about firms that were not engaged in any innovation inputs and the sample was smaller.

[12]Due to confidentiality, Apparel and Leather are reported together.

[13]All results are weighted estimates.

[14]A factor was classified as highly important to a firm if it scored 4 or 5.

[15]Dummy variables that denoted at least two, three, four or all five innovation inputs were tried. Again, similar results were obtained.

Managing Knowledge

Section Overview

Fred Gault

The two chapters in this section deal with knowledge and its management in the innovation process. Hanel continues the analysis of the 1999 Survey of Innovation data by looking at how knowledge is protected by manufacturing firms. Then, Hamdani, using the earlier 1996 Survey of Innovation, which covered service industries, looks at the relationship between skill levels and innovation. Both chapters identify themes that have already emerged in earlier parts of this book: the use of government services; and, the importance of alliances.

Hanel looks at the means of protecting the knowledge of the manufacturing firm and confirms the hypotheses that the use of intellectual property protection, such as patents, trademarks, copyrights and trade secrecy, increases with the size of firm, the doing of R&D, and engagement in innovation, especially world-first innovation. The use of intellectual property protection also varies with industry and when industries are classified as high- or low-tech, the summary finding is that small, low-tech firms have the lowest propensity to use intellectual property protection. This gives rise to two observations. The cost of learning how to use intellectual property protection and the financial burden of using such instruments are barriers for the small and medium-sized firms.

Hanel goes beyond the use of intellectual property protection to raise the importance to innovation of learning in the firm and the capacity to use knowledge drawn from internal R&D and from outside the firm, as well as knowledge gained of the market and through collaboration. Government is identified as a significant external source for firms introducing world-first innovations. In this context the use of intellectual property protection becomes a strategic decision reflecting the reaction of the firm to its economic and social environment and the way in which it manages its knowledge.

The chapter by Hamdani picks up the importance of skills in the innovation process and draws upon the Survey of Innovation 1996 of selected service industries rather than the Survey of Innovation 1999 used by most of the other authors. While these are older results, the insights gained from studying links of the firms, and their relation to innovation, are still relevant.

Hamdani uses the proportion of university graduates, college graduates, and non-graduates to classify firms as knowledge intensive, or not, and finds that knowledge intensive firms are more likely to innovate. When firms with lower knowledge intensity innovate, they tend more to contract out for expertise or to enter into alliances.

The presence in the firms of the well-educated means that the firm can absorb knowledge, from inside and outside the firm, more readily and apply it to the process of innovation and then learn from the solving of complex problems which are part of innovation. This raises questions about the learning capacity of the individual, of teams in the firm and of the firm itself. As in the work of Hanel, this leads to a question of how the knowledge of the environment of the firm and of the transformation processes within the firm are managed.

9

Protection of Intellectual Property by Manufacturing Firms in Canada

Petr Hanel

1. INTRODUCTION

Protection of intellectual property is the oldest means and one of the principal instruments of innovation policy. The objective of the study reported in this chapter was to determine how the utilization of intellectual property rights (IPRs) by Canadian manufacturing firms is related to the firms' characteristics, activities and competitive strategies, and to the industry sector in which they operate. One of the related questions that was also addressed was the extent to which Canadian firms obtain patents in Canada and abroad, and especially in the United States.

The main source of information used in the study was the Survey of Innovation 1999 conducted by Statistics Canada, which included several questions on the protection of intellectual property.

1.1 Review of Principal Findings of Previous Surveys

To put the Survey of Innovation 1999 in a proper perspective, it is useful to review the findings of major Canadian surveys of intellectual property use conducted over the last 20 years.

1.1.1 Economic Council of Canada Survey for 1980

The Economic Council of Canada survey (De Melto, McMullen and Wills 1980) examined major innovations introduced in five Canadian manufacturing industries. The study concluded that only about 32% of the 283 major innovations introduced in Canada in the preceding 20 years had been patented. Following are some of the study's findings. (1) The tendency to patent innovations increased with the size of the innovating firm. (2) Foreign-controlled firms (and particularly those under US control) patented innovations significantly more (39%) than their domestically controlled counterparts (23%). (3) Innovations based on imported technology were more often patented in Canada than innovations based on technology developed in-house. (4) There was a clear positive association between the cost of an innovation and patenting: the more costly innovations were more likely to be patented. (5) Finally, the rate of patenting declined over time, especially in the last half of the 1970s. This tendency of firms to rely less and less on the patent system to protect their major innovations was also noted in the United States and prompted the influential study by Levin, Klevorick, Nelson and Winter (1987).

1.1.2 Industry, Science and Technology Canada Report, 1989

A report commissioned by Industry, Science and Technology Canada, Consumer and Corporate Affairs Canada and the Science Council of Canada (Industry, Science and Technology Canada 1989) looked at the attitudes, practices and interests of Canadian industry with respect to IPRs. The 900 firms surveyed were broken down into four groups: top R&D performers (100); high-tech firms (300); medium- and low-tech firms (400); and major copyright users (100). The authors of the report discovered the following. (1) Even though the majority of respondents were satisfied with Canadian IPRs, there was an important variance by industry sector and size of firm. Smaller firms[1] and firms in the "new economy" sectors such as software development and biotechnology expressed the most dissatisfaction with Canadian IPRs.[2] (2) A high degree of infringement and counterfeiting was reported. Between 32%

and 40% of firms in the four groups indicated that their IPRs had been violated in the four years preceding the study. A large proportion of firms complained that litigation was too expensive, especially for smaller firms, and the penalties were insufficient to prevent infringement. (3) A significant number of firms stated that they had insufficient knowledge or expertise with respect to IPRs. (4) Finally, with the exception of copyright users, firms indicated that they had difficulties in terms of the time and cost involved in registering and obtaining IPRs.

1.1.3 Statistics Canada Survey for 1993

The study based on the 1993 Survey of Innovation and Advanced Technology conducted by Statistics Canada found that there were substantial differences in the use of trademarks, patents, trade secrets, industrial designs and copyrights between those firms that had introduced innovations in the preceding three years and those that had not. Trademarks were the most popular form of protection, followed by patents and trade secrets, industrial designs and copyrights (Baldwin 1997).[3] The survey found that firms included in the integrated portion of the Statistics Canada Business Register used IPRs as shown in Table 1.

Table 1: Use of IPRs by Innovators and All Manufacturing Firms, 1989-91 (% of Firms*)

IPR	*Patents*	*Trade-marks*	*Copy-rights*	*Trade Secrets*	*Integrated Circuit Designs*	*Industrial Designs*	*Plant Breeder's Rights*	*Other*
Innovators	24.66	31.31	9.44	17.99	1.99	13.48	0.451	1.02
All Firms	16.32	22.96	6.35	11.7	1.14	9.05	0.51	0.82

Source: Statistics Canada, Survey of Innovation and Advanced Technology 1993, special tabulation.

*This table represents firms included in the integrated portion of the Statistics Canada Business Register (typically firms with more than 20 employees). In contrast, Baldwin's (1997) results are representative of firms of all sizes — i.e., his sample also includes the smallest firms — and therefore the results in his study are not directly comparable to those in this table.

Baldwin's study corroborated earlier findings by demonstrating the following. (1) The use of IPRs increased with size of firm. (2) The use of intellectual property protection varied significantly between industries. The interindustry differences in the use of IPRs were determined at least in part by the technology sector (see Robson, Townsend and Pavitt 1988),[4] the nature of the products, their stage in the product life cycle and competitive conditions. (3) Product innovations (with or without a change in the production process) were more than twice as likely as pure process innovations to be patented. Process innovations lend themselves better to protection through trade secrecy. (4) Large firms were more likely than small firms to introduce a world-first innovation. (5) Some 15% of the innovations of large firms were world-first innovations. The firms that introduced world-first innovations made, in general, much greater use of IPRs than less innovative firms. (6) Foreign-owned firms, irrespective of size, industry or type of innovation, had recourse to instruments of intellectual property protection more often than Canadian-owned firms.

The results of the 1993 Survey of Innovation and Advanced Technology also indicated that the US findings of Levin et al. (1987), suggesting that firms tend to value alternative strategies more highly than the statutory forms of intellectual property protection, also applied to Canada. Moreover, the population of manufacturing firms as a whole ranked such strategies as patent protection as less than effective.[5] However, these rankings depended very much on the characteristics of a firm. If a firm was innovative, large and foreign-owned, and operated in one of those industries that tended to produce more innovations, the score given to the statutory forms of protection like patents increased greatly. On average, users of patents found them effective; so too did large foreign firms. Many firms used alternative strategies, such as increasing the complexity of product design to fend off imitation or being first in the market to reap the benefits from their innovations. These strategies were judged to be more effective in deriving benefits from innovation than relying on statutory IPRs.

2. ELEMENTS OF THE CONCEPTUAL FRAMEWORK USED IN THE STUDY PRESENTED HERE

The economic theory of innovation and intellectual property and earlier empirical studies suggest a conceptual framework that was used for the descriptive analysis of the Survey of Innovation 1999. Following is a summary of the rationale behind the principal relationships that were examined in the study presented in this chapter.

2.1 Relationship Between Use of IPRs and Firms' Innovation Status

The use of patents and other statutory instruments of intellectual property protection by manufacturing firms is closely, but not exclusively, related to their innovation status. Innovating firms create new knowledge and intellectual property and, in order to reap the benefits of innovation, they are more likely to protect it with various IPRs and other strategies than firms that do not innovate. The overwhelming majority (about 80%) of chief executive officers (CEOs) and head-office officials surveyed in 1999 declared that their firms had introduced an innovation in the preceding three years. The study presented here contrasted the use of IPRs by firms that introduced an innovation in the period 1997-99 with their use by firms that attempted to introduce an innovation but failed and by those that were not involved in innovation.

2.2 Type of Innovation

Patents are usually more effective in protecting product innovations than process innovations. New or significantly improved production processes are often better protected by trade secrets. Firms typically use a combination of IPRs. Their composition varies with the stage of the innovation process and the combination of the innovation's protectable elements (Cohen, Nelson and Walsh 2000). To illustrate the ways in which Canadian manufacturing firms protect their intellectual property, the use of IPRs was broken down by the type and novelty of innovation.

2.3 Novelty of Innovation

The value of intellectual property is, to an important extent, a function of its novelty. By definition, patents are granted only to producers of world-first inventions. Firms that introduce a world-first innovation are therefore more likely to use a patent than firms that introduce a Canada-first innovation or those that imitate a new process or product already in existence elsewhere in Canada. Firms that introduce the two less original types of innovation may, however, acquire or license patents or other IPRs as part of a technology transfer and may therefore also report using IPRs. The type of innovation and its novelty are therefore potentially important determinants of the use of IPRs. The results of the survey are presented here so that they may be compared by the type and novelty of innovation.

2.4 Use of IPRs and Size of Firm

The need for protection of intellectual property varies according to size of firm for at least two reasons. One is related to innovative activity, the other to financial constraints. Small firms are less likely to innovate than large ones. When they innovate, small firms introduce less frequently than larger firms innovations that contain most intellectual property worthy of protection. It is therefore likely that the use of intellectual property protection is positively related to size of firm. The cost aspects of intellectual property protection are also biased in favour of large firms. The cost of protection — including the cost of learning how to protect intellectual property and the administrative costs involved in obtaining and maintaining statutory IPRs — imposes in relative terms a greater burden on small and medium-sized enterprises (SMEs) than on large ones. SMEs face another disadvantage when it comes to enforcement of their IPRs. Since a patent is no more than a "licence to litigate," the cost of monitoring possible infringements and the prospective cost of litigation can be too high for SMEs in terms of both time and money (European Commission 2001). For all these reasons, the use of IPRs is expected to be closely related to size of firm. This pattern was found in the 1989-91 survey of Canadian manufacturing firms, and it has also been reported in Europe (European Commission 2001). To find out whether the use of IPRs remained closely related to size of firm in the period 1997-99, the use of IPRs was broken down by size of firm in the study presented here.

2.5 Interindustry Differences in Use of IPRs

Types of innovation and the mix of IPRs vary from one industry to another. These variations are related, on the one hand, to differences in technological opportunity and, on the other, to differences in the use of statutory IPRs. Technological opportunity, a term designating the potential contribution of advances in science and technology to innovation, varies from industry to industry, as do the competitive conditions. These differences explain, to a large extent, the significant interindustry differences in innovation performance that impact on the use of IPRs. Differences in the use of statutory IPRs are directly related to the particular IPR

Levin et al. (1987) found that, in chemical and pharmaceutical industries, patents were considered a more effective means of capitalizing on innovation, and they were used more frequently than in other industries. According to Cohen, Nelson and Walsh (2000), the reasons for patenting differ between industries that usually introduce a "discrete" product (a new substance introduced by a chemical or pharmaceutical firm) and those introducing a "complex" product innovation (electronic or telecommunications products typically consisting of a large number of patentable elements). Their study also showed that patenting in general and patenting to protect discrete or complex innovations in particular are motivated by reasons other than to prevent copying (negotiations and cross-licensing, licensing revenue, preventing lawsuits and enhancing reputation, for example).

2.6 Technology Sectors

A comprehensive study of innovation in the United Kingdom by Robson, Townsend and Pavitt (1988) found that different groups of industries play different roles in the innovation process. On the basis of observed patterns of innovation sources and use in manufacturing industries, the authors introduced a three-part technology sector taxonomy (core, secondary and "other" sectors). The core sector (Chemical and Pharmaceutical, Machinery, Electronics and Instruments) includes industries at the forefront of technological change whose product innovations are used in the secondary and "other" sectors and in the rest of the economy. The secondary sector (Plastics and Rubber, Nonmetallic Mineral, Primary Metal, Fabricated Metal and Transportation Equipment) is a user of product innovations from the core sector and at the same time a source of innovations used in the "other" sector and in the rest of the economy. Consumer product industries and some bulk material industries that use innovations introduced by the core and secondary sectors are included in the "other" sector. Food, Beverage and Tobacco, Textiles, Apparel and Leather, as well as Wood and Paper, belong to this sector. The usefulness of this taxonomy to synthesize the interindustry patterns of innovation and the use of IPRs is demonstrated in Baldwin and Hanel (2003). It was used in the study presented here to highlight the main differences in the use of IPRs among the three technology sectors.

2.7 Use of Combinations of IPRs

A trade secret may supplement a patent or it may be used as a substitute for patent protection. Innovators choose to use a trade secret when they believe that patent protection is too costly relative to the value of their innovation, or that it will give them a substantially lower reward than the benefit of their innovation, either because the innovation is not patentable or because the length (or other conditions) of patent protection is insufficient (Friedman, Landes and Posner 1991). Trademarks are often used along with trade secrets and/or patents and industrial designs. Trademarks are used to a certain extent by all industries, but they are used more intensively by those producing consumer products, such as leather articles and clothing, beverages or drugs.

Other IPRs, such as copyrights, are used frequently in those industries concerned with protection of printed material, recordings and software.

3. USE OF IPRS AND INNOVATION ACCORDING TO THE RESULTS OF THE SURVEY OF INNOVATION 1999

Following is a descriptive statistical analysis of the principal findings of the Survey of Innovation 1999, based on the conceptual framework introduced above. To a great extent, the use of IPRs is correlated with the basic economic characteristics of firms, their activities and the industry environment in which they operate. Many of these characteristics are interrelated. To reveal the underlying relationships, a series of multiple probabilistic regression models was set up. The probability, on the one hand, that a particular firm would use a specific instrument of intellectual property protection and, on the other hand, that it would innovate was estimated. The use of IPRs was included among the explanatory variables in the innovation equation. Since a firm's decisions to innovate and to use intellectual property protection were most likely not independent, the system of simultaneous equations was estimated by a two-stage approach to avoid the simultaneity bias. The complete results of these probabilistic models are reported elsewhere (Hanel 2001). Only the marginal effects of each variable on the probability of using patents are reported here.

The Survey of Innovation 1999 of Canadian manufacturing firms asked the CEOs or their representatives two questions on the use of IPRs.

In response to one of the questions — "... which of the following methods [Patents, Trademarks, Copyrights, Confidentiality Agreements, Trade Secrets and Other][6] have been used by your firm to protect its intellectual property during the past three years, 1997 to 1999" — about two-thirds of manufacturing firms (66.1%) responded that they had used at least one of the IPRs listed above. The question was addressed to all firms, since even those that were not involved in innovation or had not managed to introduce an innovation in the three-year period could well have used IPRs to protect innovations introduced earlier or other intangible assets not directly related to a recent innovation. Thus, the first interesting information was the overall pattern of IPR use. On the basis of their innovation status, firms were divided into three groups:

(i) firms that had innovated successfully (i.e., firms that had introduced a new or significantly improved product and/or a new or significantly improved production/ manufacturing process);

(ii) firms that had attempted to innovate but had been unsuccessful or had not completed the innovation; and

(iii) firms that were not involved in innovation in the period 1997-99.[7]

3.1 Use of IPRs by All Manufacturing Firms

As indicated in the last column of Table 2, successful innovators[8] used IPRs more often than unsuccessful innovators, and they in turn used them more frequently than firms that were not involved in innovation in the three-year period. In comparison with the period 1989-91, successful innovators in the period 1997-99 appeared to use all IPRs more frequently. For example, according to the Survey of Innovation 1999, 29.3% of successful innovators reported using patents, whereas, according to the 1993 Survey of Innovation and Advanced Technology, about one-quarter of innovators (24.7%) used patents (cf. Table 1). However, part of this apparent difference in IPR use reflects methodological differences between the two surveys.[9]

The statutory instruments of intellectual property protection require complex and costly administrative procedures. They were therefore used less frequently than trade secrets and confidentiality agreements, which are simpler to use. The choice of a particular IPR depends on the type and novelty of an innovation, and the characteristics of the firm and the industry. Trademarks were the most frequently reported statutory instrument of IPR, used by more than one-third of all manufacturing enterprises. Most firms (43.2%) protected their intellectual property by confidentiality agreements with their employees and business partners.

Table 2: Use of IPRs, by Innovation Status, 1997-99 (% of All Manufacturing Firms)

Status	***Share of Population***	*Patents*	*Trade-marks*	*Copy-rights*	*Trade Secrets*	*Confiden-tiality Agreemts.*	*Other*	*Any IPR*
Successful Innovators	**80.7**	29.3	39.8	13.6	28.4	48.4	2.7	72.6
Unsuccess. Innovators	**7.2**	14.[illegible]	25.3	6.4	14.4	32.6	1.8	49.7
Not Involved	**12.1**	8.3	19.1	4.5	7.5	16.9	2.3	35.9
All	**100.0**	25.7	36.0	12.0	24.7	43.2	2.5	66.1

Source: Statistics Canada, Survey of Innovation 1999, preliminary results.

Note: The statistics from the Survey of Innovation 1999 presented in this and all other tables and figures are weighted by gross business income and are representative of the population of Canadian manufacturing "provincial enterprises."

Only some products and processes satisfy patenting criteria of novelty, usefulness and non-obvious improvement. About one-quarter of all manufacturing firms used patents to protect their innovations. When a firm patents an innovation, it reveals the substance of the innovation in exchange for statutory temporary monopoly protection. Since, as indicated above, the efficacy of patent protection is far from perfect, some innovations, particularly new or significantly improved production processes, are often better protected by trade secrecy.[10] Trade secrets were used about as frequently as patents (24.7%). Copyrights are used mainly to protect such products as works of art and software. The use of copyrights is therefore distributed rather unevenly among industries. Only 12% of all manufacturing firms reported using copyrights.

3.2 Use of IPRs by Innovating Firms

About 80% of firms innovated in the period 1997-99. Firms that were involved in an innovation process protected their intellectual property more frequently than the overall population of manufacturing enterprises: almost three-quarters (72.6%) of successful innovators reported the use of at least one of the IPR instruments. About one-half of those firms that attempted unsuccessfully to introduce an innovation in the three-year period reported using an IPR. The lowest proportion of users of any IPR (35.9%) was found among firms that were not involved in innovation over the period 1997-99. A roughly similar pattern was observed for each particular IPR with the exception of the category "Other," which is insignificant.

3.2.1 Use of IPRs According to Type of Innovation: Product or Process

As indicated in the Introduction, the choice of an IPR instrument depends on the type of innovation. Patents are considered to be more suitable for the protection of product innovations, and trade secrecy is considered to be a more effective means of protection for process innovations. One should not, however, read too much into the difference between product and process innovations, because, in the 1999 survey, two-thirds (67%) of all firms reported introducing a combination of product and process innovations. Firms that introduced a combination of innovations used IPRs more frequently than creators of a product- or process-only innovation. The most pronounced difference between product and process innovators was with regard to patent protection. Innovators that introduced only a new or significantly improved product were three times as likely to protect them with patents than were process innovators (see Table 3).

Table 3: Use of IPRs, by Type of Innovation, 1997-99 (% of Innovating Firms)

IPR	*Type of Innovation*			
	Product and Process	*Product*	*Process*	*All Firms*
% of All Innovations	**67.0**	**17.6**	**15.4**	**100.0**
Patents	32.7	34.1	11.1	29.3
Trademarks	43.4	44.7	18.6	39.8
Copyrights	15.0	14.8	6.7	13.6
Trade Secrets	32.1	24.4	17.5	28.4
Confidentiality Agreements	52.6	46.6	36.3	48.4
Other	2.9	2.2	2.8	2.7
At Least One IPR	77.0	75.1	52.7	72.6

Source: Statistics Canada, Survey of Innovation 1999. Tabulated by author.

3.2.2 Use of IPRs According to Novelty of Innovation

Firms were asked to classify their most important innovation according to its novelty: world-first, Canada-first or firm-first. This question elicited a response from 88.3% of innovators.

A minority of innovators providing information on the novelty of their most important innovation introduced a world-first innovation (14.5%). Almost twice as many firms (24.5%) introduced a Canada-first innovation. The largest proportion of firms (61%) introduced a firm-first innovation (see Table 4).

Table 4: Use of IPRs, by Novelty of Innovation, 1997-99 (% of Innovating Firms)

	Novelty of Innovation		
IPR	*World-First*	*Canada-First*	*Firm-First*
% of All Innovations	**14.5**	**24.5**	**61.0**
Patents	62.8	40.7	20.8
Trademarks	57.9	49.1	32.9
Copyrights	26.7	18.7	9.6
Trade Secrets	44.3	37.5	23.3
Confidentiality Agreements	72.1	60.1	42.7
Other	2.9	2.6	3.1
Any IPR	93.3	83.9	66.8

Source: Statistics Canada, Survey of Innovation 1999. Tabulated by author.

Note: This table presents the distribution of innovators that responded to the question on novelty and were able to classify their innovation in one of the three categories. The classification of an innovation in a particular category was exclusive and not cumulative — i.e., if an innovation was classified as world-first, it was not classified at the same time as Canada-first or firm-first.

As expected, firms that introduced a world-first innovation used all IPR instruments more frequently than firms that introduced a Canada-first innovation. Almost two out of three used patents. Firms that introduced a firm-first innovation obviously did not have to fear imitation, and they used all means of protecting intellectual property less than the first two groups. Even though this pattern applied to all IPR instruments, it was most pronounced for patents (Table 4).

The novelty-related differences in the use of IPR instruments other than patents were less pronounced, but the pattern was similar. Less original innovators had less intellectual property to protect and therefore used all IPR instruments less often than world-first innovators. This confirms Baldwin's (1997) results.

3.2.3 Use of IPRs According to Size of Firm

Large firms use IPRs more frequently than small firms. There are two reasons for this. Small firms are less likely to innovate than large ones and, when they do, they are less likely to introduce world-first innovations, which are most often protected by IPRs. The cost of intellectual property protection also imposes a greater burden on SMEs than on large firms. Small firms are therefore less likely to use IPRs than large firms. The data from the 1999 survey confirm this pattern for Canada.

The data presented in Table 5 show that the use of IPRs was positively correlated with size of firm. The use of at least one IPR increased monotonically with size of firm, from 65.2% for the smallest firms to 87.3% for the largest ones. This pattern was replicated for each of the IPR instruments and, with some minor exceptions, for each innovation type (product, process and both). Larger firms were more likely to use each IPR instrument than smaller ones.

Table 5: Use of IPRs, by Size of Firm, 1997-99 (% of Innovating Firms)

	Number of Employees			
IPR	*20-49*	*50-99*	*100-499*	*500+*
Patents	21.2	25.7	36.8	50.6
Trademarks	32.8	35.5	47.8	52.4
Copyrights	9.9	11.4	17.1	24.2
Trade Secrets	25.2	26.5	31.2	39.8
Confidentiality Agreements	39.5	44.3	58.3	67.0
At Least One IPR	65.2	69.2	80.4	87.3

Source: Statistics Canada, Survey of Innovation 1999. Tabulated by author.

Note: The category "Other" is not included.

3.2.4 Sectoral Differences

As expected, firms in the core sector with high technological opportunity, supplying innovations to other manufacturing firms and, for that matter, to the rest of the economy, had in general a greater stake in protecting knowledge assets than firms in technologically less progressive industries. When firms were regrouped, according to the taxonomy of Robson, Townsend and Pavitt (1988),[11] into the three technology sectors (core, secondary and "other"), the frequency of use of all IPRs, except for trademarks, unequivocally declined from the former to the latter group. Trademarks were used least frequently in the secondary sector.

Technological competition is most intense among industries in the core sector. Core-sector firms, especially those that introduced world-first innovations, sought patent protection more often than firms in the secondary sector, and they in turn sought it more often than those in the "other" sector. Core-sector firms also used all other statutory instruments of IPR protection more often than firms in the other two sectors.

3.2.5 Interindustry Differences in Use of IPRs by Innovating Firms

Even though the IPR strategies of firms belonging to the same technology sector are fairly similar, significant interindustry differences still exist within each sector. These differences are mainly due to differences in the perceived effectiveness of various IPRs (Cohen, Nelson and Walsh 2000). Some features of the interindustry pattern are presented in Table A1.2 in Appendix 1. As in other countries, pharmaceutical firms in Canada protected their intellectual property most intensively. Almost 19 out of 20 pharmaceutical and communications equipment firms used at least one of the IPRs. The most frequently used IPR instruments were confidentiality agreements, trademarks and patents.

The use of patents varied significantly from one industry to another. The top users of patents were, somewhat surprisingly, not pharmaceutical and chemical industries where, according to other studies (Baldwin 1997; Levin et al. 1987; and Cohen et al. 2000), patents are perceived as being the most effective means of protection. According to the 1999 survey, the highest proportion of innovating firms using patents was found in Agricultural, Construction and Mining Machinery and Industrial Machinery Manufacturing (72.3%) and Electrical Equipment, Appliance and Component Manufacturing (66.1%), followed by Pharmaceutical and Medicine Manufacturing (59.4%).

In Computer and Peripheral Equipment Manufacturing, almost all innovating IPR users (94.9%) protected their intellectual property with confidentiality agreements, but less than half of them used patents and/

or trade secrets. Since software is often an integral part of computer and peripheral equipment, and since software can be protected by copyright, the computer industry was the most frequent user of copyrights (47.7%).

The top user of trade secrets was Semiconductor and Other Electronic Component Manufacturing, followed by Petroleum and Coal and Beverage and Tobacco.

On the low end of the spectrum for IPR use were firms in the "other" sector, most notably those fabricating and transforming wood products. This industry mainly produces standard industrial materials and components for further transformation in downstream industries and services (construction), which rely more on price competition than on product differentiation and technological characteristics. Less than half of the firms in this sector used any IPRs. When they did, they relied on trademarks, trade secrets and, to a lesser extent, patents.

3.3 R&D Activity and Use of IPRs

Innovative ideas and solutions come from various sources, from both within and outside a firm. Even though R&D is not always the most important source of innovative ideas, the majority of innovative firms that protected their intellectual property acknowledged that R&D played an important role in their innovation process. An overview of all firms (innovators and non-innovators) shows that, indeed, firms carrying out R&D[12] used all IPR instruments more often than firms that did not carry out R&D. To illustrate this, Figure 1 shows the relationship, for the subpopulation of innovating firms, between the use of IPRs by firms that carried out R&D and their use by those that did not. This pattern remained true for each firm size and each technology sector. Thus, it appears that firms that pursued active innovation strategies based on R&D felt the need to protect their intellectual property and developed the competence to do so. This was particularly notable for firms that often collaborated with universities and colleges. These firms used IPRs, especially patents, more often than other firms.[13]

R&D activity is, however, not a prerequisite for innovation. Less than two-thirds (about 59%) of all manufacturing firms carried out R&D activities (Table 6). Almost two-thirds of firms that did not carry out

Figure 1: Use of IPRs by Performers and Non-Performers of R&D, 1997-99

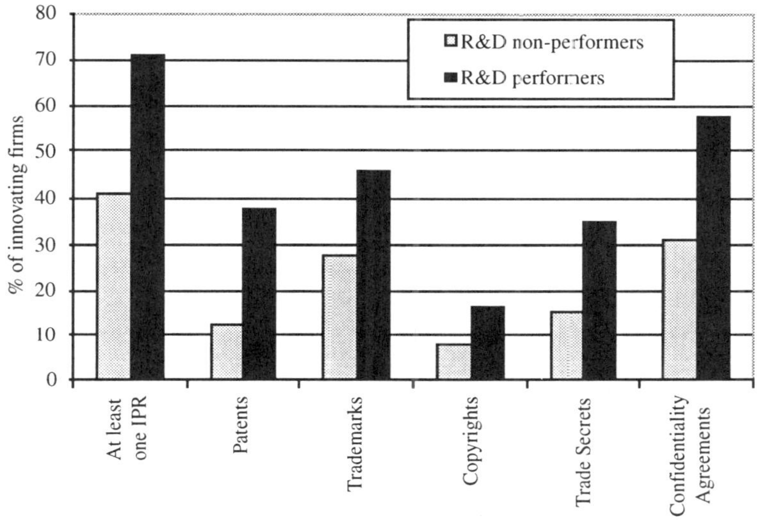

Source: Statistics Canada, Survey of Innovation 1999. Tabulated by author.
Note: The category "Other" is not included.

Table 6: Relationship Between Innovation and R&D, 1997-99

Innovation	*R&D*		*Total (%)*
	No (%)	*Yes (%)*	
No (%)	14.8 (76.9) (36.6)	4.5 (23.1) (7.5)	19.3 (100.0)
Yes (%)	25.7 (31.8) (63.4)	55 (68.16) (92.5)	80.7(100.0)
Total (%)	40.6 (100.0)	59.4 (100.0)	100.0

Source: Statistics Canada, Survey of Innovation 1999. Tabulated by author.

R&D nevertheless innovated successfully.[14] As innovation surveys show, R&D is only one of the important sources of innovative ideas and technical solutions. Firms frequently use ideas from management and production departments, from customers, suppliers, affiliated firms and competitors, and from public institutions such as government laboratories.

Firms that did not carry out R&D activities had much less use for all forms of intellectual property protection than those that did. This suggests that innovators that did not carry out R&D introduced mostly incremental, imitative innovations with a lower intellectual property value. World-first innovations depend on R&D more than firm-first innovations. The difference between performers and non-performers of R&D in their use of IPRs was most evident in the use of patents.

3.4 Extent of Patenting

The temporary monopoly protection granted by a patent may encourage firms to apply for as many patents as possible. The accelerating pace of technological change and innovation, as well as the recent introduction in Canada of the "first-to-file" patent system, have increased the propensity of firms to file patent applications (Rafiquzzaman and Whewell 1999). Firms adopt various strategies with respect to patenting, and the number of patentable innovations varies significantly from firm to firm. Given the high degree of economic integration between Canada and the United States, many firms patent their innovations in both countries. Some apply for patents in the United States only, and a small minority files patent applications in other foreign countries.

In addition to the question regarding use of IPRs, firms were asked to provide information on (1) the extent and geographical pattern of patent applications, and (2) the number of patent applications filed in Canada and the United States.

3.4.1 Extent and Geographical Pattern of Patent Applications

As far as the extent of patent application was concerned, almost one in five manufacturing firms (19%) applied for at least one patent in the period 1997-99 (see Table 7). As expected, the proportion of firms that

Table 7: Applications for Patents in Canada and the United States, 1997-99

	Applied for At Least One Patent (%)	*Of These, % That Applied for Patents in:* Canada	US	Both Canada and US	Canada Only	US Only	Neither Canada nor US
Manufacturing Industries							
All Firms	19.0	85.1	75.6	65.8	19.3	9.8	5.1
Innovators	22.4	85.2	75.4	65.8	19.5	9.6	5.1

Source: Statistics Canada, Survey of Innovation 1999, preliminary results.

applied for patents was larger in the group of innovating manufacturers (22.4%).[15] With regard to the geographical pattern of patent application, the table shows that, even though two-thirds of Canadian firms applied for patents in both Canada and the United States, the majority (about 85%) filed patent applications in Canada, and about 20% of firms filed them in Canada only. Only about 10% of firms that applied for patents did not bother to file an application in Canada and applied for patents only in the United States. It is probable that these were mainly US affiliates.[16] A small group of firms (5.1%) applied for patents in countries other than Canada and the United States.

3.4.2 Number of Patent Applications Filed in Canada and the United States

Firms were asked to indicate not only whether they applied for patents in Canada and/or the United States, but also the number of applications in each country. Details on industry distribution of patent applications are presented in Table A1.3 in Appendix 1.

Most firms applied for one patent only (41.6% in Canada and 34.4% in the United States) (see Table 8). The percentage of firms that applied for more than one patent declined rapidly with the number of applications. Those, mainly larger, firms that applied for patents most frequently applied for patents more in the United States than in Canada. For

Table 8: Number of Patent Applications in Canada and the United States, 1997-99

		Of These, Number of Patents Applied for:						
		1	*2*	*3*	*4*	*5-9*	*10 or More*	*Unknown*
					%			
All Manufacturing Firms								
Applied for At Least One Patent	**19.0**							
Of These, % That Applied for Patents in:								
Canada	85.1	41.6	17.7	11.3	4.2	9.4	9.6	6.2
United States	75.6	34.4	15.7	13.4	3.7	11.8	13.4	7.6
Innovating Manufacturers								
Applied for At Least One Patent	**22.4**							
Of These, % That Applied for Patents in:								
Canada	85.2	40.3	18.1	11.8	4.4	9.6	9.6	6.3
United States	75.4	32.9	16.1	13.7	3.9	12.0	13.7	7.7

Source: Statistics Canada, Survey of Innovation 1999, preliminary results.

instance, a larger proportion of those firms that applied for more than 10 patents did so in the United States (13.4%) than in Canada (9.6%).

Further analysis of the propensity of firms to file patent applications in Canada and in the United States shows that firms that were found to be more likely to use patents (large firms performing R&D, active in the core sector and having introduced world-first innovations) also tended to apply for more than one patent. Detailed tables and figures are available on request, but the main conclusions can be summarized as follows.

(i) The number of patent applications filed by firms increased from the "other" sector to the secondary and to the core sector.

(ii) The frequency distribution of the number of patent applications in Canada and in the United States by size of firm shows that, in each size category, firms that rarely filed patent applications (for instance, fewer than three patents) tended to do so more often in Canada than in the United States. However, firms that applied for a larger number of patents did so more frequently in the United

States than in Canada. This trend was most notable for the largest firms that applied for a large number of patents. Forty per cent of these firms applied for more than 10 patents in the United States and only 31% in Canada.

(iii) The preference for patent applications in the United States was most notable among firms that applied for a large number of patents in the technologically most progressive core sector. The observed pattern suggests that the leading-edge firms that patented innovations most frequently applied for patents in both countries, but in the United States more often than in Canada.

4. RESULTS OF MULTIVARIATE MODELS

The joint effect of a firm's economic characteristics, the type of innovation activities it pursues and the competitive strategies it adopts in the industrial and institutional environment in which it evolves is closely associated with the probability of using a particular type of intellectual property. To reveal the complex relationships involved, a series of multivariate logit regression models was estimated. The probability of a firm using a particular type of intellectual property protection was regressed on a list of explanatory variables that included not only those firm and industry characteristics used in the tabulations presented above, but also strategies, attitudes and activities not included in the descriptive statistical analysis. The regression results[17] indicate whether a given characteristic increased or reduced the probability that a firm would use a given instrument of intellectual property protection. The marginal effects of each explanatory variable on the probability of a firm using patent protection are presented in Table 9.

4.1 Model Findings

The results of the regression models confirmed in a multivariate context the findings from the tabulations.

(i) Smaller firms were less likely than larger ones to use any or all IPR instruments. The relationship was statistically significant for the use of all IPRs except trade secrets, and it was most notable

Table 9: Marginal Effects of Explanatory Variables on the Probability of Using Patents, 1997-99

Explanatory Variables	*Marginal Effect (%)*
Product Innovation	24
Both Product and Process Innovation	18
World-First Innovation	17
Located in Ontario	16
Performs R&D in a Separate Division	12
Uses R&D Subsidies	12
Canada-First Innovation	12
Performs R&D	11
Belongs to Core Sector	11
Located in Alberta	11
Size (500+ Employees)	10
Contracts Out R&D	9
Uses R&D Credits	7
Highly Competitive in the Product Market	5
Promoting Firm Reputation	4
Developing Export Market	3
Located in Quebec	1
Developing New Markets	0
Size (50-99 Employees)	-7
Size (20-49 Employees)	-8
Difficulty Hiring or Retaining Staff	-9
Belongs to "Other" Sector	-14

Source: Statistics Canada, Survey of Innovation 1999. Tabulated by author.

Note: The marginal effects were estimated by a single-step logit regression equation evaluated at the mean values of independent variables with respect to a medium-sized firm in the secondary sector employing 100-499 people, not performing R&D and having introduced a firm-first innovation. Thus, for example, all things being equal, the probability that a small firm employing 20-49 people used patents is 8% less than the probability that a medium-sized firm used patents.

for patents. In contrast, small firms did not use trade secrets less frequently than medium-sized firms. This finding corroborates the hypothesis that, for SMEs, cost considerations may discourage the use of IPRs other than trade secrets.

(ii) Firms carrying out R&D, and especially those doing so in a separate R&D unit and/or contracting out R&D, were also more likely than firms not involved in R&D to use any or all IPR instruments.

(iii) Innovating firms, especially those that introduced world-first innovations and, to a slightly lesser extent, those that introduced Canada-first innovations, were likely users of IPRs. However, world-first innovators relied less on trade secrets and more on patents, copyrights and confidentiality agreements.

(iv) The probability of a firm using IPRs increased notably when the firm introduced product innovations or both product and process innovations rather than process innovations alone. Again, these characteristics of innovation were more closely associated with the use of patents than with the use of other IPRs.

(v) Firms that based their competitive strategy on the development of new markets were likely to protect their intellectual property using trademarks, trade secrets and confidentiality agreements. Curiously, they were unlikely to use patents for this purpose. Equally surprising was the absence of a statistical association between export strategy and the use of any IPRs. As expected, the regression results confirmed that trademarks were used as a means of enhancing a firm's reputation.

(vi) Firms that reported receiving a government subsidy for their R&D activities were more likely to use patents, trademarks and/or confidentiality agreements than firms that received the more generally available government tax credit. Both groups were likely to use those IPRs more frequently than firms that did not receive any form of government assistance. However, the estimated effects of the use of government R&D support were probably influenced by a simultaneity bias. The importance of the effects of R&D support policies decreased sharply when the simultaneous nature of a firm's decision to use IPRs and to innovate was taken into consideration.

(vii) Paradoxically, firms that reported using government information services through the Internet or in other ways were more likely than other firms to use trade secrets, confidentiality agreements and copyrights.

(viii) Firms located in Ontario and Alberta were more likely to use most of the IPRs than firms in other provinces. Firms in Quebec were more likely than firms in other provinces to protect their products with trademarks, but were less likely to use copyrights, trade secrets and confidentiality agreements.

(ix) Firms in the core sector were more likely to use all IPRs except trade secrets than firms in the secondary sector. Patents were used less frequently in the "other" sector than in both the core and secondary sectors. In keeping with their consumer product orientation, firms in the "other" sector were more likely than those in the secondary sector to use trademarks and copyrights.

4.2 Use of Trade Secrets, Patents and/or Trademarks

As predicted by the economic theory of innovation, those firms that protected their intellectual property with patents, trademarks and, most notably, trade secrets were more likely to innovate than other firms. However, the decisions to innovate and to protect the intellectual property that the innovation was expected to produce may well not have been independent. When the possible interdependence of the decisions to innovate and to use patents was taken into consideration, the positive correlation between the probability that a firm would innovate and that it would use patents remained unchanged. On the other hand, the use of patents had less effect on a firm's decision to innovate than the single-equation estimates would suggest. The simultaneous equations estimates also cast doubt on the importance of the effect of government support programs on the use of IPRs, derived from estimates of the ordinary (single-step) logit model that did not properly take into account the interdependent character of a firm's decision to protect intellectual property and to innovate.[18]

4.3 Relationship Between Protection of Intellectual Property and Profitability of Innovation

Presumably, firms use IPRs to appropriate benefits from innovation. If this hypothesis is true, one would expect innovators that protect their intellectual property to be more profitable than those that do not. This hypothesis was tested using the survey information on the impact of innovation and on the use of IPRs. Respondents indicated on a scale of 1 to 5 their strong disagreement (1) or strong agreement (5) with a series of statements regarding the impact of new or significantly improved products or processes that they introduced during the period 1997-99. Two of the statements concerned an innovation's impact on profits: the innovation increased the firm's profitability; and the innovation allowed the firm to maintain its profit margins. Scores of 1 and 2 were considered to be an indication that the innovation had not contributed to profitability, and scores of 4 and 5 as evidence of a positive contribution to profitability.

Answers were tabulated in a series of two-way contingency tables, classifying firms according to the impact of innovation on profitability and as users and non-users of each IPR. The chi-square statistics reject for every IPR the test of homogeneity, i.e., the hypothesis that the impact of innovation on profitability is independent of the use of the particular IPR. The statistical tests show that the users of IPRs reported more often than the non-users that their innovation allowed them to maintain or to increase profitability.[19]

This positive relationship was not very strong, but it was statistically very significant. It held for all innovating firms, but there were some sectoral differences. The positive relationship between the use of patents and profitability obtained for the core and secondary sectors, but not for the "other" sector. In contrast, firms in the "other" sector that found innovation profitable used trade secrets and confidentiality agreements more frequently. Trademark users reported maintained or increased profitability more frequently than non-users in all three sectors. Overall, these results provide convincing evidence that innovators that protected their intellectual property found that their innovations contributed to their firm's profitability.

5. CONCLUSION

Even though the protection of intellectual property is far from being a perfect means of appropriating benefits from innovation and new technology, its importance, especially for world-first innovations, is well documented. To innovate successfully, firms must not only learn how to conduct R&D in order to absorb, create and adapt new knowledge, how to find and process market information and how to collaborate; they must also learn to use IPRs efficiently and to combine them with other appropriation strategies. Further research is needed to evaluate the use and efficiency of these other strategies and to compare them with statutory IPRs.

The results of the study presented here suggest several policy measures. The evidence shows that SMEs use any or all IPR instruments less frequently than large firms. This suggests that the cost of obtaining, maintaining and enforcing IPRs by litigation imposes a heavier burden on SMEs. One solution would be improved information and training for SMEs in how to use IPRs effectively. Another would be to consider making SMEs eligible for tax credits or grants to cover the cost of applying for, or renewing, statutory IPR instruments (primarily, patents). A further study is needed to determine whether SMEs would benefit from the introduction of a specialized court for litigation of IPR-related cases.

Appendix 1: Technology Sectors, Intellectual Property Protection and Patents Applications

Table A1.1: Author's Conversion from NAICS to Robson, Townsend and Pavitt's Technology Sectors

NAICS	*Industry*	*Sector*			*Sector*
		Other			**Secondary**
311	Food	Other	326	Plastics-Rubber	Secondary
312	Beverages-Tobacco	Other	327	Non-Metallic	Secondary
313	Primary Textiles	Other	331	Primary Metals	Secondary
314	Textiles	Other	332	Fabricated Metals	Secondary
315	Clothing	Other	3335	Fabricated Metals	Secondary
316	Leather	Other	3361	Transportation	Secondary
3211	Wood	Other	3362	Transportation	Secondary
3212	Wood	Other	3363	Transportation	Secondary
3219	Wood	Other	3364	Transportation	Secondary
322	Paper	Other	3365	Transportation	Secondary
323	Printing	Other	3366	Transportation	Secondary
337	Furniture	Other	3369	Transportation	Secondary
339	Other Man.	Other			
3346	Other Man.	Other			
		Core			
324	Refined Petroleum	Core			
3251	Chemicals	Core			
3252	Chemicals	Core			
3253	Chemicals	Core			
3254	Chemicals	Core			
3255	Chemicals	Core			
3256	Chemicals	Core			
3259	Chemicals	Core			
3331	Machinery	Core			
3332	Machinery	Core			
3333	Electrical	Core			
3334	Machinery	Core			
3339	Machinery	Core			
3341	Electrical	Core			
3342	Electrical	Core			
3343	Electrical	Core			
3344	Electrical	Core			
3345	Scientific Instruments	Core			
335	Electrical	Core			

Table A1.2: Methods Used to Protect Intellectual Property, by Industry, 1997-99 (% of Innovating Firms)

Industry	*Method*						
	Firms That Protected Intellectual Property	*Of These, % That Used:*					
		Patents	*Trademarks*	*Copyrights*	*Confident. Agreements*	*Trade Secrets*	*Other*
				(%)			
Total Manufacturing Industries	**72.9**	40.3	54.8	18.8	66.7	39.1	3.7
Food Manufacturing	**78.4**	21.7	73.3	17.7	69.6	45.6	2.7
Beverage and Tobacco Product Manufacturing	**87.8**	36.3	86.3	35.2	77.3	55.8	3.4
Textile Mills/Textile Product Mills	**68.4**	35.0	65.3	21.6	52.4	40.5	2.3
Apparel Manufacturing	**71.7**	12.1	83.2	10.4	35.6	27.2	1.3
Leather and Allied Product Manufacturing	**81.4**	24.0	64.6	10.3	21.8	39.5	0.0
Sawmills and Wood Preservation, Veneer, Plywood and Engineered Wood Product Manufacturing, Other Wood Product Manufacturing	**46.2**	22.9	43.8	6.2	55.0	32.6	6.1
Paper Manufacturing	**67.4**	35.1	42.9	12.4	76.1	34.1	2.1
Printing and Related Support Activities	**60.6**	21.5	39.2	28.6	70.0	33.4	4.4
Petroleum and Coal Products Manufacturing	**92.7**	31.6	57.9	13.2	71.1	57.9	5.3
Chemical Manufacturing (excluding 3254)	**87.1**	47.9	67.4	18.9	80.9	55.0	2.4
Pharmaceutical and Medicine Manufacturing (3254)	**94.2**	59.4	74.9	29.5	96.0	55.7	0.0
Plastics and Rubber Products Manufacturing	**82.1**	51.4	51.8	13.9	67.7	41.5	3.9
Non-metallic Mineral Product Manufacturing	**68.7**	48.5	51.3	17.3	58.0	45.3	8.9
Primary Metal Manufacturing	**66.3**	33.7	23.8	13.6	74.6	54.1	0.9
Fabricated Metal Product Manufacturing	**70.9**	37.4	41.3	13.6	68.0	32.9	4.4
Agricultural, Construction and Mining Machinery & Industrial Machinery Manuf. (3331 & 3332)	**84.6**	72.3	55.2	20.5	71.6	29.3	0.0
Machinery Manuf. (excluding 3331 & 3332)	**79.3**	54.6	47.7	18.8	68.4	35.0	4.0
Computer and Peripheral Equipment Man.	**91.9**	44.5	66.9	47.7	94.9	43.9	7.6
Communications Equipment Manufacturing	**94.6**	53.6	48.4	29.5	91.0	42.3	1.8
Semiconductor and Other Electronic Component Manufacturing	**84.8**	50.7	49.0	32.5	92.2	66.4	5.2
Navigational, Measuring, Medical and Control Instruments Manufacturing & Manufacturing and Reproducing Magnetic and Optical Media	**88.5**	55.9	54.0	33.3	81.5	39.1	3.3
Electrical Equipment, Appliance and Component Manufacturing	**76.2**	66.1	56.0	21.7	70.2	33.9	6.0
Motor Vehicle Manufacturing & Motor Vehicle Body and Trailer Manufacturing & Motor Vehicle Parts Manufacturing	**77.6**	51.5	39.9	21.1	74.5	41.0	1.2
Aerospace Product and Parts Manufacturing	**85.6**	39.8	28.4	28.6	81.9	47.7	13.8
Railroad Rolling Stock Manufacturing & Ship and Boat Building & Other Transport. Eqpt. Manuf.	**71.3**	39.3	7.0	23.6	58.4	32.1	4.9
Furniture and Related Product Manuf.	**61.6**	43.0	63.4	16.3	53.1	26.4	3.6
Miscellaneous Manufacturing	**75.0**	48.9	63.4	30.7	60.5	42.9	3.9

Source: Statistics Canada, Survey of Innovation 1999, preliminary results.

Table A1.3: Applications for Patents in Canada and the United States by Innovating Firms in Manufacturing, 1997-99

Industry	***Applied for At Least One Patent***	*Of These, % That Applied for Patents in:*					
		Canada	*US*	*Both Canada and US*	*Canada Only*	*US Only*	*Neither Canada nor US*
				(%)			
Total Manufacturing Industries	**22.4**	85.2	75.4	65.8	19.5	9.6	5.1
Food Manufacturing	**10.5**	80.9	63.6	50.5	30.4	13.2	6.0
Beverage and Tobacco Product Manufacturing	**23.8**	100.0	41.9	41.9	58.1	0.0	0.0
Textile Mills /Textile Product Mills	**20.7**	91.8	79.9	71.7	20.2	8.2	0.0
Apparel Manufacturing	**3.9**	100.0	67.9	67.9	32.1	0.0	0.0
Leather and Allied Product Manufacturing	**16.2**	88.9	77.8	66.7	22.2	11.1	0.0
Sawmills and Wood Preservation, Veneer, Plywood and Engineered Wood Product Manufacturing, Other Wood Product Manufacturing	**6.8**	91.9	67.3	61.8	30.1	5.5	2.6
Paper Manufacturing, Printing and Related Support Activities, Petroleum and Coal Products Manufacturing	**19.2**	76.4	64.8	55.1	21.4	9.7	13.8
Chemical Manufacturing (excluding 3254)	**29.4**	77.3	71.1	62.0	15.3	9.0	13.6
Pharmaceutical and Medicine Manufacturing (3254)	**30.0**	93.8	67.2	61.0	32.8	6.2	0.0
Plastics and Rubber Products Manufacturing	**30.6**	86.4	81.1	73.5	12.8	7.5	6.1
Non-metallic Mineral Product Manufacturing	**20.8**	80.0	78.7	67.9	12.1	10.8	9.2
Primary Metal Manufacturing	**17.3**	100.0	74.9	74.9	25.1	0.0	0.0
Fabricated Metal Product Manufacturing	**20.8**	80.0	72.6	57.3	22.7	15.3	4.7
Agricultural, Construction and Mining Machinery & Industrial Machinery Manufacturing (3331 & 3332)	**54.1**	94.1	81.3	75.4	18.7	5.9	0.0
Machinery Manufacturing (excluding 3331 & 3332)	**33.8**	82.8	84.6	73.0	9.8	11.6	5.6
Computer and Peripheral Equipment Manufacturing	**36.2**	67.9	65.2	46.0	21.9	19.3	12.8
Communications Equipment Manufacturing	**48.2**	96.6	81.7	78.3	18.3	3.4	0.0
Audio and Video Equipment Manufacturing	**40.0**	100.0	100.0	100.0	0.0	0.0	0.0
Semiconductor and Other Electronic Component Manufacturing	**40.9**	89.5	88.8	78.3	11.2	10.5	0.0
Navigational, Measuring, Medical and Control Instruments Manufacturing & Manufacturing and Reproducing Magnetic and Optical Media	**39.3**	75.6	88.6	68.8	6.8	19.8	4.6
Electrical Equipment, Appliance and Component Manufacturing	**40.6**	91.4	73.3	68.6	22.8	4.7	3.9
Motor Vehicle Manufacturing & Motor Vehicle Body and Trailer Manufacturing & Motor Vehicle Parts Manufacturing	**34.5**	84.4	77.9	70.2	14.3	7.7	7.8
Aerospace Product and Parts Manufacturing	**20.1**	73.6	82.0	64.0	9.6	18.0	8.4
Railroad Rolling Stock Manufacturing & Ship and Boat Building & Other Transportation Equipment Manufacturing	**20.8**	...	80.4	...	19.6	...	0.0
Furniture and Related Product Manufacturing	**20.7**	84.9	52.7	44.1	40.8	8.6	6.5
Miscellaneous Manufacturing	**28.4**	95.7	80.2	75.9	19.8	4.3	0.0

Source: Statistics Canada, Survey of Innovation 1999, preliminary results.

Notes

The author sincerely thanks Fred Gault, Frances Anderson, Susan Schaan and Brian Nemes from Statistics Canada and Pierre Therrien from Industry Canada for access to and help with the data. He is grateful to his students M. St-Pierre and C. Trudeau for help with the computer work, and he also acknowledges the useful comments received from J. Putnam. The author alone is responsible for the opinions expressed in this chapter and for any errors and omissions that may remain.

[1]Smaller firms — those with sales of less than $5 million — used IPRs less than larger firms and were less satisfied with Canadian IPRs.

[2]These responses have to be considered today in a proper perspective. Amendments to the *Copyright Act* introduced in June 1988 extended copyright protection to computer programs, strengthened the right of artists to control the use of their work and improved systems to collect copyright payments. The new act also increased penalties for infringement of copyright up to a maximum of $1 million, with prison terms ranging from six months to five years. The *Canadian Patent Act* also underwent significant changes in 1989. Thus, the criticism that Canadian IPRs did not provide sufficient protection and had not kept pace with technological developments may no longer be valid.

[3]Baldwin's study was based on the 1993 Survey of Innovation and Advanced Technology, which surveyed both small and large manufacturing firms. One of the features of the survey was that, in the case of larger firms, different sections of the questionnaire were addressed to those in a firm most likely to be intimately involved with the subject at hand. Thus, questions relating to IPRs were answered by the person in charge of intellectual property protection in the firm. These and other methodological differences make it difficult to compare the results of this survey with those of the Survey of Innovation 1999.

[4]See the classification of industries into core, secondary and "other" technology sectors in Appendix 1.

[5]It should be noted that firms may not have confidence in the effectiveness of patents to protect their inventions from imitation (or other IPRs), but may use them anyway for other strategic reasons (Cohen, Nelson and Walsh 2000), such as to signal their technological prowess on the stock markets or their ability to attract highly qualified workers on the labour market.

[6]The questionnaire did not include such statutory IPRs as industrial designs, integrated circuit designs and plant breeder's rights. Widely used strategies other than the statutory IPRs, such as being first in the market or increasing the complexity of product design, were not available options in the questionnaire either.

[7]Firms that innovated successfully were called "innovators" in accordance with the Oslo definition (OECD/Eurostat 1997). It should be noted that firms that had not completed an innovation in the period 1997-99, but that might have completed it successfully later were classified as unsuccessful innovators in the Survey of Innovation 1999. The survey does not allow a distinction to be made between incomplete innovations and those that failed for technical or commercial reasons (Therrien 2002).

[8]The definition of innovation used, for example, in the Survey of Innovation 1999 was as follows:

"A new product (good or service) is a product which is new to your firm whose characteristics or intended uses differ significantly from those of your firm's previously produced products. A significantly improved product (good or service) is an existing product whose performance has been significantly enhanced or upgraded. . . . New production/manufacturing processes are processes which are new to your firm. They involve the introduction into your firm of new production/manufacturing methods, procedures, systems, machinery or equipment which differs significantly from your firm's previous production/manufacturing processes. Significantly improved production/ manufacturing processes involve significant changes to your existing processes which may be intended to produce new or significantly improved products (goods or services) or production/manufacturing processes."

[9]Owing to methodological differences, the comparison of survey results is only approximate.

[10]Patents are often used by competitors to obtain valuable technical information. For example, 38% of top R&D performers reported using patents "quite a bit" to obtain information (Industry, Science and Technology Canada 1989).

[11]See the description of the taxonomy in section 2, and the classification of industries by technology sector in Appendix 1.

[12]Firms can carry out R&D themselves in a separate R&D department or in other departments, and they can also contract it out to other firms.

[13]The chi-square test rejects the hypothesis of independence between R&D collaboration with universities and use of all IPRs beyond the 0.001 level of significance. Firms that collaborated with universities on R&D were almost twice as likely to use patents as those that did not collaborate. The relationship was strongest for the use of patents. This was true for firms of all sizes and in all three technology sectors. The contingency tables are not presented here, but they are available on request.

[14]R&D is only one of the innovation inputs, and the cost of R&D represents less than half of the total cost of innovation in Canada. Basic and applied research accounted for only 17% and development expenditures (defined more liberally than in official R&D statistics) for 30% of total innovation cost in the period 1989-91. Recent statistics on officially defined R&D expenditures as a share of innovation cost in Canada are not available. However, in the European Community officially defined R&D expenditures range from 25% of total innovation expenditures in electrical products to just 10% in the pulp and paper sector (European Commission 2001).

[15]The number of respondents that indicated that they had applied for patents in the period 1997-99 was smaller than the proportion that indicated in the previous question that they had used patents to protect their intellectual property in the same period. The correlation of the two responses was not particularly strong ($r = 0.76$). It suggests that the respondents gave an approximate answer to the more general first question that asked them to enumerate the various means their firm used to protect its intellectual property. The response could, in the case of patents, include patents licensed from suppliers of technology. In this case, a firm would not have applied for the patent and would have responded "no" to the second question, which asked specifically whether the firm had applied for a patent in Canada, the United States or elsewhere in the period 1997-99. This explains at least part of the difference in the response rate to the two questions; their loose wording is probably responsible for the rest.

[16]The lack of information on the country of ownership makes it impossible to determine to what extent firms that filed patent applications in the United States only were affiliates of US companies.

[17]Presented in detail in Hanel (2003).

[18]The marginal effects computed using the regression coefficients estimated by a simultaneous two-stage logit were very similar for most variables, except for the effect of R&D subsidies (decreased to +2% instead of 12%) and R&D tax credits (+4% instead of +7%).

[19]The test *z* shows that the positive relationship between the use of IPRs and increased profitability was significant well beyond the 1% level. The values of the statistic *phi* ranged between 4% and 9% (*phi* can be interpreted as a correlation coefficient: *phi* = 0 — no relationship; *phi* = 1 — perfect correlation).

10

Innovation and Labour Skills

Daood Hamdani

1. INTRODUCTION

The role of knowledge and skills, as inputs to the production process, has become progressively more important in economic theory. Marshall noted the importance of human capital in his *Principles of Economics* (1890). While knowledge appeared as an extraneous factor in neo-classical models, it was nevertheless a factor in economic growth. The emerging theories of economic growth treat it as an unknown variable whose evolution is determined by economic factors (Romer 1986, 1990; Lucas 1988). Parallel to the developments in economic theory, knowledge creation has become to corporate strategy what excellence was in the 1980s. The resource-based approach to organization and management has shifted the focus from products and market shares to the development of competencies, capabilities and skills as the source of sustainable competitive advantage (Prahalad and Hamel 1990; Quinn, Dooley and Paquette 1991; Stalk, Evans and Shulman 1992). According to this view, the process by which firms learn and create competitively valuable organizational knowledge has emerged as the primary function of the firm (Nonanka and Takeuchi 1995). Consequently, organizations are paying an increasing amount of attention to their knowledge assets.

However, analysis of the statistical relationship between innovation and knowledge and skills has been constrained by lack of data.

While macroeconomic analyses and case studies have provided useful insights, particular industry characteristics and their effect on this relationship remain to be studied. Moreover, in the absence of relevant data, these studies rely on proxies for innovation. For example, investment in scientific research or in high-tech machinery is employed as an indicator of innovation rather than more relevant indicators such as the number of new products, the contribution of new products and processes to sales and the number of innovating firms.

The study presented in this chapter advances the discussion in two directions. First, it developed an indicator of skill levels, or knowledge intensity, appropriate for the study of the relationship between innovation and skills in professional service industries. The indicators of skill levels used in macro-level studies are too broad to capture particular industry characteristics, and those employed in case studies are too specific to be relevant or easily measurable at the industry level. As an alternative, the study presented here distinguished between conceptual and implementation skills within an occupational group. Conceptual skills enable workers to challenge the premises of existing knowledge and develop new mental models, or hypotheses, whereas implementation skills enable workers to make use of, or implement, existing knowledge contained in knowledge objects such as instruction manuals. Second, it used more direct indicators of innovation, made available by the innovation surveys of the 1990s, but innovators were identified by criteria different from those used in the innovation surveys. Instead of novelty being defined by spatial factors, newness was considered from the firm's point of view, so that innovations that were duplications or replications of existing products were distinguished from those that gave a firm sustainable competitive advantage and global leadership.

The engineering services industry was the focus of the study, since it exemplifies knowledge-intensive industries. Unlike most of the other studies presented in this book, which were based on the Survey of Innovation 1999 in goods-producing industries, the study described here used data from the Survey of Innovation 1996, which covered service industries. Conducted in the spring of 1997, it was a natural choice, as it contains considerably more detail on human capital than is found in other Canadian surveys. Although the data are somewhat dated, the types of relationships under review are stable over time.

2. RELEVANT FEATURES OF THE ENGINEERING SERVICES INDUSTRY

The engineering services industry is composed of firms engaged primarily in applying the principles of engineering in the design, development and utilization of machines, materials, structures, processes and systems. It offers services ranging from feasibility studies to the design and commissioning of the operation of projects. Notwithstanding the variety of services, most firms are highly specialized in pure engineering services.

The industry earned $5,456 million in fees[1] in 1996 (Statistics Canada 1996a). It was export-oriented and ranked fourth largest in the world, with the United States dominating. The other major players with which it had to compete in international markets were the United Kingdom, the Netherlands, Germany and France (Industry Canada 1997a). Canadian engineering services firms exported $1,109 million worth of the value of their services (or 20%) in 1996, but the actual exposure to international markets was much greater when the work performed by their foreign subsidiaries, which were excluded by the export data, is included. The volume of business performed by foreign subsidiaries was significant, as they were estimated to account for 21% of the total workforce working for the Canadian engineering services firms in Canada and foreign countries (ACEC 1998, 1999).

The industry is very intensive in human capital. This is illustrated in a number of ways. First, workers are its principal asset and the main source of value added. In 1996, 54% of current expenditure consisted of payments to employees for their services (Statistics Canada 1996a). Payment for expertise contributed by consultants engaged for short-term assignments and by partners whose remuneration was not designated as salary was over and above this. By contrast, physical capital plays a substantially smaller role in production, compared with its role in goods-producing industries. In 1996, only 7% of current expenditure was related to machines and buildings (Statistics Canada 1996a).

Second, the industry stands out as knowledge-intensive among similar consulting industries and in the economy. Measures of knowledge intensity at a fine industrial disaggregation level are not readily available.

However, average salary is a reasonable proxy. Although it is determined by many factors, it can reasonably be expected to be higher in knowledge-intensive industries if society accords knowledge more value. Average salary in this industry (broadly defined to include architectural, engineering and scientific services) at $44,000 in 1996 was 30% higher than in the business services sector, which includes various professional service industries. It was 44% above the average for the economy as a whole (Statistics Canada 1998c).

Finally, as providers of inputs in the production of capital goods, firms in this industry play a key role in the diffusion of up-to-date technologies, whether they develop or borrow them for use in the projects they undertake. As examples, computer-aided design and drafting and computer-assisted engineering design developed by consulting engineers are widely used in the manufacturing sector. Forty-four per cent of the manufacturing establishments used CAD/CAE technology and 36% used CAD/CAM technology. Their use was twice as prevalent among the large establishments (Sabourin and Beckstead 1999).

3. KNOWLEDGE INTENSITY AND PRODUCT LIFE CYCLE

The education and skill levels of workers in an industry vary from firm to firm, reflecting a firm's product and competitive strategies. Interfirm differences in skill levels depend, in part, on whether firms engage in innovative activities and on the phase of the product life cycle. Firms that pursue technological innovation as a strategy are likely to be more intensive in knowledge or skills compared with those for whom innovation is a matter of chance. In some firms, knowledge intensity responds to the product life cycle. It tends to be higher in the early phases of the conception of a product. When the product is marketed, its production can be automated, and the knowledge accumulated during its design and development can be codified for repeated use. Consequently, the demand for people who are good at implementing the codified knowledge contained in objects such as manuals, machines or software increases relative to the demand for people who develop new concepts. As a result, knowledge intensity declines, because the product can be produced in large batches through the use of new machines that have a higher codified knowledge content (Nelson and Phelps 1966; Bartel and Lichtenberg 1987). As the product reaches maturity, the firm's focus begins to shift towards research to find new products or more efficient methods of production, and therefore knowledge intensity begins to increase.

However, firms that pursue innovation as a regular activity and keep replenishing their product portfolio need a continuous stream of innovations. They require a core of in-house personnel who are able to assimilate existing knowledge and extend its boundaries. Therefore, these firms are less likely to adjust their knowledge intensity to product cycles; if they do, the adjustments are subtle.

4. CONCEPTS

Knowledge and innovation are difficult concepts to measure. The aim of the following discussion is to clarify some of the issues that need to be tackled in order for measurable definitions of knowledge intensity and novelty of innovation to be developed.

4.1 Knowledge Intensity

There are few statistical indicators that appropriately capture the knowledge intensity or skill mix of a firm. With the emphasis in the 1990s on developing competencies and skills in corporate strategy, the tracking of intellectual capital has started to attract more attention, but the development, and in particular the reporting, of indicators is still in a very early stage. The innovation surveys have filled some of the gaps in recent years. However, a number of other options have been suggested to calculate knowledge intensity.

A simple approach assumes that intangible assets are a reasonable proxy for intellectual capital and defines it as the difference between the market valuation of a company and the book value of its physical assets. The ratio between the value of intangible assets and the company's total market value provides an indicator of knowledge intensity (Stewart 1998). While this approach has the advantage of covering all the various capabilities of a firm, including organizational changes, which are of particular importance for service sector innovation, it raises a number of conceptual issues. First, market valuation fluctuates over time as a result of many factors, and it may therefore give false signals. For example, does a decrease in market valuation when the stock markets are in a general decline signal a reduction in knowledge assets? Although in times of declining stock markets a firm's successful launch of a new product (good or service) may be jeopardized, the firm's intrinsic capabilities are nevertheless unchanged unless it divests itself of its capital or reduces its workforce. In fact, a portion of a firm's market value could be attributed to the perception of the firm's ability, rather than its actual capability, to produce innovations. Second, all intangible assets are not relevant to the production of technological innovations. For example, marketing skills are an essential component of a firm's product development strategy and are often a deciding factor in the success or failure of a new product; however, technological innovations initially depend on the availability of human capital to invent and to convert inventions into marketable and commercially viable products.

In addition, this approach raises a number of practical issues with respect to firm-level analysis in general and its application to service industries in particular. The market valuation approach can be applied only to public companies whose shares trade on the stock exchanges. It means that wholly owned subsidiaries are excluded. Exclusions are more serious in the case of the service sector: many service firms are unincorporated organizations, and many firms providing professional services are partnerships that are not listed on the stock exchanges and therefore their market valuations are unavailable.

An alternative approach to calculating knowledge intensity is to use the skill mix as an indicator, because most of the intellectual capital of professional service firms rests in their personnel. According to Rothwell (1977), one-third of all the knowledge a firm uses in innovation comes from external sources; the remainder is the organizational knowledge available or generated within the firm. According to one estimate, a firm's staff holds one-half of the firm's internal knowledge as a result of previous education and experience (Gibbons and Johnston 1974). This figure is higher for professional service industries because of their greater reliance on internal human capital. Furthermore, knowledge generated within a firm matters more in innovation than that acquired externally, as firms usually outsource only complementary competencies. Firm size may be a factor in this type of decision. While large firms keep control of their core competencies because they act as strategic barriers to the entry of new competitors, small firms may rely on consultancies to acquire specialized knowledge that they may not be able to afford on a full-time basis. While Gibbons and Johnston (1974) note that the remaining one-half of internally generated organizational knowledge comes from research and the exchange of ideas among staff, the depth and effectiveness of this activity depend on the knowledge that the staff already possesses. When a firm hires an employee, it acquires not only a body of accumulated knowledge and skills, but also the ability to access and utilize knowledge generated elsewhere and to generate new knowledge by engaging in relevant R&D.

To create an index of skill mix or knowledge intensity, various skills need to be taken into account. The most common approach makes use of the traditional classification of workers into those in white-collar jobs and those in blue-collar occupations. An increase in the ratio implies an increase in skills and knowledge and hence in the economy's capacity to invent and innovate. This approach is used by Berman, Bound and Griliches (1994), among others, to analyze the demand for labour skills. While many macro-level studies show that an increase in the number of white-collar workers relative to blue-collar workers is positively correlated with technological advances (Collecchia and Papaconstantinou 1996), further refinements are needed to take into account particular industry characteristics. This classification of workers is unique to goods-producing industries. Its application to the service sector is questionable, and it is clearly irrelevant for professional service industries.

A refinement of the use of a classification of occupations, with each occupation characterized by its knowledge intensity, is to examine the characteristics of the actual population of the occupational classification, including the level of education and the specialization (Gault and Hansen 1999). In Canada, Census of Population data can be used to do this and, as the census is done every five years, it is possible to examine change over time. Hansen (1997) has compared the educational levels of people in particular occupations over time and for different industries. This has allowed her to draw inferences about human capital and changes in skill level. Of course, a quinquennial measure cannot identify the rapid changes that occur in some industries, especially those that occurred in response to the economic conditions prevailing in 2002. An alternative is to use studies of firms to capture some of the history of change in human capital.

Firm-level case studies provide useful insights. A recent study of management consulting firms (Hansen, Nohria and Tierney 1999) is of particular interest in this respect. Although specifically concerned with the relationship between growth strategies and knowledge management practices, it provides valuable insights into the development of indicators for the analysis of the interaction between knowledge and innovation. Among the many distinctive characteristics of highly innovative firms, according to the study, is the high ratio of partners to consultants. Firms that build a reputation as providers of cutting-edge solutions to unique and complex problems tend to concentrate knowledge in humans. They maintain a high ratio of partners (who, according to the authors, have more knowledge) to consultants (who have relatively less knowledge), and hire staff for their conceptual skills. They specialize in solving problems that defy the capabilities of firms that treat knowledge as a commodity by storing it in tangible objects for repeated use and that therefore have a relatively lower partner/consultant ratio. Firms in the latter category typically hire people who are good at implementing rather than conceptualizing. Physical capital is important to organizations that treat knowledge as a commodity and are less innovative, and whose innovations have less new technological content. Its importance in the more innovative firms stems from its ability to facilitate routine tasks, because codification of subtle knowledge is not possible.

The macro approach is easier to observe but less meaningful, while the case studies' finer distinction is more meaningful but more difficult to implement from a statistical point of view. There are few objective criteria to distinguish the knowledge content of a partner from that of a consultant in the same occupation. Differences in remuneration commanded by partners and consultants may provide a yardstick. An examination of the hiring practices of firms to determine whether they hire people primarily for their conceptual or for their implementation skills can also shed light on the issue. But the differentiation of skills and abilities on the basis of remuneration or detailed analysis of hiring practices is time-consuming and fraught with difficulties for analysis involving thousands of firms.

While the distinction between conceptual and implementation skills is a useful one for the study of innovation and changes in skill level, the skills being compared must be different enough to be distinguishable and similar enough to be substitutable to some degree. For example, a change in the ratio of engineers to legal secretaries indicates a structural change in the economy. It says little about the ability of the engineering or legal services industries to innovate. Only when the occupations are substitutable can the conclusions about the relationship between innovation and skills be more precise.

In the study reported in this chapter, the differentiation between conceptual and implementation skills was maintained, but more distinct occupational groups were used. The workforce of engineering firms that was directly engaged in engineering work was divided into two groups. The first group was made up of workers who were trained and whose primary function was to use standard procedures set out in manuals. They made optimum use of knowledge embodied in objects and machines. This group was estimated on the basis of the number of technicians and technologists in a firm. The second group consisted of workers who enhanced knowledge, which subsequently led to the production of smarter machines, which, in turn, required technologists to upgrade their skills. Engineers constituted the second group. Some engineers may not have possessed well-developed conceptual skills, but their ability to conceptualize was greater than that of the technologists. The ratio of engineers to technologists was used as an indicator of skill level. Since engineers and technologists constituted most of the personnel of an engineering services firm, the ratio was representative of its competencies and skills. The higher this ratio was, the more knowledge-intensive the firm was deemed to be for the purpose of analysis.

4.2 Novelty of Innovation

The measurable definitions of innovation start with the lowest grade of innovation, which is any product or process new to the firm offering it on the market. Then, innovations are distinguished by degree of newness defined by the geographical boundaries of the market — for example, new to the country or new to the world (OECD/Eurostat 1997). Since a vast majority of innovators are of the firm-first or country-first type, it is useful to probe the novelty of their innovations to subclassify them further. Issues concerning measurement of novelty of innovation in the service sector are discussed in Gault and Hamdani (2000). Their taxonomy of novelty formed the basis of calculations in the study presented in this chapter.

5. DATA

The principal source of data for the study reported here was the 1996 Survey of Innovation, carried out in the spring of 1997 for the period 1994-96. All firms that had at least one employee in 1996 were randomly sampled using strata that related to firm size, region and four-digit industry, using the Canadian 1980 Standard Industrial Classification. In addition to questions about innovation, respondents were asked to provide information on skill requirements. There were also questions directly linking innovation to factor substitution. The survey was consistent with international standards as set out in the *Oslo Manual* (OECD/Eurostat 1997) in both content and methods used, and therefore its findings can be compared with those for other countries for further validation of the results.

However, the innovation surveys do not adequately measure human capital in professional service industries. For example, employment as measured by the innovation surveys does not cover partners because they are owners and not employees. Also excluded is outsourced expertise (Hamdani 1998, 1999). The workforce strategies that firms are adopting in pursuit of their objectives are changing. It is not only the boundaries of firms that are changing; the concepts that revolutionized inventory management are being applied to factors of production. Just as machines are leased rather than purchased, firms are resorting to just-in-time labour or the "Hollywood model" of employment in which workers arrive when there is work to be done and leave when the task is completed (Hamdani 1996). This is compounded by the practice of leasing rather than hiring employees; this means that the firm where these workers provide services is not an employer of record — in other words, it has no employees.

Some of the gaps can be filled by manipulating information from other sources (Statistics Canada 1996a). For example, the Annual Survey of Engineering Services collects data on consulting fees paid to other firms. There is also some information on the number of working owners and principals of unincorporated firms, as well as the occupational distribution of the workforce.

6. ANALYSIS

The analysis reported here addressed three issues. First, the relationship between a firm's ability to innovate and knowledge intensity was analyzed. Next, novelty of innovation was introduced as an additional element so that the degree to which innovation is skill enhancing could be examined. Finally, level of education was incorporated into the relationship between innovation and skills in order to validate the hypothesis that knowledge augmentation and diffusion in the process of innovation also depend upon the stock of knowledge that a firm already possesses.

6.1 Skills and Innovation

The results show that firms with higher skill levels were more likely to innovate than firms with a lower engineer/technologist ratio. They were in a better position to develop new premises and paradigms to supersede the existing ones because they had the capability to probe concepts. The makeup of their workforce was a reflection of their strategy to invent and innovate. Firms with a lower engineer/technologist ratio, on the other hand, emphasized codifying and "commoditizing" knowledge for repeated use, and therefore preferred employees who were good at working with knowledge objects such as manuals, work schedules and procedures.

While firms that reported no innovation employed, on average, as many engineers as technologists, innovating firms had, on average, 19%

more engineers than technologists (Table 1). Where firms with a low engineer/technologist ratio were able to innovate, it was often because they either outsourced expertise or entered into alliances with other firms to maximize their limited resources.

Table 1: Knowledge Intensity of Innovating Versus Non-Innovating Firms: Canadian Engineering Services Industry, 1994-96

Firm Characteristics	*Number of Engineers per 100 Technologists*	
	Firms Reporting an Innovation	*Firms Not Reporting an Innovation*
All Firms	119	101
Firm Size in Terms of Income:		
Less than $2 million	93	89
$2 million to $5 million	119	83
$5 million to $10 million	130	86
More than $10 million	125	121
Partners and Principals as % of Employees	2.3	1.9
Payments to Consultancies as % of Expenditures	9.6	4.6

Source: Statistics Canada 1996a, 1996b.

It should be noted that skills obtained through consultants and the contribution of partners and principals of firms, neither of which the innovation surveys include in the resources of the firm actually receiving their services, played a significant role in the innovation capacity of a firm. Several firms that had a low ratio of engineers to technologists compensated by hiring consultants. A firm's decision as to whether to hire a consultant depends on several factors, including the need for full-time, highly specialized expertise and concern about the loss of its intellectual property through consultants who may be advising many firms. The latter could be a major concern in situations where new products or processes have a high technological content. However, firms introducing such breakthroughs are few. Often, the innovation's technological sophistication, which makes it difficult to copy, becomes an effective way of protecting it. In a list of eight methods to protect intellectual property, which included legal instruments, technological sophistication and business strategies, the innovators chose complexity of product design as the second most effective method. Being first on the market was rated as the most effective way of reaping the benefits of investment in intellectual property, and more than twice as many firms chose it as those selecting such legal instruments as patents and copyrights (Hamdani 2001).

6.2 Novelty of Innovation and Skill Enhancement

The findings indicate the emergence of a circular relationship between innovation and knowledge intensity. Just as high knowledge intensity helps firms to innovate, the process of innovation itself enriches their skill levels and knowledge. To determine the effect of innovation on skills, innovators were asked whether their experience with innovation was skill enhancing, deskilling or neutral. In general, the innovations introduced over the three-year period 1994-96 were skill enhancing or neutral. Only a very small number of firms reported that their skill requirements had declined, probably because they had automated their processes and therefore needed more technologists and fewer engineers. Skill requirements increased in 39% of firms, while 59% reported no change in the skill mix (Table 2). This result is replicated for other industries in the 1996 survey of selected service industries (Baldwin, Gellatly, Johnson and Peters 1998).

The finding that a relatively large number of innovators (59%) experienced no change in skill requirements might, on the surface, appear somewhat unexpected, but it is not surprising because the definition of innovation employed in the analysis was very broad.[2] It may be argued that, while the first commercial application of an idea represents the true innovation, subsequent applications of that same technology in a different setting involve a certain degree of adaptation, whether the new setting is a different country, a different location within the same country or even a different firm. Each stage of adaptation involves some learning and improvement in skills. However, at each stage the

Table 2: Impact of Innovation on Skills of Innovating Firms (% Distribution): Canadian Engineering Services Industry, 1994-96

Degree of Novelty	*Skill Enhancement*	*Deskilling*	*No Change*
		(%)	
All Innovators	39	2	59
Increasing Grade of Novelty:			
Quality Improvement/Cost Reduction	44	3	53
Adding New Product to Existing Line	49	1	49
Adding New Product Line	55	1	44
Breakthrough (World-First)	59	1	40

Source: Statistics Canada 1996b.

innovation content or novelty diminishes, and additions to learning and knowledge occur at a declining rate.

Therefore, the extent to which a firm acquires more knowledge and needs improved skills in the process of innovation is directly related to the technological content of the new product or process, or, in other words, the novelty of the innovation. Breakthroughs that involve a considerable amount of R&D are obviously the source of original knowledge and enhance the learning of everyone involved. Innovations that are of a lower grade of novelty are less likely to be the source of additional learning and knowledge.

The statistics corroborate the notion that skill enhancement increases with the innovation content of new products and processes. Firms introducing mainly lower grades of innovation, defined as quality improvement and cost reduction, were split 44% to 53% between skill enhancement and no change in skills. It is noteworthy that this was the group with the largest number of firms that could manage with less-skilled staff. Cost reduction, which is typically associated with new or improved methods of production or delivery, could mean substitution of smarter machines for skills provided by humans.

Innovations aimed at augmenting a firm's existing product line are taken to be the next highest grade of novelty. Such an innovation can often be achieved by the simple adaptation of a technology that the firm already possesses and uses. For this type of innovation, firms reporting skill enhancement and firms experiencing no change in skill requirements were divided equally (49%), with a very small number (1%) reporting deskilling. It is noteworthy that the percentage of firms reporting skill enhancement was higher for this type of innovation than it was for quality improvement.

The third category of innovators included firms that intended to add a new product line, allowing them to bring it to an established market for the first time. The results were as expected, with firms that reported skill enhancement outnumbering those that reported no change in skill requirements.

In the highest grade of novelty — the introduction of products and processes that are new to the world market — the relationship between innovation content and skill enhancement reached its culmination. More firms reported skill enhancement for this type of innovation than for any other.

The self-assessment of the impact of innovation on respondents' skill levels was corroborated by the longitudinal information on the engineer/technologist ratio. While firms reporting innovation during the period 1994-96 typically increased their employment of both engineers and technologists in subsequent years, the number of engineers grew relatively faster.

6.3 Novelty of Innovation, Level of Education and Skill Enhancement

To this point, no account has been taken of the role of workforce members' previous education in the interaction between innovation and skill enhancement. Given that level of education is a factor in how quickly the workforce can assimilate existing knowledge, the question of whether skill enhancement in the process of innovation is related to educational level was examined. Three groups of workers were identified: university graduates, college graduates and others designated as non-graduates (Table 3). The table shows that university graduates received the most skill enhancement (44% of the innovators reported skill enhancement for this group) and non-graduates received the least (22% of the firms reported skill enhancement for them). The gap widened as

Table 3: Novelty of Innovation, Education and Skill Enhancement, Innovating Firms (% Distribution): Canadian Engineering Services Industry, 1994-96

	Skill Enhancement	*Deskilling*	*No Change*
		(%)	
All Innovators:			
Overall	39	2	59
Non-Graduates	22	3	74
College Graduates	35	2	63
University Graduates	44	2	54
Increasing Grade of Novelty:			
Quality Improvement/Cost Reduction			
Overall	44	3	53
Non-Graduates	26	4	70
College Graduates	42	2	56
University Graduates	47	3	50
Adding New Product to Existing Line:			
Overall	49	1	49
Non-Graduates	20	2	78
College Graduates	36	1	63
University Graduates	48	2	50
Adding New Product Line:			
Overall	55	1	44
Non-Graduates	22	1	77
College Graduates	24	1	75
University Graduates	61	2	37
Breakthrough (World-First):			
Overall	59	1	40
Non-Graduates	19	2	79
College Graduates	28	1	71
University Graduates	68	1	31

Source: Statistics Canada 1996b.

innovation became more complex. Since introducing higher grades of novelty often requires complex technologies and the right combination of humans and machines, improvement in higher-level skills, as represented by the number of university graduates, far outweighed enhancement in lower-level skills, as denoted by the number of college graduates and non-graduates.

7. CONCLUSION

Firms that are more knowledge-intensive are more likely to innovate. Where firms with lower knowledge intensity are able to innovate, it is often because they outsource expertise or enter into alliances with other firms.

Learning and innovation are linked. While knowledge intensity is closely related to innovation, the process of innovation, in turn, sets in motion a new cycle of learning through insights gained from experimentation to find solutions to complex problems. The link between innovation and learning is directly related to the novelty of innovation.

The educational levels of the workforce are of importance. Members of the workforce with higher education tend to receive more skill enhancement than employees without a college or university degree. The gap between the two groups widens with an increase in the technological content or complexity of the innovation.

Notes

[1]The total revenue, estimated at $6,928 million, included reimbursable expenses that the engineering services firms incurred on behalf of their clients. As reimbursable expenses do not represent value added, fee income is a more accurate measure of the size of the industry.

[2]Innovation was defined as any product or process new to a firm. This is the lowest grade of novelty. The other extreme is a product or process new to the world. For a detailed discussion, see Booz, Allen and Hamilton (1982); OECD/Eurostat (1997); and Gault and Hamdani (2000).

Canada and Europe

11

Comparing the Innovation Performance of Manufacturing Firms in Canada and in Selected European Countries: An Econometric Analysis

Pierre Mohnen and Pierre Therrien

1. BACKGROUND

Innovation is seen by many as one of the major engines of growth, and yet many studies show that innovation rates vary considerably from one country to another. Using patenting rates (international patents/million of population) as a measure of innovation, (Stern, Porter and Furman 2000) have shown that, from 1975 to 1995, only Switzerland managed to keep pace with US inventors with more than 150 international patents per million of population. Japan caught up with the two leaders towards the end of the 1980s and even outpaced them in the 1990s. Following these three countries in the top group, Sweden, Finland, Germany and Canada were in the second group with fewer than 100 international patents per million of population. Finally, a third group, with fewer than 50 international patents per million of population, contained France and Spain. These findings have been corroborated by (Trajtenberg 2000), using the same database but for slightly different years.

Other studies have used indicators derived from innovation surveys and have also found substantial differences in realized innovation performance among OECD countries. For instance, Guellec and Pattinson (2001) looked at the results of the second European Community Innovation Survey (CIS 2). They found that the overall level of innovation (percentage of innovators) for European manufacturing firms was 49%, ranging from 73% in Ireland to 26% in Portugal. Therrien and Mohnen (2003) compared the results for some of these European Union (EU) countries (Germany, France, Ireland and Spain) with those for Canada. They found that Canada, at least in terms of the percentage of innovators, led the pack with more than 80% of innovative firms, followed by Ireland (73%), Germany (68%), France (44%) and Spain (30%).

What is the explanation for these differences in innovation or patenting rates? The literature on innovation proposes several determinants of innovation, involving both macro- and microeconomic variables. Trajtenberg (2000) found a strong correlation between civilian R&D (with a three-year lag) and the number of international patents. Focusing on the Canadian situation, he showed that Canada was located in the middle of the pack with regard to patenting. However, he predicted that, because of weak investment in civilian R&D, Canada would hardly be able to improve, or even maintain, its position as a producer of relevant technology in the immediate future.

Stern, Porter and Furman (2000) created an innovation index of "national innovation capacity," providing the expected level of international patents by country. Their index combined factors derived from three major areas of research on innovation: the ideas-driven endogenous growth theory (Romer 1990); the cluster-based theory (Porter 1990); and the national system of innovation concept (Nelson 1993). This index, based on macroeconomic and political variables, drew a picture of an "economy's potential ... to produce a stream of commercially relevant innovation" and, to some extent, explained differences in innovation or patenting rates among countries. The authors found that variables related to the ideas-driven endogenous growth and national system of innovation models — stock of knowledge (GDP per capita or stock of patents), human resources devoted to innovation (full-time science and technology personnel or R&D expenditures), political institutions or policies (intellectual property, openness of the market) — are important determinants of innovation. Private R&D funding, R&D performed by universities, and the linkages between institutions and industrial clusters — central variables in the cluster-based theory of innovation — are also important determinants of innovation.

While macro-level policies and institutions establish the general context for innovation, it is ultimately firms that introduce innovations. Therefore, it is important to understand the process leading to innovation within firms. Firms must develop an internal capability to generate novel ideas and develop them into innovative products or processes. It is firms' innovation capability that will shape a country's overall innovation performance.

Several studies have looked at the determinants of innovation at the level of the firm (see Kleinknecht 1996). From these studies, firms' general characteristics (size, industrial sector), as well as their environment (proximity to the knowledge sector, competitive pressure, demand-pull effect), emerge as important determinants of innovation. Moreover, a firm's internal decisions regarding its involvement in innovation (activities linked to innovation, continuous R&D, co-operation, skills, etc.) are also important in determining the firm's innovation capability. For instance, Crépon, Duget and Mairesse (1998), using a French database, employed a system of equations to examine the process from the decision to engage in innovation activities to the impact of the innovation output on firms' performance. They found that the size of a firm has a positive impact on the probability of engaging in innovation activities, but not on the innovation output (measured by the share of sales from innovative products).[1] However, sales from innovative products increase with the research effort (measured by R&D intensity), demand-pull and technology-push variables. Using a similar model, Klomp and van Leeuwen (2001) have obtained similar results for the Netherlands and Lööf and Heshmati (2003) for Sweden.

The studies reviewed above focus on firms in a particular country. Other authors have compared firms' innovation performance in different countries. For instance, Mohnen and Dagenais (2002) created an innovation index, defined as the expected share of sales from innovation, for Ireland and Denmark, controlling for the usual determinants of innovation (size, industry, R&D intensity, continuous R&D, cooperative R&D). From the innovation index for one country (say, Ireland), they took the structure of the other country (Denmark) to compute what the expected share of sales from innovation would be using the industrial structure of the other country. Using the estimates obtained for each country, they found that Denmark was slightly more innovative than Ireland.

In a more recent study, Mairesse and Mohnen (2001) refined the comparison of innovation in different countries, using a framework that accounted for innovation and distinguishing two different but related measures of innovation: the expected share of sales from innovation; and innovativeness (defined as the difference between the observed and the expected share of sales from innovation). Pooling data from seven European countries, they computed the expected intensity of sales from innovation for a hypothetical European country (the means of the seven countries studied), and deduced the impact of selected variables on the difference between the hypothetical country and each European country. As noted by the authors, "The interest of the expected innovation indicator (and the underlying accounting framework) is that it goes beyond merely reporting the observed share of innovative sales, and attempts to explicitly assess the differences which are imputable to the differences in industry, size, R&D effort, economic environment, and so on." Innovativeness, on the other hand, acts as the residual of the innovation function, and therefore measures what is still unknown or what cannot be estimated from the model. The authors found that, even though the model could explain some of the differences between countries, there was still much that needed explanation.

The aim of the study reported in this chapter was to compare the innovation performance of manufacturing firms in Canada and in four European countries — Germany, France, Ireland and Spain. Using, as much as possible, the potential determinants of innovation suggested in the literature, a dichotomous probit determining the probability of innovating was estimated, followed by an ordered probit on the intensity of innovation. Due to administrative constraints, data from Europe and Canada could not be pooled. Therefore, estimates from the model described above were used to compute the expected innovation intensities in Canada and in Europe; to compare these, a decomposition of the differences was then undertaken, explaining which factors caused the difference in innovation performance between Canada and the four European countries.

2. DATA[2]

The comparison of the innovation performance of manufacturing firms in Canada and the four European countries was based on two separate databases. The Canadian data came from Statistics Canada's Survey of Innovation 1999, while the European data came from CIS 2, the second European Community Innovation Survey. These two surveys shared several characteristics, since both of them followed the guidelines set out in the *Oslo Manual* (OECD/Eurostat 1997). They used the same definition of innovation and contained the same questions regarding the innovation process (obstacles to innovation, activities related to innovation, etc.) and its outcomes.

However, even though international comparability was at the core of both survey designs, there were subtle and more substantial differences between the two surveys. Some adjustments regarding industry classifications, categories of magnitude of sales from innovation, and the definition of the novelty of innovation to allow a fair comparison of inter-country innovation performance were suggested by Therrien and Mohnen (2003). Adjustments for some discrepancies could not be made, however. The first discrepancy concerned the statistical unit: the European survey was based on a sample of enterprises, whereas the Canadian survey was based on a sample of "provincial enterprises."[3] After some data analysis, the conclusion was that using the provincial enterprise rather than the enterprise as the statistical unit did not seem to lead to any serious bias. The second discrepancy concerned the years covered by the surveys: the Canadian survey covered the period 1997-99, while the European survey covered the period 1994-96. It was suspected that the later period of the Canadian survey may have favoured small provincial enterprises in Canada and those in the low-tech intensive industries.[4] These firms had a higher innovation rate than their European counterparts, although this was no longer the case if novelty was restricted to the stronger definition of first innovation (meaning first introduction of a product innovation to the market for European firms and the aggregation of world-first and Canada-first product or process innovations for Canadian firms). The third discrepancy concerned the data themselves: microaggregated data for the four European countries and micro-data at the level of the provincial enterprise for Canada. However, Mairesse and Mohnen (2001) showed that, for an econometric model similar to the one used in the study reported here, the microaggregated nature of the data had little bearing on the results.

The core question in the questionnaires asked whether new or significantly improved products or processes had been introduced into a firm during the three-year period of reference (1994-96 and 1997-99 for Europe and Canada, respectively). Firms that indicated that they had introduced an innovation (be it product or process) were asked to answer the other questions regarding the innovation process and its outcome. The variable that was considered particularly interesting for the purpose of comparing the innovation performance of firms in different countries, and that was unique to the innovation surveys, was the share of sales resulting from innovative products. In Canada, only product innovators (which included product and process innovators and product innovators only, but excluded process innovators only) were asked to answer this question. Also in Canada, the respondents were given a choice of six size categories (1-5%, 6-15%, 16-25%, 26-50%, 51-75% and 76-100%) within which to indicate the percentage of sales from new or significantly improved products, while in Europe the choice was unrestricted. So that this variable could be used in the same way for Canada and Europe, two steps were taken. First, all innovators that were process innovators only — identified as those that declared themselves to be process innovators and that reported no sales of innovative products — were excluded from the samples. Second, the categories given to indicate the percentage of sales from innovative products in the Canadian survey were created for the European countries.

Both surveys provided data on a set of variables that could explain the share of sales resulting from innovative products. For the study reported here, size of firm was measured by the number of employees (expressed in logarithms). Industries were divided into three groups according to their technological intensity (see the Appendix). A dummy indicated the incidence of internal R&D. For Canada, the dummy was set to 1 if a firm carried out R&D internally in a separate and distinct R&D department or if it performed R&D and did not contract it out to other firms. The data sets provided information on the pursuit of a number of innovation activities. For Canada, these innovation activities were R&D, acquisition of machinery and equipment, industrial engineering and industrial design, tooling up and training related to innovation. The European survey, in addition, divided R&D into internal and external R&D and included the activity of introducing innovations to the market. The number of innovation activities was considered to be an indicator of innovation efforts. Therefore, firms with more than the median number of innovation activities were distinguished from those with less than the median number of activities. Other explanatory variables were the use of some kind of government support for innovation, involvement in co-operation or collaboration with other firms, and the presence of a first innovation as an indication of innovation strength. Finally, a measure of proximity to basic research and a measure of competitive pressure, both of which are expected to foster innovation, were constructed. The former was constructed as a dummy set to 1 if the respondent used universities or government laboratories as a source of information for innovation (Canada) or gave a score greater than the median score on a Likert scale to the use of universities or government laboratories as a source of information for innovation (Europe). For Europe, only national universities and national government laboratories were taken into account, while for Canada no such distinction was made. The variable competitive pressure was constructed as a dummy set to 1 if the objective of "opening up new markets or increasing market share" was given a score above the median. For Canada, the same procedure was applied to the success factor "seeking new markets."

The original data sets were cleaned to remove outliers and to harmonize the Canadian and European data. To obtain comparable data, all firms with a gross business income of less than €165,000 (or C$250,000) or fewer than 20 employees,[5] and all those from the printing and publishing industry were removed.[6] For the European data, missing values for the criterion defining innovators (i.e., firms that had introduced a new product or process) were replaced by zeros — in other words, these firms were treated as non-innovators — and missing values for the share of sales from innovative products were also replaced by zeros (with the possible elimination of these firms if they declared themselves to be process innovators). Missing values for the dummies underlying the measures of proximity to basic research, competitive pressure and co-operation were assigned a value of zero. This was not an unreasonable assumption and in any case involved less than 3% of European firms. To remove potential outliers in the European data, all firms for which the logarithm of labour productivity was located outside the interval defined by the country mean plus or minus four times the country standard error were excluded. Finally, for Europe, some 30 firms with R&D/sales or R&D personnel/total personnel above 50%, which could represent research units, were also excluded.[7] That left 4,404 observations for Canada and 10,407 observations for Europe (constituted, respectively, of 4,434, 1,537, 4,054 and 382 for France, Germany, Spain and Ireland). Weights were applied to the data to cover the total population.

Tables 1a and 1b show that industry composition was quite similar for Canada and the four European countries. One-third of the firms were located in the low-tech sector, and more than one-half were located in the medium-tech sector, with the remainder in the high-tech sector. However, the percentage of innovative firms was higher in Canada (78%) than in Europe (49%). One factor that might explain the difference between the Canadian innovation performance and that of the European countries is the size composition of the two samples. While small firms represented only 31% of the Canadian population of manufacturing firms, they represented 53% of the European population.[8] As smaller firms tend to be less innovative, the size composition of the sample would certainly have affected the overall European innovation performance. This is demonstrated in Tables 1a and 1b, where the proportion of innovative small firms in each case is shown to have been less than the proportion of small firms in the whole population (29% vs. 31% for Canada and 40% vs. 53% for Europe). The use of an econometric model helped to separate the effect of size from other potential effects on innovation performance.

3. MODEL

The comparison of the innovation performance of manufacturing firms in Canada and in the four European countries was based on the output

Table 1a: CANADA – Distribution of Innovators and Non-Innovators Across Sectors and Size Categories

	Non-Innovators	*Innovators*	*All Firms*
Industrial Sector			
Low-Tech	39%	32%	33%
Medium-Tech	54%	55%	55%
High-Tech	7%	13%	12%
Size of Firm			
Small	38%	29%	31%
Medium	53%	55%	55%
Large	9%	16%	14%
Observations (weighted)			
Percentage	22%	78%	100%
N	1,566	5.464	7,030

Source: Statistics Canada, Survey of Innovation 1999. Tabulated by authors

Note: Pure process innovators were excluded from the sample.

Table 1b: EUROPE – Distribution of Innovators and Non-Innovators Across Sectors and Size Categories

	Non-Innovators	*Innovators*	*All Firms*
Industrial Sector			
Low-Tech	40%	24%	32%
Medium-Tech	52%	50%	56%
High-Tech	8%	16%	12%
Size of Firm			
Small	65%	40%	53%
Medium	31%	43%	37%
Large	4%	17%	10%
Observations (weighted)			
Percentage	51%	49%	100%
N	35,730	34,652	70,382

Source: CIS 2, Eurostat. Tabulated by authors.

Note: Pure process innovators were excluded from the sample.

measure of innovation that was unique to the innovation surveys — namely, the share of sales resulting from innovative products. In determining the appropriate econometric model to handle these data, three issues concerning this variable had to be addressed.

First, the way in which the survey questionnaire was set up raised a censoring problem. Before the question on the percentage of sales resulting from innovative products, there was a filtering question requiring firms to declare whether they had introduced a new product or a new process in the previous three years. If they answered "no" to both questions, they then had to provide only a minimal amount of information, essentially their size (in terms of turnover and employment) and their main sector of activity. If they answered "yes" to either of the questions, they declared themselves to be innovative and only then were they invited to respond to another set of questions regarding the inputs, the outputs and the organization of their innovation. Second, some firms identified themselves as innovative but reported no percentage of sales from their innovative products. Either these firms were process innovators only or, in the case of European firms, they were innovative but had not yet been able to introduce their product to the market. All pure process innovators were eliminated, and the European product innovators with no sales from innovative products were put in the category of innovators with less than 5% of sales from innovative products. Third, as mentioned before, the data on the percentage of sales from innovative products were categorical and not continuous, because the Canadian questionnaire reported only categorical data.

The purpose of the study reported here was not only to compare countries in terms of their firms' distribution across categories of percentages of sales from innovative products, but also to explain what caused the difference in innovation performance. This also affected the choice of model. Given the structure of the data, the natural choice was a probit model distinguishing between innovative and non-innovative firms, followed by and linked to a multinomial ordered probit model (with known bounds) determining the importance of sales from innovative products. The econometric model was thus as follows. To simplify notation, the enterprise index was omitted whenever possible.

INNO denotes the binary variable of whether a firm innovated or not and *INNO** the latent variable that underlay this decision, such that

$$\begin{aligned} INNO = 1 \quad & \text{if} \quad INNO^* = X\beta + \varepsilon \geq 0 \\ INNO = 0 \quad & \text{if} \quad INNO^* = X\beta + \varepsilon < 0 \end{aligned} \tag{1}$$

where X is a matrix of explanatory variables, β represents the coefficients to be estimated and ε is a random error term with mean zero and unit variance.[9] A negative value for *INNO** was associated with a zero response to *INNO*, and a positive value for *INNO** was associated with a unity response to *INNO*. *INNO** is like the threshold beyond which it is worth innovating.

INNO_S denotes the share of sales from innovative products. With the observed ordered responses to the variable *INNO_S* was associated a latent variable

$$ZINNO\_S^* = \ln(INNO\_S/(1 - INNO\_S)) = Z\gamma + \sigma_\eta \eta \tag{2}$$

where Z is a matrix of explanatory variables, γ represents the corresponding coefficients to be estimated and η is a random error term with mean zero and variance 1. The ordered responses to *INNO_S* correspond to defined intervals of realization of the latent variable:

$$INNO\_S = 1 \quad \text{if } INNO^* = 0 \text{ and } 0\% < INNO\_S^* \leq 5\% \text{ or } -\infty < ZINNO\_S^* \leq -2.94$$

$$INNO\_S = 2 \quad \text{if } INNO^* = 0 \text{ and } 5\% < INNO\_S^* \leq 15\% \text{ or } -2.94 < ZINNO\_S^* \leq -1.73$$

$$INNO\_S = 3 \quad \text{if } INNO^* = 0 \text{ and } 15\% < INNO\_S^* \leq 25\% \text{ or } -1.73 < ZINNO\_S^* \leq -1.10$$

$$INNO\_S = 4 \quad \text{if } INNO^* = 0 \text{ and } 25\% < INNO\_S^* \leq 50\% \text{ or } -1.10 < ZINNO\_S^* \leq 0$$

$$INNO\_S = 5 \quad \text{if } INNO^* = 0 \text{ and } 50\% < INNO\_S^* \leq 75\% \text{ or } 0 < ZINNO\_S^* \leq 1.90$$

$$INNO\_S = 6 \quad \text{if } INNO^* = 0 \text{ and } INNO\_S^* > 75\% \text{ or } 1.90 < ZINNO\_S^* \leq \infty \tag{3}$$

ε and η are assumed to be independently and identically distributed according to a standard bivariate normal distribution with correlation coefficient ρ. The logit transformation of the latent variable *INNO_S** into $ZINNO\_S^* = \ln(INNO\_S^*/(1-INNO\_S^*))$ stretches its domain of definition from [0,1] to [-∞,+∞]. The thresholds defining the categories

t_i (i = 1,…,5) are transformed accordingly: e.g., $t_i = 0.05$ becomes $t_i' = \ln(0.05/0.95) = -2.94$. The β, γ and ρ parameters are estimated by maximum likelihood — i.e., the likelihood of observing the 0/1 responses to *INNO* and the categorical responses to *INNO_S* that were observed in innovators is maximized. The log-likelihood function is given by

$$
\begin{aligned}
\ln L = {} & \sum_{INNO=0} \ln\left(\Phi_1[-X\beta]\right) + \sum_{INNO=1,\, INNO\_S=1} \ln\left(\Phi_2\left[t_1^* - Z\gamma^*, X\beta, \rho\right]\right) + \\
& \sum_{INNO=1,\, INNO\_S=2} \ln\left(\Phi_2\left[t_2^* - Z\gamma^*, X\beta, \rho\right] - \Phi_2\left[t_1^* - Z\gamma^*, X\beta, \rho\right]\right) + \\
& \sum_{INNO=1,\, INNO\_S=3} \ln\left(\Phi_2\left[t_3^* - Z\gamma^*, X\beta, \rho\right] - \Phi_2\left[t_2^* - Z\gamma^*, X\beta, \rho\right]\right) + \\
& \sum_{INNO=1,\, INNO\_S=4} \ln\left(\Phi_2\left[t_4^* - Z\gamma^*, X\beta, \rho\right] - \Phi_2\left[t_3^* - Z\gamma^*, X\beta, \rho\right]\right) + \\
& \sum_{INNO=1,\, INNO\_S=5} \ln\left(\Phi_2\left[t_5^* - Z\gamma^*, X\beta, \rho\right] - \Phi_2\left[t_4^* - Z\gamma^*, X\beta, \rho\right]\right) + \\
& \sum_{INNO=1,\, INNO\_S=6} \ln\left(\Phi_2\left[\infty, X\beta, \rho\right] - \Phi_2\left[t_5^* - Z\gamma^*, X\beta, \rho\right]\right)
\end{aligned} \tag{4}
$$

where Φ_1 is the cumulative univariate normal distribution, Φ_2 is the cumulative bivariate normal distribution, and the indices under the summation signs indicate the observations over which the sums are taken, $t_i^* = t_i' / \sigma_\eta$(i = 1,…,5) and $\gamma^* = \gamma / \sigma_\eta$.

For reasons of statistical confidentiality, the European and Canadian data could not be pooled. Thus, the data for Canada and for the four European countries were estimated separately. Any country heterogeneity in the European data that was not accounted for by the explanatory variables introduced in the model was captured by country-specific effects introduced in the specification of (1) and (2). Dummy variables were introduced to capture specific industry-group effects. The explanatory variables were industry group and size (expressed in logarithms) for the probit equation. For the ordered probit equation, dummies were constructed for the following: incidence of internal R&D; firms with more than the median number of innovation activities; co-operation or collaboration with other firms; proximity to basic research; competitive pressure; novelty of innovation; and use of government support for innovation. Each sample unit in each database was assigned a weight based on the number of sample units it represented in a given stratum for the population. This weight (the inverse of the sampling rate) was applied to each term in the log-likelihood function. In this way, inferences could be drawn about the whole population.

4. RESULTS

Before turning to an analysis of the model described in the previous section, it would be instructive to look at the distribution of the dependent variable — the percentage of sales from innovative products — by the selected variables of interest. Table 2 shows that a larger proportion of European firms than of Canadian firms was able to derive a substantial part of their revenue from product innovation. The percentage of European firms deriving more than 50% of their revenue from innovation was almost three times as high as the percentage of Canadian firms (44% vs. 15%). Although firms in Canada had a better innovation performance than those in Europe in terms of the percentage of innovators (Tables 1a and 1b), that advantage did not seem to hold when intensity of innovation, as measured by the share of sales from innovative products, was considered.

These results are consistent with those obtained by Trajtenberg (2000) on international patenting. With regard to the rate of patenting (international patents/million of population), Canada trailed Germany but was ahead of France, Ireland and Spain. These results were in line with Canada's strong innovation performance in terms of the percentage of innovative firms. However, according to other patent indicators — percentage of patents owned by foreign interests and unassigned patents — Canada lagged behind the rest of the world.[10] As Trajtenberg stated: ["There is reason for concern ... in that a full half of Canadian inventions may not fully benefit the Canadian economy, either because they are done by individuals that may have a hard time commercializing them, or because they are owned by foreign assignees."[11]] Foreign owners of patents could decide to commercialize the innovative product in their home country instead of Canada, which might partly explain why Canada lagged behind Europe in terms of sales from innovative products.[12]

Table 2: Percentage of Sales from Innovative Products, by Selected Variables of Interest, Canada and Europe

	<5%	6-15%	16-25%	26-50%	51-75%	>75%	All Firms
				Canada			
Industrial Sector							
Low-Tech	38%	34%	32%	26%	32%	31%	32%
Medium-Tech	52%	56%	55%	58%	46%	53%	55%
High-Tech	10%	11%	13%	15%	22%	16%	13%
Size of Firm							
Small	37%	32%	29%	28%	27%	23%	29%
Medium	48%	53%	58%	54%	57%	59%	55%
Large	15%	16%	14%	17%	17%	18%	16%
Innovation Activities							
Below Median	63%	56%	49%	40%	45%	38%	48%
Above Median	37%	44%	51%	60%	55%	62%	52%
Other Activities Related to Innovation							
Internal R&D	60%	55%	67%	71%	64%	76%	64%
Co-operation	33%	31%	37%	42%	39%	42%	36%
Govt. Support	62%	59%	62%	68%	70%	67%	63%
Firm and Industry Characteristics							
Proximity to Basic Research	11%	15%	16%	17%	16%	18%	16%
Competitive Pressure	44%	42%	43%	49%	44%	49%	45%
First Innovators	19%	31%	24%	36%	35%	37%	31%
Observations (weighted)							
Percentage	5%	33%	21%	26%	7%	8%	100%
N	247	1,817	1,168	1,397	377	458	5,464
				Europe			
Industrial Sector							
Low-Tech	34%	33%	26%	19%	22%	25%	24%
Medium-Tech	49%	52%	59%	65%	59%	61%	60%
High-Tech	17%	15%	15%	16%	18%	14%	16%
Size of Firm							
Small	43%	40%	34%	36%	46%	45%	40%
Medium	46%	42%	47%	46%	36%	40%	43%
Large	11%	18%	19%	18%	18%	15%	17%
Innovation Activities							
Below Median	68%	51%	40%	30%	35%	41%	40%
Above Median	32%	49%	60%	70%	65%	59%	60%
Other Activities Related to Innovation							
Internal R&D	58%	66%	74%	78%	77%	71%	73%
Co-operation	26%	32%	29%	28%	32%	28%	29%
Govt. Support	12%	19%	21%	19%	25%	25%	21%
Firm and Industry Characteristics							
Proximity to Basic Research	27%	24%	32%	33%	35%	34%	32%
Competitive Pressure	46%	65%	66%	54%	49%	56%	56%
First Innovators	26%	39%	46%	50%	48%	43%	45%
Observations (weighted)							
Percentage	6%	12%	13%	26%	15%	29%	100%
N	2,120	3,992	4,550	8,878	5,126	9,986	34,652

Sources: Canada: Statistics Canada, Survey of Innovation 1999; Europe: CIS 2, Eurostat. Tabulated by authors.

Note: Totals may not sum due to rounding.

Both in Canada and in Europe, firms with more than 15% of sales from innovative products tended to have more than the median number of innovation activities; the opposite was true for small innovative firms. The distribution of innovators across size categories was not greatly affected by the intensity of innovation. The latter seemed to be correlated with the incidence of R&D activities. Firms with a large share of sales from innovative products seemed to be first innovators. Proximity to basic research and, to a lesser extent, pressure of competition also seemed to boost innovation. The three major differences between the Canadian and European firms were the level of government support, the proximity to basic research and the percentage of first innovators.[13] In the Canadian sample, 63% of firms reported using some kind of government support compared with only 21% in Europe. In the European sample, on the other hand, twice as many firms received information needed for innovation from basic research institutions (i.e., universities and government laboratories) (32% vs. 16%), and 45% of European firms were first innovators compared with only 31% in Canada. This once again reinforces the impression that Canadian firms, while more often innovative, were in fact less strongly innovative than their European counterparts.

Table 3 presents the econometric estimates of the model described in the previous section. The estimates for the European data are more precise, which is not surprising given the greater number of data points in the European sample. For the European countries, there were additional dummies to account for missing explanatory variables that were country-specific. The reference country was Germany. Both in Europe and in Canada, the probability of innovating and the intensity of innovation increased with sectoral technological intensity. The country dummy coefficients reproduced the relative rankings reported in Therrien and Mohnen (2003). The incidence of innovation increased with the number of employees more in Europe than in Canada. According to the European data, intensity of innovation decreased with size of firm, whereas the Canadian data indicated that the effect of size was positive. This finding is not out of line with those from previous studies.[14] The incidence of internal R&D had a negative effect in Europe, whereas in Canada, as would be expected from a knowledge production function, the coefficient was positive. Since in Europe the proportion of internal R&D performers and first innovators varied in

Table 3: Estimations of Probit on Probability of Innovating and Ordered Probit on Share of Sales from Innovative Products, Europe and Canada

Explanatory Variables	*Europe*		*Canada*	
	Probit on Probability of Innovating			
Industrial Sector				
Low-Tech	-1.311	(-175.84)	0.030	(0.47)
Medium-Tech	-1.022	(-149.08)	0.197	(3.41)
High-Tech	-0.755	(-82.85)	0.472	(6.15)
Country				
Ireland	0.234	(64.26)	—	—
France	-0.604	(-79.54)	—	—
Spain	-0.995	(-163.37)	—	—
Log of Number of Employees	0.355	(241.59)	0.139	(10.28)
	Ordered Probit on Share of Sales from Innovative Products			
Industrial Sector				
Low-Tech	-0.090	(-0.81)	-2.135	(-13.71)
Medium-Tech	0.261	(2.68)	-2.066	(-14.60)
High-Tech	0.296	(3.21)	-1.782	(-13.51)
Country				
Ireland	-0.824	(-90.25)	—	—
France	-1.676	(-58.10)	—	—
Spain	-0.422	(-7.94)	—	—
Log of Number of Employees	-0.127	(-8.98)	0.052	(2.85)
Activities Related to Innovation				
Internal R&D	-0.080	(-10.71)	0.252	(8.98)
No. of Innovation Activities	0.113	(9.83)	0.238	(8.58)
Co-operation	0.068	(10.02)	0.119	(4.14)
Government Support	0.312	(55.31)	0.060	(1.12)
Firm and Industry Characteristics				
Proximity to Basic Research	0.026	(3.00)	-0.085	(-1.23)
Competitive Pressure	0.044	(2.69)	0.071	(2.34)
First Innovators	0.435	(82.53)	0.105	(3.45)
Sigma	1.742	(206.63)	1.128	(34.01)
Rho	-0.080	(-1.92)	-0.432	(-1.84)
Log-likelihood	-99364.99		-12568.07	
N (weighted)	34,652		5,464	
N (unweighted)	10,407		4,404	

Sources: Canada: Statistics Canada, Survey of Innovation 1999; Europe: CIS 2, Eurostat. Tabulated by authors.

Note: Bold = significant at the 5% level; asymptotic t-stat in parentheses.

the same way across increasing ranges of sales from innovative products (see Table 2), there might have been an identification problem between these two variables. Perhaps only the sum of the two effects could be identified for Europe with the data at hand. All the other effects followed the same direction, at least when they were significant. The number of innovation activities, co-operation or collaboration with other firms, use of government support programs, proximity to basic research, pressure of competition and novelty of innovation all contributed favourably to innovation. These findings corroborate the assumption that to be innovative (and to profit from innovation) firms must develop their innovation capability, and that such capability cannot be taken for granted.[15] Therefore, firms must invest in R&D as well as in other innovation activities, such as acquisition of existing technologies, training and so on, to increase their innovation capability. Moreover, proximity to basic research and co-operation with other firms are other means of strengthening that capability.

Table 4 reports the marginal effects of the main explanatory variables, taken at their mean values, on the various modalities of innovation: the propensity not to innovate; or, if innovative, the propensity to derive a percentage of sales — less than 15%, 15-50% or more than 50% — from innovative products. The sum of the marginal effects across the different possibilities (i.e., on each line) is equal to zero. The results have to be interpreted in the following way. The propensity to be innovative and to fall into one of the three sales categories of innovation behaviour was computed at the mean values of the explanatory variables — i.e., at the observed proportion of firms in the various industrial sectors, of firms carrying out internal R&D, of firms receiving government support and so on. The marginal effects were the changes in those probabilities if a firm increased its size by 1%, or shifted from one industry to another, or moved from non-collaboration to co-operation with other firms, etc. In other words, what was the additional likelihood of being in one of the four categories of innovation behaviour shown in Table 4 as the result of a change (*ceteris paribus*) in one selected variable? For example, if a firm in one of the European countries switched from a low-tech to a high-tech industry, the following changes would be likely: its propensity not to innovate would fall by 21% (0.50 to 0.29); its probability of being a low-intensive innovator (deriving

Table 4: Marginal Effects of Explanatory Variables (in Percentage Points) on Various Modalities of Innovation, Europe and Canada

	Propensity Not to Innovate		*Propensity to Derive Sales from Innovative Products*					
			<15%		*15-50%*		*>50%*	
	Europe	*Canada*	*Europe*	*Canada*	*Europe*	*Canada*	*Europe*	*Canada*
Industrial Sector								
Low-Tech	0.50	-0.01	-0.12	0.49	-0.19	-0.14	-0.20	-0.34
Medium-Tech	0.39	-0.06	-0.13	0.50	-0.16	-0.12	-0.11	-0.33
High-Tech	0.29	-0.14	-0.11	0.48	-0.12	-0.07	-0.07	-0.28
Country								
Ireland	-0.09	—	0.11	—	0.06	—	-0.08	—
France	0.23	—	0.12	—	-0.04	—	-0.31	—
Spain	0.38	—	-0.05	—	-0.13	—	-0.20	—
Log (No. Empl.)	-0.14	-0.04	0.05	0.01	0.05	0.02	0.03	0.01
Competitive Pressure	—	—	-0.01	-0.02	0.00	0.00	0.01	0.01
Activities Related to Innovation								
Internal R&D	—	—	0.01	-0.06	0.00	0.02	-0.01	0.04
No. of Innovation Activities	—	—	-0.01	-0.05	0.00	0.02	0.01	0.04
Co-operation	—	—	-0.01	-0.03	0.00	0.01	0.01	0.02
Govt. Support	—	—	-0.03	-0.01	-0.01	0.00	0.04	0.01
Firm and Industry Characteristics								
Proximity to Basic Research	—	—	0.00	0.02	0.00	-0.01	0.00	-0.01
First Innovators	—	—	-0.03	-0.02	-0.02	0.01	0.06	0.02

Sources: Canada: Statistics Canada, Survey of Innovation 1999; Europe: CIS 2, Eurostat. Tabulated by authors.

Note: Evaluated at means of the variables for innovative firms.

less than 15% of sales from innovative products) would increase by 1% (–0.12 to –0.11); its probability of being a medium-intensive innovator (sales from innovative products of between 15% and 50%) would increase by 7% (–0.19 to –0.12); and its propensity to be an intensive innovator (more than 50% of sales from innovative products) would increase by 13% (–0.20 to –0.07). The same change in a Canadian firm would result in a 13% higher probability of innovating, a 1% lesser chance of being a low-intensive innovator, a 7% greater chance of being a medium-intensive innovator and a 6% greater chance of being an intensive innovator.

Size of firm increased the probability of innovating in both regions, but much more in Europe than in Canada, with an almost equal distribution across the categories denoting percentage of sales from innovative products. Size had two effects on the intensity of innovation: an indirect effect on the probability of innovating and a direct effect on the intensity of innovation. The net effect was positive for both Europe and Canada. The European firms were more sensitive than the Canadian firms to government support and novelty of innovation, whereas the characteristics of carrying out internal R&D, co-operation with other firms and number of innovation activities had more of an effect on sales from innovation in Canada than in Europe.

Table 5 presents an analysis of the difference in innovation performance between Europe and Canada. The analysis was performed once using the estimates obtained from the European data, and once using the estimates obtained from the Canadian data. In both cases, the data and the estimates of the model parameters allowed the prediction of the expected intensity of innovation for every firm in the sample, innovative or not, given by

$$E(INNO\_S \mid X, Z = \int_{-X\beta}^{\infty}\int_{-\infty}^{\infty} \exp(Z\gamma + \sigma_\eta \eta) / \left(1 + \exp(Z\gamma + \sigma_\eta \eta)\right) f(\varepsilon, \eta)\, d\varepsilon d\eta \tag{5}$$

where $f(\varepsilon, \eta)$ is the bivariate standard normal distribution with correlation coefficient ρ. The conditional mean was evaluated at the estimated values of β, γ, ρ and σ_η using a Gauss-Legendre quadrature to compute the integrals.

The quadrature is actually a convolution of two effects, the probability of innovating and the expected intensity of innovation for innovative firms, respectively given by

$$E\left(INNO^* \mid X\right) = \int_{-X\beta}^{\infty} f_1(\varepsilon) d\varepsilon \tag{6}$$

$$E\left(INNO\_S \mid Z, INNO^* \geq 0\right) = \int_{-\infty}^{\infty} \exp\left(Z\gamma + \sigma_\eta \eta\right) / \left(1 + \exp\left(Z\gamma + \sigma_\eta \eta\right)\right) f_2(\eta) d\eta \tag{7}$$

Table 5: Comparison of Predicted Probability of Innovating, and Intensity of Innovation for Innovators and for All Firms, Europe and Canada

	Using European Estimates			*Using Canadian Estimates*		
	Prob. of Innovating All Firms	*Intensity of Innovation Innovators*	*Intensity of Innovation All Firms*	*Prob. of Innovating All Firms*	*Intensity of Innovation Innovators*	*Intensity of Innovation All Firms*
Predictions of ...						
at the Averages of the Variables in ...						
Canada	70.51%	49.51%	34.12%	79.25%	25.82%	17.83%
Europe	69.73%	49.04%	33.40%	79.18%	26.13%	18.04%
Canada–Europe	0.78%	0.47%	0.72%	0.07%	-0.31%	-0.21%
Decomposition of the Predicted Difference (Canada–Europe)						
Industrial Sector						
Low-Tech	-3.76%	-0.70%	-2.38%	0.06%	-2.13%	-1.57%
Medium-Tech	1.98%	0.13%	1.08%	-0.27%	1.48%	1.01%
High-Tech	0.84%	0.12%	0.50%	-0.26%	0.51%	0.29%
Firm and Industry Characteristics						
Log (No. Empl.)	1.67%	-0.28%	0.64%	0.54%	0.10%	0.25%
Competitive Pressure		-0.08%	-0.06%		-0.12%	-0.09%
Internal Innovation Capacity						
Internal R&D		0.11%	0.08%		-0.35%	-0.26%
No. of Innovation Activities		-0.16%	-0.10%		-0.30%	-0.22%
Co-operation		0.08%	0.06%		0.13%	0.10%
Proximity to Basic Research		-0.07%	-0.05%		0.21%	0.15%
Govt. Support		2.15%	1.50%		0.37%	0.28%
First Innovators		-0.88%	-0.61%		-0.20%	-0.15%
Sum of Struct. Effects	0.74%	0.42%	0.66%	0.07%	-0.31%	-0.21%
Observed Averages of ...						
in ...						
Canada	78.59%	29.60%	23.26%	78.59%	29.60%	23.26%
Europe	51.58%	39.38%	20.31%	51.58%	39.38%	20.31%
Canada–Europe	27.01%	-9.78%	2.95%	27.01%	-9.78%	2.95%
Innovativeness	26.27%	-10.20%	2.29%	26.94%	-9.47%	3.16%

Sources: Canada: Statistics Canada, Survey of Innovation 1999; Europe: CIS 2, Eurostat. Tabulated by authors.

where $f_1(\varepsilon)$ and $f_2(\eta)$ are, respectively, the marginal normal distribution functions of ε and η. It should be noted that (5) is the product of (6) and (7) when ε and η are independent.

The predicted values of these three expressions were evaluated at the mean values (for innovative firms) for the variables in the respective countries. To decompose the difference in prediction into its various components, a linear expansion of the predictions was performed — once around the European mean with the gradients evaluated with the European estimates, and once around the Canadian mean with the gradients evaluated with the Canadian estimates.

Not surprisingly, Table 5 shows that the estimates for the European firms, evaluated at the averages of the variables, more closely predicted the average European performance and the estimates for the Canadian firms the average Canadian performance. If the difference is evaluated with respect to the European estimates, firms in the Canadian sample were predicted to have a greater probability of innovating (column 1) and, for innovative and all firms, a higher intensity of innovation (columns 2 and 3). If the difference is evaluated with respect to the Canadian estimates, firms in the Canadian sample were still predicted to have a greater probability of innovating (column 4), but these firms, whether innovative or not, would no longer have a higher intensity of innovation (columns 5 and 6). The model allowed the difference in predicted innovation performance to be split into what could be attributed to structural effects, such as differences in size, R&D efforts and pressure of competition, and what, for lack of a better explanation, could be attributed to innovativeness. If it had been possible to pool the data for all five countries, a unique set of estimates would have been obtained, and any difference in predicted innovation performance between Canada and the European countries could have been attributed to differences in the levels of the variables. With two sets of estimates, there were inevitably differences in predictions and explanations for them. The proper course to take, then, was to ask: If innovation works as it seems to work in Canada, how does Canada compare to Europe, and, conversely, how do the two regions compare if innovation works as it seems to work in Europe?

Both estimates revealed that the sectoral composition of manufacturing output favoured innovation in Europe. Europe had a lower proportion of output in what was classified as the low-tech sector, where the occurrence and intensity of innovation were lower. Unambiguous, too, were the effects of size, co-operation with other firms and use of government support. These factors were more prominent in Canada than in Europe. There were sometimes differences in order of magnitude between the two sets of estimates. For example, the use of government support for innovation as a factor in predicted intensity of innovation for all firms accounted for 0.28 percentage points according to the Canadian estimates compared with 1.50 points according to the European estimates. Differences in competitive pressure, in the scope of innovation, measured by the number of innovation activities undertaken, and in the novelty of innovation (i.e., the occurrence of first-to-the-market innovations) were favourable to innovation in Europe — that is, they were more prevalent in European firms than in Canadian firms. The only two factors on which the two decompositions disagreed were proximity to basic research and incidence of internal R&D. If, as mentioned earlier, the effects of engaging in internal R&D and being a first innovator could not be properly dissociated and the respective contributions of these variables were therefore added to the decomposition, almost identical contributions from the sum of their effects were obtained with both sets of estimates. According to the Canadian estimates (as opposed to the European estimates), proximity to basic research had a negative effect, but the estimated coefficient was not significant. The structural effects — i.e., of all the predicted differences that could be attributed to explanatory variables — almost added up to the differences in prediction. The small discrepancy was due to the linear approximation error in the decomposition of a non-linear function. Table 5 shows that, on average, almost 80% of Canadian firms were likely to innovate compared with barely 50% of European firms, but that innovative firms in Europe were predicted to have a higher intensity of innovation than those in Canada. If the two effects were combined, a Canadian firm was, on average, predicted to be slightly more innovative. Whatever difference in observed innovation performance was not explained by the model was what Mairesse and Mohnen (2001) call "innovativeness." Innovativeness in all three dimensions (the probability of innovating, and the intensity of innovation for innovators and for all firms) was thus measured as the difference in observed innovation unexplained by the structural effects in the model. Here again, the two estimates were consistent: Canadian firms, despite all the differences in size, competitive pressure and output composition in comparison with the European firms, had a greater tendency to innovate, but the typical innovative firm in Canada had a lower share of sales from innovative products. Because the former effect was more prevalent than the latter, a Canadian firm was expected to be more innovative than a European firm.

5. CONCLUSION

The comparison of the innovation performance of manufacturing firms in Canada and selected European countries on the basis of information provided by the innovation surveys of the late 1990s reveals that Canada had a higher proportion of innovative firms, but a lower share of sales from innovative products for its innovating firms. On average, the Canadian firms in the sample had a slightly higher share of sales from innovative products. The econometric analysis attempted to explain part of this difference as the result of the influence of the sectoral composition on output, and the effects of size, competitive pressure and innovation capacity. The estimates from both surveys concurred that the Canadian firms had a slight advantage in terms of size, tendency to co-operate with other firms and use of government support, whereas the European firms had a slight advantage in terms of sectoral composition of manufacturing output, pressure of competition, scope of innovation activities and novelty of innovation. These structural effects combined, while informative, were not enough to explain the basic pattern of innovation revealed by the raw data. Canada appeared more innovative than the four European countries combined.

It would be useful for future studies to examine in more detail how representative national samples are. Do non-respondents or non-sampled firms behave in the same way as respondents — an implicit assumption of the analysis presented here? Another possible source of error lies in the definition of innovation. Although efforts were made in the questionnaires to define the concept of innovation, there was still a serious potential measurement error due to different interpretations of the meaning of innovation and of its order of magnitude. To some extent, a more nuanced picture of inter-country differences in innovation performance could be obtained if the share of sales from innovative products was divided into incremental innovations, first-to-the-market innovations or innovations at different stages of the product life cycle. It would also be useful to get an idea of the selection bias resulting from the mandatory/voluntary nature of the surveys. To enhance the understanding of innovation, an attempt should be made to increase the set of explanatory variables, which would imply the merging of innovation survey data with other information from firms. Similarly, the model used in the study reported here would have benefited from the use of quantitative, rather than qualitative, variables (i.e., the use of R&D expenses instead of the dichotomous variable of performing R&D or not) for refining the estimates. Finally, the explanation for differences in innovation performance might be found not at the firm level, but at the macro level of the national organization of innovation.

Appendix 1: Concordance Between NAICS and NACE (Rev. 1) Industry Classifications by Technological Intensity*

NAICS Code	*NACE Code (Rev. 1)*	*Corresponding Economic Activities*
		Low-Tech
311-312	15-16	Food, Beverage and Tobacco Products
313-31617	19	Textile Mills, Textile Product Mills, Clothing, Leather and Allied Products
321-322	20-21	Wood Products and Paper
		Medium-Tech
324	23	Petroleum and Coal Products
326-327	25-26	Rubber and Other Non-metallic Products
331-332	27-28	Basic and Fabricated Metal Products
333	29	Machinery and Equipment
3345-3346	33	Navigational, Medical, Medial and Optical Equipment
3361-3363	34	Motor Vehicles, Trailers and Semi-Trailers
337 & 339	36	Furniture and Related Products and Miscellaneous Manufacturing
		High-Tech
325	24	Chemicals and Chemical Products
3341	30	Computers and Peripheral Equipment
3344 & 335	31	Electrical and Electronic Machinery and Equipment
3342-3343	32	Radio, Television and Communications Equipment and Apparatus
3364-3369	35	Aerospace Products and Parts, and Other Transportation Equipment

*Taxonomy drawn from Hatzichronoglou (1997).

Notes

The authors would like to thank Jacques Mairesse and Anthony Arundel for valuable comments, and Brian Nemes, Frances Anderson and Susan Schaan for guiding them through the data from Statistics Canada's Survey of Innovation 1999. They also wish to thank Louis-Philippe Bergeron for his research assistance. The authors alone are responsible for the opinions expressed in this chapter and for any errors and omissions that may remain.

[1]At least, size has no direct impact on innovation output, but, as innovation output is a function of innovation input, which in turn is a function of size, size has an indirect impact on innovation output.

[2]For a complete description and comparison of CIS 2 and Statistics Canada's Survey of Innovation 1999, and the tests performed and assumptions made to harmonize the two databases, see Therrien and Mohnen (2003). For official reports regarding the surveys, see Schaan and Nemes (2002) for Canada and Foyn (1999, 2000) for Europe.

[3]A "provincial enterprise" included all establishments of a given enterprise in the same province and industry at the four-digit NAICS level. Using this definition, a firm could be represented more than once in the sample.

[4]Therrien and Mohnen (2003) reported that 77% of Canadian firms in the low-tech sector were innovative compared with 41% of European firms. Similarly, 75% of small firms in Canada reported an innovation compared with 42% in Europe. In contrast, 88% of large Canadian firms were innovative compared with 82% of large European firms.

[5]The target population of both surveys was firms with a gross business income of at least €165,000 (C$250,000) and more than 19 employees.

[6]The publishing industry is classified differently according to the European industry classification (NACE) or the Canadian industry classification (NAICS). The NAICS classifies the publishing industry outside the manufacturing sector, but the NACE classifies it as a subgroup of printing industries in the manufacturing sector. Because publishing activities constitute an important share of the printing industry, the whole Printing and Related Support Activities industry (NAICS 323 and NACE 22) was excluded.

[7]It was decided not to eliminate observations on the basis of outliers in the turnover or labour productivity growth rates, as these data were not used in the econometric model.

[8]Small firms had fewer than 50 employees, medium-sized firms had between 50 and 250 employees, and large firms had more than 250 employees.

[9]As the variance of ε was not identifiable, it was set as equal to 1.

[10]Here, Trajtenberg compared Canada to the G7 countries and to some other countries like Israel and Finland. He had no information regarding Spain and Ireland.

[11]Analyzing Canada and the United States in greater depth, Trajtenberg drew a dark picture of the future for Canadian innovation. For instance, he found that Canada trailed the United States in terms of patent quality (measured by citations). Even more troubling is the fact that Canada patented more in traditional fields (such as transportation and agriculture) than in the upcoming field of computer and communications, which is the core of the new general-purpose technologies (GPT).

[12]Unfortunately, with the databases it was not possible to control for the status of a firm (foreign or national).

[13]For Canada, government support programs for innovation included financial as well as non-financial programs, while for Europe only financial support programs were included in the definition.

[14]In Crépon, Duget and Mairesse (1998) and in Mairesse and Mohnen (2001), size was not significant, while in Lööf and Heshmati (2003) size was negatively correlated with innovation output (sales from innovation per employee).

[15]See Cohen and Levinthal (1989) and Oerlemans, Meeus and Boekema (1998) for a literature review of innovation capability.

Beyond Manufacturing

Section Overview

Fred Gault

Most chapters in this book have focused on innovation in manufacturing firms, and one, by Hamdani, drew inferences from firms in services industries. This section goes beyond manufacturing, and services, to look at the forest sector and construction. What the papers by Schaan and Anderson find is that innovation in these sectors is quite different from that in manufacturing and that there is no simple answer to the question of what are the characteristics of a world-first innovative firm. There is a second finding of significance; firms in these sectors cannot be examined in isolation.

The systems approach to analysing innovation has already been discussed. It stresses the importance, not just of the activities carried out by the actors in the system, but also the linkages to the activities of other actors. There are the outcomes of the activities which can occur at the level of the firm (new products or processes) and, more broadly, in the economy and society as a result of the network that forms part of the system (collaboration, industry standards, regulation, and framework conditions are examples).

As Schaan points out, in logging, there is process innovation but little product innovation, given that the purpose of the industry is to deliver logs to market. The process innovation results from the acquisition of machinery and equipment, which may not be seen by the firm as an "innovation" in process development, but more as a means of survival. Even if the capital investment is viewed as part of a process innovation, the lifetime of the machinery and equipment often exceeds the three-year period used to identify the activity. A major change put in place four years ago, rather than three, makes the firm "non-innovative." Understanding innovation in logging, then, requires an understanding of what is going on in the firms that supply the machinery and equipment, some of which will be outside of Canada. If the system is to be improved so that there is more process innovation, better quality of product, and increased productivity, the role of all components has to be understood. That requires a mix of survey data, case studies, administrative data on trade, and knowledge of the regulatory framework in which logging is done.

Anderson makes the case for the importance of understanding the innovation system and its effect on the construction industry and she uses the concept of "the construction cluster." This is motivated by the argument that observation only of the industries found on a construction site presents a misleading view of the complexity of the construction production system. The observations have to be expanded to include the owners and managers of the buildings, the suppliers of capital equipment and services, the suppliers of raw material and building components, as well as the regulators, suppliers of infrastructure, and the education and research institutions. Such an analysis includes the public and private sector, and, within the private sector, the extraction, manufacturing, and service sectors.

12

An Innovation System for the Forest Sector

Susan Schaan

1. INTRODUCTION

Innovation is vital to economic growth and development. Through innovation, new products are introduced to the market, new production processes are developed and introduced, and organizational changes are made.

The Survey of Innovation 1999 conducted by Statistics Canada surveyed firms in the manufacturing sector and was the first innovation survey of selected natural resource industries in Canada.[1] It employed the framework and guidelines detailed in the *Oslo Manual* (OECD/Eurostat 1997) to collect innovation data. Questions were also included in the Survey of Innovation 1999 to explore the relationship between manufacturing suppliers and selected natural resource industries.

The aim of the study presented in this chapter was to examine innovation in the forest sector. A systems approach was used to analyze the characteristics of and relationships between innovative firms in this system.

2. BUILDING A SYSTEM OF INNOVATION

There are several elements that influence how industries operate and that accordingly should be incorporated into a description of a system of innovation. These include infrastructure and macroeconomic conditions, the characteristics of the production process and the firm's role in innovation activities (Freeman 1987; Lundvall 1992b).

The cultural and institutional context and economic structure of a country form the infrastructure within which firms operate. These underlying conditions influence all activities within a firm, including innovation. Lundvall (1992b, 1998) has proposed that innovation is rooted in knowledge and that learning is a prerequisite of knowledge. Furthermore, learning is predominantly interactive, making it a "socially embedded process." He hypothesizes that, if research institutions define how things are done and how learning takes place, the economic structure affects "what is done" and, as a result, what is learned. These institutions can have their origins far back in social history. Accordingly, for an understanding of innovation within an economy it is necessary not only to consider the macroeconomic conditions, but also to examine the institutional and cultural context of that economy. This then leads to an understanding of what is being done within the firm.

Innis's (1956) "staple theory" provides a description of the institutional and cultural context of the Canadian economy and the cumulative effects of the exploitation of staple products such as lumber, wheat, fish (cod), furs and gold. These staple products constitute an important component of Canadian economic history. Migrants to Canada produced goods in demand in their home countries that would yield large profits. Harvesting of raw materials stimulated the manufacture of finished products in demand in the European mother countries and in the colony. The scale of production of agricultural products such as wheat and lumber and the extraction of minerals such as gold, nickel and other metals grew with demand as production techniques (machinery and equipment), marketing and transportation infrastructure improved. These improvements affected both the harvesters and the manufacturers. Greater demand stimulated technological advances to increase the supply of raw materials and to accommodate the influx of supply to manufacturers.

Technological change in the production process is only one aspect of innovation, yet it has historically played a large role in firm innovation in the logging industry that continues today. Utterback and Abernathy (1975) and Abernathy and Clark (1985) have explored the influence of the production process on firm innovation. Characteristics of innovation correspond to the stage of development of production process technology. Incremental technical change during the evolution of an industry is emphasized. The environment in which a firm operates and the competitive significance of an innovation, which includes the objective of the innovation and factors for firm success, have an impact on the type of innovation a firm attempts.

How is innovation measured within a firm? Models of firm innovation explore interactions involving knowledge flows both within the firm and from external sources from the perspective of the firm (Kline and Rosenberg 1986; Padmore, Schuetze and Gibson 1998). The *Oslo Manual* provides guidelines for collecting and interpreting innovation data within a system of innovation. Technological changes are dealt with at the level of the individual firm or business enterprise existing within a complex system composed of many interacting components. Application of the *Oslo Manual* guidelines allows the production of internationally comparable, meaningful indicators of innovation. In the case of product innovation, the product must have been introduced to the market. An innovative firm is one that has offered a new or significantly improved product or introduced a new or significantly improved production/manufacturing process during the previous three years. The term "product" includes both goods and services as innovation outputs. A process innovation must have been used within the production process.

In this firmcentric conceptual framework there are four domains, each of which plays a role in business innovation. The firm is referred to as an "innovation dynamo," as it is through the firm that innovations are introduced. Innovation in the firm is affected by a series of "transfer factors" based on learning. These are the human, social and cultural factors that affect the transmission of information to firms and the learning that occurs as a consequence. Firms exist and carry out business under a series of established conditions and institutions that affect the range of opportunities for innovation. These "framework conditions" include the basic education system, the communications infrastructure, financial institutions, the legislative and macroeconomic settings, market accessibility, industry structure and the competitive environment.

The various institutions that provide the technological training and scientific knowledge for this system form the "science and engineering base."

Understanding the ways in which new knowledge and information are generated and diffused is another important issue in the measurement of technological change. They can be embedded in new products or processes, as described by Rothwell's (1994) "system integration and networking" model of innovation which provides for the transmission of technological change through changes in technologies. New knowledge and information can also exist in the form of organizational practices that are adopted by a given industry. They can take the form of tacit knowledge possessed by people with specialized expertise gained through education and experience or they can be codified in manuals, other forms of publication, computer software or product design. Diffusion of knowledge requires a user who can find, understand, absorb and then apply science and technology (S&T) knowledge in the form that it is transmitted (Statistics Canada 1998b).

Foray and David (1995) have explored the diffusion of S&T knowledge, describing three basic activities in the S&T system: generation; transmission; and use. Statistics Canada's Science, Innovation and Electronic Information Division developed a framework for the organization of data as a means of understanding the S&T system (Statistics Canada 1998b). Recognizing that innovation is just one of the activities involved in the use of S&T knowledge,[2] the approach considers interactions with society, the economy and the political order as a system of information flows. It promotes the examination of an industry within this system. "Actors" undertake product or process innovation, with various indicators such as objectives and outcomes providing a comprehensive statistical description of firm innovation.

This chapter presents evidence that logging is part of a system of innovation, which encompasses many industry sectors and knowledge flows between sectors.

3. AN INNOVATION SYSTEM FOR THE FOREST SECTOR

It can be argued that innovation for natural resources takes place in a larger interactive system. No part of this system should be viewed in isolation if the innovation process is to be understood. At Statistics Canada, the NAICS or North American Industry Classification System (Statistics Canada 1998a) allows for the comparable collection and dissemination of industry data. Forest or forestry is listed in the NAICS's alphabetical index under 10 entries encompassing industries from a wide range of sectors (two-digit NAICS codes). These include: Agriculture, Forestry, Fishing and Hunting (11); Wholesale Trade (41); Transportation (48); Real Estate and Rental and Leasing (53); Professional, Scientific and Technical Services (54); and Educational Services (61).[3] Only one of these industry sectors has the word "forestry" in its title: Agriculture, Forestry, Fishing and Hunting. Logging is included in this industry sector.

A systems approach to describing the forest sector combines concepts from the "innovation policy terrain" described in the *Oslo Manual* and concepts from Statistics Canada's *Framework for a Statistical Information System* (Statistics Canada 1998b). To reduce complexity in the study reported here, the construction sector and manufacturing industries that produce products such as furniture, windows, doors and cabinets were not included, but it is recognized that they are users of outputs from logging (either directly or as outputs from another industry).

Preliminary analytical work focused on the innovation activities of the logging industry. The scope of the industries examined was quickly expanded to include industries that had a role in the activities of the logging firms. The forest sector system proposed in the study presented here includes the following elements: industries involved in resource management; harvesters of raw materials; manufacturers of products from raw materials; providers of services to support the other actors in the system; and suppliers of machinery, equipment and instruments used by the other actors in the system. Harvesting industries served as a starting point in the development of the system. Groups of actors were defined by the industries they include (Appendix 1). Aspects of the forest sector system are presented in Figure 1. Between each group of actors is a series of transfer factors that influence the diffusion of S&T knowledge (depicted by the various arrows in Figure 1). Following is a description of the six main groups of actors.

Figure 1: The Forest Sector System

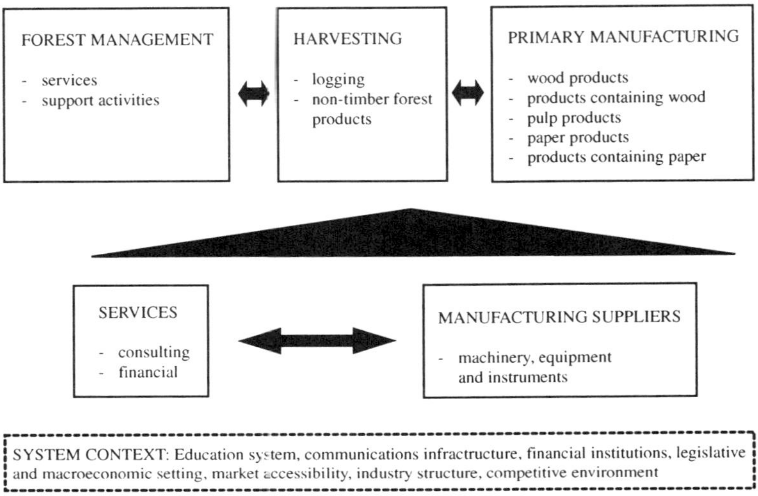

3.1 Harvesting

- Forest products (raw materials) are harvested by industries such as logging.

3.2 Primary Manufacturers

- They transform raw materials into products composed of wood, pulp or paper, or products containing wood or paper.
- Products are output for use by consumers or are input into other industries for further processing.

3.3 Forest Management

- This includes a variety of surveying and mapping services, environmental services, engineering, support services and consulting industries.

- Services are directly related to forests and to management of the resource so that continued harvesting is assured and competitiveness is maintained.
- It has a large role in natural resource industries, but it is not an issue for manufacturing industries that have a factory production system.

3.4 Services

- Services can be provided to a variety of industries or specialization in one sector can be developed.
- Providers of services include private R&D laboratories, banks and investment counsellors.

3.5 Manufacturing Suppliers

- They produce machinery, equipment and other instruments.
- They can be used by any industry.

3.6 Context

- This is the underlying social and economic context in which all industries operate.
- It includes public R&D facilities, communications infrastructure (roads, other transportation networks, etc.), operational regulations set by government, and industry protocols.
- Conditions and institutions determine the range of opportunities for innovation and can also affect the diffusion of S&T knowledge.

4. INNOVATION IN THE FOREST SECTOR

4.1 Methodology

The target population for the Survey of Innovation 1999 was all firms operating in Canada in the manufacturing sector or in selected natural resource industries that had completed existing production surveys. Five natural resource industries, including logging, were selected for sampling; the total sample size was 674 firms, of which 168 were logging firms. All 31 manufacturing industries were sampled, for a total sample size of 5,944 firms.[4] The survey results provide innovation data for only some of the industries in the various groups of actors identified in Figure 1. Data for the following were used for the study presented here: one harvesting industry — Logging (NAICS 1133); four primary manufacturing industries — Sawmills and Wood Preservation (NAICS 3211), Veneer, Plywood and Engineered Wood Product Manufacturing (NAICS 3212), Other Wood Product Manufacturing (NAICS 3219) and Paper Manufacturing (NAICS 322); and all manufacturing industries (NAICS 31-33).

4.2 Innovation Activities in Forest Sector Industries

Understanding why a firm might innovate can help to predict the type of innovation it will undertake: product or process. The primary manufacturing and harvesting industries in the forest sector distinguished themselves by the type of innovations, either product or process, they produced in the period 1997-99 (Figure 2).

Figure 2: Percentage of Innovators, by Type of Innovation (Product or Process), for Logging, All Manufacturing and the Primary Manufacturing Industries in the Forest Sector, 1997-99

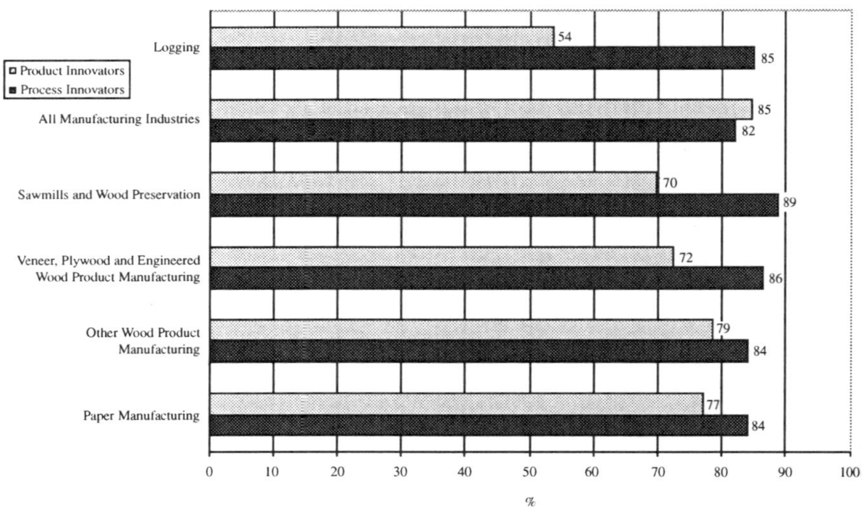

Source: Statistics Canada, Survey of Innovation 1999.

Overall, innovative manufacturing industries showed a stronger affinity for product innovation than for process innovation. For industries in the forest sector, however, the relationship was reversed. Both logging and the primary manufacturing industries favoured process innovation over product innovation. In addition, the difference between the percentage of product innovators and the percentage of process innovators within the logging industry was greater than in the primary manufacturing industries, although the more raw materials were processed within the primary manufacturing industries the narrower this gap became — for example, between the outputs of sawmills and engineered wood products (Figure 2).

Natural resource industries produce raw materials with long life cycles. Only 11% of innovative logging firms indicated that replacing products being phased out was an important objective of innovation compared to over one-third (35%) of innovative manufacturing firms. The innovation indicator is influenced by the length of a product's life cycle (OECD/Eurostat 21: 39). Only slightly over one-third of logging firms felt that developing new products and processes was an important factor for firm success compared to one-half to two-thirds of manufacturing industries. Satisfying existing clients was the factor to which the highest percentage of the primary manufacturing and harvesting industries attributed their success, yet less than one-quarter (22%) of logging firms introduced new products during the period 1997-99 compared to over two-thirds (68%) of manufacturing firms. If logging firms did not feel that they needed new products in spite of the opinion that satisfying existing clients was important to firm success, why would one expect that they would develop new products?

Industries that produce products with shorter life cycles have a more frequent need for innovation than industries that produce products with longer life cycles. Considering the nature of the products produced by the logging industry, there would not be a strong need for it to be innovative in the three-year cycle stipulated by the *Oslo Manual*. This industry should show a stronger affinity for process than for product innovation. Are there other motivators of innovation in logging?

The macroeconomic and cultural context of the Canadian logging industry was established in the country's early economic history, when markets for raw materials were developed and interdependencies with manufacturers were forged. Only one-half of innovative logging firms considered seeking new markets to be an important factor for firm success compared to more than three-quarters of innovative manufacturing firms. An even smaller percentage of logging firms felt that developing export markets was a factor for firm success (Table 1). A major market for the outputs of the logging industry is the manufacturing industry. Changes in production processes allowed harvesting industries to meet the demand for raw materials imposed by the manufacturing sector.

Table 1: Factors for Firm Success for Innovative Logging, Manufacturing and Selected Primary Manufacturing Industries in the Forest Sector, 1997-99

	% of Innovative Firms Indicating Factor Is Important[5]					
Factor for Firm Success	*Logging*	*All Manufacturing Industries*	*Sawmills and Wood Preservation*	*Veneer, Plywood and Engineered Wood Product Manufacturing*	*Other Wood Product Manufacturing*	*Paper Manufacturing*
Satisfying existing clients	93.3	97.0	95.2	95.6	95.6	98.2
Developing new products (goods or services) and processes	37.6	72.2	63.7	59.3	51.2	61.0
Promoting firm or product (good or service) reputation	60.5	83.8	78.6	87.9	80.3	79.3
Developing export markets	27.9	62.3	68.3	65.8	69.6	59.3
Seeking new markets	49.8	77.0	78.9	80.6	83.0	75.0

Source: Statistics Canada, Survey of Innovation 1999.

The percentage of process innovators in logging was comparable to the percentage of process innovators in the primary manufacturing industries (Figure 2). This is compatible with the findings of MacDonald and Clow (1999), who studied the industrialization of tree harvesting systems. They believe that technology must be systematically located in the production or labour process as part of the division of labour that is characteristic of this system of harvesting.

Other natural resource harvesting industries have shown similar trends. The prevalence of process innovation in mining is supported by studies undertaken by the Australian Bureau of Statistics (1998). They found that 39% of all mining businesses[6] had undertaken process innovation between July 1994 and June 1997, compared to 11% of all mining businesses that had undertaken product innovation.

4.3 Why Do Harvesting and Primary Manufacturing Firms Innovate?

Improving product quality was the objective of innovation chosen by the largest percentage of innovative logging firms and by three out of the four innovative primary manufacturing industries (Table 2).[7]

Table 2: Objectives of Innovation for Innovative Logging, Manufacturing and Selected Primary Manufacturing Industries in the Forest Sector, 1997-99

Objective	*% of Innovative Firms Choosing Listed Objective*					
	Logging	*All Manufacturing Industries*	*Sawmills and Wood Preservation*	*Veneer, Plywood and Engineered Wood Product Manufacturing*	*Other Wood Product Manufacturing*	*Paper Manufacturing*
To improve product quality	90.6	82.5	81.8	79.3	78.9	87.0
To reduce environmental damage	56.8	24.8	35.1	37.2	17.5	29.9
To deal with or to respond to new government regulations	50.0	17.9	25.8	10.1	12.6	19.2
To improve production flexibility	46.7	66.5	66.7	61.3	66.2	68.2

Source: Statistics Canada, Survey of Innovation 1999.

The objectives listed in Table 2 would focus attention on changes to the harvesting or manufacturing process rather than on the development of new products. This helps to explain why there was more process innovation than product innovation in these industries.

The Conference Board of Canada (2001a) also found that innovation in the natural resource industries focuses more on process innovation than on the development of value-added products. It found that the emphasis is often on process innovation as a way to drive down costs, especially when commodity prices have a significant influence on profit margins.

Data from the Survey of Innovation 1999 show that the logging industry undertook innovation to respond to environmental issues and government regulations, whereas these objectives were less common for the primary manufacturing industries. For instance, reducing environmental damage was the third most common objective for logging, chosen by well over half of the firms (56.8%), but this objective was chosen much less frequently by the primary manufacturers. Half of the firms in logging chose dealing with or responding to new government regulations as an objective of innovation compared to a maximum of one-quarter of the primary manufacturing industries. Changes to the harvesting processes in logging could result in benefits that would help to meet the objectives mentioned in this paragraph.

4.4 How Is Knowledge Transmitted?

How is the knowledge that facilitates technological change in the harvesting process transmitted? Laestadius (1998) argues that, in the case of the pulp and paper industry, process technology developed along a trajectory already chosen in the mid-1800s during Canada's early economic development. How does innovation take place in this system? He suggests that the technologies based on traditional processes are now part of very sophisticated technological systems. These systems develop to a large extent through relatively minor incremental changes and, on occasion, in response to radical shifts in technology. He cites small-scale technologies and radical biotechnological breakthroughs as part of innovation in the pulp and paper industry. These technological changes do not occur in the logging industry proper as identified by the NAICS, but rather become part of the harvesting system when new machinery and equipment are purchased by logging firms. For industries involved in the harvesting or extraction of natural resources, innovation rates tend to be low. In these cases, an assessment of innovation rates cannot be made by considering the activities in the harvesting industries in isolation, since part of the innovation takes place in industries supplying machinery and equipment to the harvesting industries.

This is evidence that the concept of a production system is not as clear in these harvesting industries where there is no factory per se such as exists in manufacturing. Just over one-third (37.6%) of harvesters felt that developing new products and processes was a factor for firm success compared to about one-half to just under two-thirds of the primary manufacturers. Logging firms see themselves as harvesters of a raw material. When asked to indicate why they did not innovate, the majority gave responses that related to this issue. The second most common reason for non-innovation related to the belief that the techniques employed in the industry were standard and therefore there was no need for change. Less than half (46.7%) of logging firms indicated that an objective of innovation was to improve production flexibility compared to two-thirds (66.5%) of manufacturing firms. This suggests that the logging industry operates under a different regime from that of manufacturing industries — most notably, one in which there is no production factory. There is an issue with production concepts and with the notion of a production process within harvesting industries.

There are many ways to acquire the knowledge that leads to innovation. Such knowledge need not be acquired exclusively through innovation activities carried out within a firm, such as internal R&D. Technology can be acquired in the form of patents, non-patented inventions, licences, know-how, trademarks, services with a technological content, consultancy services, the contracting out of R&D, the purchase or sale of an enterprise or the mobility of skilled personnel, to name just a few ways.

New technologies can be acquired through the purchase of equipment. The top innovation activity for logging, chosen by over 80% of innovative firms, was acquisition of machinery, equipment or other technology linked to new or significantly improved products or production/manufacturing processes. In fact, close to one in five (17%) manufacturing firms had products used by logging and forestry industries.

Another innovation activity is R&D. Just over one-quarter (28.3%) of innovative logging firms indicated that performing R&D within the firm was an important factor for firm success (Table 3). Among firms responding to the question on sources of information for innovation, only 15.7% of innovative logging firms indicated that R&D staff were an important internal source of information — not surprising when only one-quarter (26.2%) of innovative logging firms undertook R&D.

The percentage of logging firms that only contracted out R&D to other firms was more than twice the percentage of firms in which R&D was only carried out in an in-house R&D department. The outcome of this innovation activity would be attributed to the firms undertaking the R&D in other industries and not to logging.

Table 3: R&D in Innovative Logging, Manufacturing and Selected Primary Manufacturing Industries in the Forest Sector, 1997-99

	% of Innovative Firms in Given Category					
R&D	*Logging*	*All Manufacturing Industries*	*Sawmills and Wood Preservation*	*Veneer, Plywood and Engineered Wood Product Manufacturing*	*Other Wood Product Manufacturing*	*Paper Manufacturing*
Performing R&D within firm important[8] to firm success	28.3	59.0	41.9	54.1	43.2	50.1
R&D staff as an important source of information for innovation	15.7	51.3	24.0	37.8	27.4	51.4
Undertook R&D[9]	26.2	67.5	52.6	63.9	54.8	68.9
For those firms that undertook R&D, % for which R&D was:						
Carried out by a separate and distinct R&D department	37.7	45.2	30.7	40.6	25.4	53.2
Contracted out to other firms	59.8	28.5	44.6	46.3	26.6	37.4
Only carried out by a separate and distinct R&D department	21.6	29.6	16.2	18.3	16.7	33.4
Only contracted out to other firms	43.8	12.9	30.1	23.9	17.9	17.6
Neither carried out by a separate and distinct R&D department nor contracted out to other firms	18.6	41.9	39.2	35.4	56.7	29.2

Source: Statistics Canada, Survey of Innovation 1999.

Mohnen, Romain and Gallant (1996) discuss the attempts by the wood industry to devise electronic scanners to check the length, taper and straightness of logs, to design new saw teeth for more efficient cutting and to develop a lumber-grading method based on physical properties instead of visible characteristics. The electronic scanner would contribute to the manufacturer's innovation rate. Whether it would be recognized as a process innovation by logging would depend on how long the scanner was in place, as the survey asked for changes to processes in the previous three years.

4.5 What Is the Provenance of Ideas for Innovation?

Manufacturing industries overall and the primary manufacturers relied more heavily on internal sources of information for innovation than did the logging industry. The top three sources of information for logging in the period 1997-99 were all external. They included suppliers of equipment, material and components (66%), trade fairs and exhibitions (63%) and professional conferences, meetings and publications (52%). The top two sources of information for innovation for manufacturing and for the primary manufacturers were the same: management staff and production staff within the firm. Firms in the primary manufacturing industries gave suppliers of equipment, material and components as their third most common source of information. Trade fairs and exhibitions were the third most common source for innovative manufacturing firms overall.

There are several institutes that carry out R&D activities specifically for industries in the forest sector. The Forest Engineering Research Institute of Canada (FERIC) conducts research to improve machinery and equipment used for logging and transport. Forintek Canada Corporation performs research in the solid wood sector. The Pulp and Paper Research Institute of Canada (PAPRICAN), the largest of the three institutes, is devoted to improving pulp and paper products and processes. FORCAST is a coalition for the advancement of science and technology in the forest sector. Its primary goal is to promote innovation activities as tools for resource management and global competition in the forest sector.

The reliance of the natural resource industry on outsourcing for research was one of the findings of the Conference Board of Canada (2001a). Research is conducted in collaboration with universities, government laboratories and other research institutions. Work with suppliers, contractors, consultants, service industries and even customers helps to generate new product and process innovations. Even co-operation with competitors to share the costs of R&D or new technologies can occur.

4.6 How Is S&T Knowledge Transmitted?

The acquisition of technologies is one mechanism for the transmission of S&T knowledge, facilitating technological change. Five of the manufacturing industries containing the highest percentage of firms with products used by selected natural resource industries (more than 30% of firms) were among the top 10 innovators out of 35 industries (Table 4). Within this group of five industries, four were producers of machinery and equipment or instruments.

In addition, 13.6% of the firms in Agricultural, Construction and Mining Machinery and Industrial Machinery Manufacturing (3331 & 3332) indicated that 50% or more of their revenues were derived from products supplied to logging and forestry. These firms can be considered as dedicated suppliers to logging and forestry. Innovations in the machinery manufacturing industry would benefit the harvesting process in the logging and forestry industries. Logging relies on the machinery manufacturing industry for process innovation. Suppliers of equipment, material and components were chosen as the most important source of information for innovation by two-thirds (66%) of innovative logging firms. Over three-quarters of machinery manufacturers undertook R&D activities during the period 1997-99.

Activity in natural resource harvesting industries experiences boom-and-bust cycles that are subject to the upturns and downturns in the economy. The reference period for the Survey of Innovation was 1997-99. There is evidence that, within this period, the harvesting industries were not at the same point with regard to innovation as manufacturing industries overall. Several non-innovators in logging indicated that they had made changes to their production system outside the survey's

Table 4: Selected Manufacturing Industries with Products Used by Selected Natural Resource Industries, 1997-99

Manufacturing Industry	*% of All Firms with Products Used by Selected Natural Resource Industries*	*% of All Firms with Products Used by Logging*	*% of Innovative Firms in Selected Industry*	*Relative Innovation Rank*	*% of All Firms That Undertook R&D*
Electrical Equipment, Appliance and Component Manufacturing	47.5	55.4	89.8	5	74.2
Agricultural, Construction and Mining Machinery & Industrial Machinery Manufacturing (3331 & 3332)	46.6	37.8	38.0	6	80.4
Machinery Manufacturing (excluding 3331 & 3332) [Commercial and Service Industry Machinery Manufacturing & Ventilation, Heating, Air-Conditioning and Commercial Refrigeration Equipment Manufacturing & Metalworking Machinery Manufacturing & Engine, Turbine and Power Transmission Equipment Manufacturing & Other General Purpose Machinery Manufacturing]	35.3	26.9	86.9	8	75.9
Navigational, Measuring, Medical and Control Instruments Manufacturing & Manufacturing and Reproducing Magnetic and Optical Media	32.0	14.9	91.3	4	77.3
Chemical Manufacturing (excluding 3254)	31.2	23.9	87.7	7	75.2

Source: Statistics Canada, Survey of Innovation 1999.

reference period. Other responses indicated that changes to the production system were slow and methods remained essentially unchanged.

Concurrent with growth cycles in an industry there will be cycles of capital investment in machinery and equipment. Forest sector industries, including logging and the primary manufacturing industries, invested heavily in machinery and equipment but relatively little in R&D in 1997[10] (Figure 3). The top three R&D investors — Communications Equipment, Aerospace Product and Parts, and Pharmaceutical and Medicine manufacturers — invested only about one-fifth as much in machinery and equipment in 1997.[11]

Figure 3: A Comparison of R&D Expenditures and Capital Expenditures on Machinery and Equipment for Selected Forest Sector Industries and the Top R&D Investors, 1997

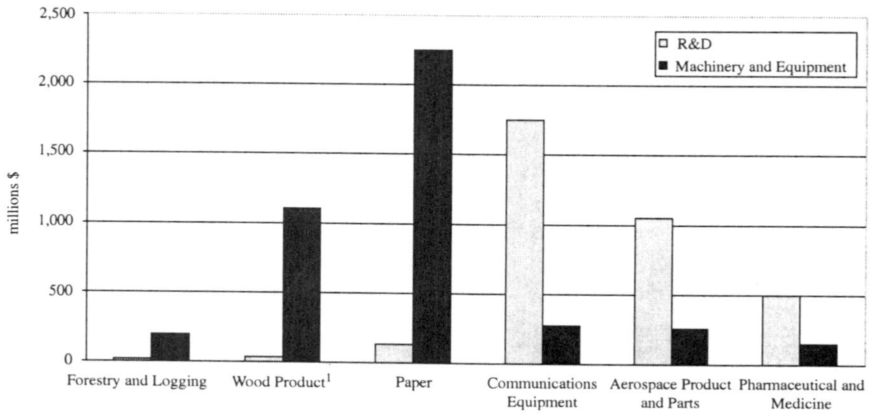

Source: Statistics Canada, Capital Expenditures on Machinery, 1997 actual (CANSIM matrix 11501 and 11504) and Industrial R&D, 1997-2001 (Catalogue No. 88-001-XIB).

Note: 1. Includes Sawmills and Wood Preservation; Veneer, Plywood and Engineered Wood Product Manufacturing; and Other Wood Product Manufacturing.

Although the forest sector industries showed far less R&D expenditure than the top three industries, forest sector capital expenditures on machinery and equipment were more than five times the R&D expenditure made by the top three R&D-performing industries. This makes the point that forest sector industries are technology adopters and they are using the knowledge developed in other industries through capital investment.

5. CONCLUSION

Logging, a harvesting industry, is part of a larger system of innovation that includes resource management industries, primary manufacturers, service industries, manufacturers of machinery and equipment and the context in which a firm operates. For a complete picture of innovation within a system, components of the system must include all actors that are developing or participating in the innovation process.

Logging is characterized by a high degree of process innovation but a much lower rate of product innovation. In an industry that produces commodities (raw materials) with long life cycles, high rates of product innovation are not expected. Innovation in logging occurs predominantly through the acquisition of machinery and equipment as embodied technologies. The manufacturers of machinery and equipment have high rates of innovation. Using the *Oslo Manual*'s innovation indicator for one industry in isolation would result in the attribution of this innovation activity to the manufacturers of the machinery and equipment as a product innovation. There is evidence that the logging industry does not acknowledge the adoption of technologies to be part of process innovation. If the innovation activity of the supplier of machinery and equipment is not taken into consideration, this component of process innovation may not be recognized.

Appendix 1: The Forest Sector System, Industry Details

Forest Management

1153 Support Activities for Forestry
- Forest fire fighting services
- Log hauling in the bush
- Pest control services, forestry
- Reforestation services
- Timber cruising
- Timber valuation
- Forest & timber land management

54136 Geophysical surveying and mapping services
- Surveying services, geophysical
- Gravimetric surveying services, geophysical
- Seismic geophysical surveying services

54137 Surveying & Mapping (except Geophysical) Services
- Aerial surveying
- Cadastral surveying services
- Geographic information system (GIS), base mapping services
- Land surveying services
- Map making services
- Oceanic surveying services
- Photogrammetric mapping services
- Production of topographic materials & maps
- Informatics/Computer Modeling/Decision-Making Technologies

541510 CAD, CAE

54 Professional, Scientific & Technical Services
- **54133** Engineering Services
  - Logging engineering services
- **54162** Environmental Consulting Services
- **541710** Forestry R&D Laboratories

Harvesting

Non-timber forest products

113 Forestry & Logging
- **1132** Forest Nurseries & Gathering of Forest Products

111 Crop Production
- **111421** Nursery & Tree Production
- **111999** All Other Miscellaneous Crop Farming
  - Maple products production

Logging

113 Forestry & Logging
- **1131** Timber Tract Operations
- **1133** Logging

Primary Manufacturers

321 Wood Product Manufacturing
- **3211** Sawmills & Wood Preservation
- **3212** Veneer, Plywood & Engineered Wood Product Manufacturing
- **3219** Other Wood Product Manufacturing

322 Paper Manufacturing
- **3221** Pulp, Paper and Paperboard Mills
- **3222** Converted Paper Product Manufacturing

Services

52 Financial and Insurance

Manufacturing Suppliers

333 Machinery Manufacturing
- **3332** Industrial Machinery Manufacturing

Notes

The innovation system for the forest sector presented in this chapter is the result of work carried out by staff from the Science, Innovation and Electronic Information Division of Statistics Canada under the direction of Fred Gault, in collaboration with staff at Natural Resources Canada. In particular, the author would like to thank Frances Anderson for her continued direction of research into innovation in the natural resource industries and for sharing the findings of work carried out with John Hector at several forest industry workshops. In addition, this chapter has benefited from the contributions of Rob Dunn, Tim Norris, John Hector, Jai Persaud and Hugh Deng at Natural Resources Canada. The author alone is responsible for the opinions expressed in this chapter and for any errors and omissions that may remain.

[1]Statistics Canada has conducted several surveys of innovation since 1993 to gain a better understanding of innovation in Canada. The Survey of Innovation and Advanced Technology 1993 surveyed manufacturing firms. The Survey of Innovation 1996 surveyed the communications, financial services and technical business services industries. The Survey of Innovation, Advanced Technologies and Practices in the Construction and Related Industries 1999 was the first survey of advanced technologies and practices in the construction sector.

[2]Other S&T activities within the framework include: R&D; invention; and technology diffusion.

[3]Industries listed include: Timber Tract Operations (113110); Support Activities for Forestry (115310); Construction and Forestry Machinery, Equipment and Supplies Wholesalers-Distributers (417210); Forest Products Trucking: Local (484223) and Long Distance (484233); Lessors of Other Real Estate Property (the forestry component) (531190); Research and Development in the Physical, Engineering and Life Sciences (the forestry component) (541710); Community Colleges and C.E.G.E.P.s (the forest technology schools) (611210); and Technical and Trade Schools (for forestry) (611510).

[4]See Schaan and Nemes (2002) for details on the methodology for the Survey of Innovation 1999.

[5]Firms were asked to indicate the importance of a series of success factors on a scale of 1 to 5, where 1 was low importance and 5 was high importance. Firms giving a score of 4 or 5 to a particular success factor were considered to regard this factor as important.

[6]This included coal mining, oil and gas extraction, metal ore mining, other mining and services to mining.

[7]Veneer, Plywood and Engineered Wood Product Manufacturing firms had improving product quality as their second most frequently chosen objective.

[8]Firms were asked to indicate the importance of a series of success factors on a scale of 1 to 5, where 1 was low importance and 5 was high importance. Firms giving a score of 4 or 5 to a particular success factor were considered to regard this factor as important.

[9]Data on R&D performance and on how this R&D was carried out came from question 24 of the Survey of Innovation 1999, where respondents were asked if their firm undertook R&D activities.

[10]R&D expenditures are reported by Standard Industrial Classification (SIC) codes. The correspondence to these industries in the NAICS classification is not direct. For instance, R&D expenditures in Forestry and Logging include support activities for forestry (SIC 0411, 0412 and 0511), but capital expenditures on machinery and equipment for support activities are not included in the figure. R&D expenditures in Wood Product Manufacturing include those of some industries from Furniture and Related Product Manufacturing, but exclude those of some industries that are included in the NAICS industry code.

[11]Communications Equipment, Aerospace Product and Parts, and Pharmaceutical and Medicine manufacturers invested a total of $3,308 million in R&D and $683.7 million in machinery and equipment in 1997.

13

The Flow of Innovative Products from Manufacturing Industries to Construction Industries

Frances Anderson

1. INTRODUCTION

Firms can gain a competitive advantage through the introduction of new or significantly improved products to the market.[1] New or significantly improved products confer upon a firm a temporary monopoly until other firms are able to catch up and develop the same or similar products for the market. Intellectual property rights can extend the period of time in which firms have an effective monopoly in their market.

Similarly, the implementation of new or significantly improved production processes can also provide firms with a competitive advantage. New or significantly improved processes can increase the productivity of the workforce and can reduce the cost of producing products, allowing a firm to arrive in the marketplace with products at a lower cost than their competitors. New or significantly improved processes are often a necessary component of introducing new products. They can also result in improvements in the quality of products, providing the firm with a competitive advantage in the marketplace.

A competitive advantage can also be gained by the use of new or significantly improved products. They can be bought from a supplier and used by a firm. The user firm would not have contributed directly to the R&D for the product and yet it could gain a competitive advantage from being a user. This is the case for much of the improvement in process technologies. Process technologies, in the form of machinery, equipment and tools, are sold to other firms that can gain a competitive advantage in the ways discussed above. Firms that buy new products and use them can also gain a competitive advantage.

Construction industries in particular can derive a further advantage from new and significantly improved products and processes. The built structures[2] produced by construction industries tend to have long life cycles. Reducing the initial cost of built structures by the incorporation of innovative products can benefit their owners. The initial construction costs of a building are but one component of the total cost of maintaining the built structure over the time that it is in use. Benefits can accrue to the owner of the built structure if the products used in its construction can reduce or delay the amount of repair and maintenance required over time.

It is argued in the study presented here that an understanding of the impact of innovation on firms and on the economy as a whole requires an understanding of the generation of knowledge, the transmission of this knowledge and its use.[3] A focus only on the development of new or significantly improved products and processes provides a partial picture and neglects the component of the use of innovative products and processes, which constitutes one of the important impacts of an innovation. In the study presented in this chapter, this general framework was applied to construction industries; the flow of building products from manufacturing to construction industries was examined, with specific attention paid to the creation of new products by construction products manufacturers, the means of transmission of knowledge embodied in construction products and the use of these new products.

2. THE CONSTRUCTION CLUSTER

A number of recent studies have adopted a cluster approach to analyzing construction activities (see den Hertog and Brouwer 2001; Vock 2001; Dahl and Dalum 2001; and Anderson and Manseau 1999). They argue that an analysis based only on those industries that are found on the construction site presents a misleading view of the complexity of the construction production system. Vock (2001: 245) in his analysis of the Swiss construction sector, defines a cluster as "a set of organisations either characterized by their interdependence (buyer-supplier, input-output or value chain relationships) or by their commonalities (geographical location, the generation or use of a common technology, the use of common distribution channels, or a common labour pool)."

A key issue for authors using the cluster approach is to develop a taxonomy of the industries that comprise the construction cluster that goes beyond the taxonomy of the construction sector found in the industrial classification systems used by national statistical agencies. Dahl and Dalum (2001) using the industry classification system based on the Statistical Classification of Economic Activities in the European Community (NACE), provide a definition of the industry components of what they call the "construction mega-cluster." Their mega-cluster includes industries from the primary sector, the manufacturing sector and the service sector. Vock (2001) in his analysis of the Swiss construction cluster, and den Hertog and Brouwer (2001) in their analysis of the Dutch construction cluster, adopt similar approaches using NACE industry coding. Den Hertog and Brouwer (2001) include industries from the wholesale sector.

An analysis of the construction cluster has been underway for a number of years in a joint project of the National Research Council of Canada's Institute for Research in Construction and Statistics Canada. Preliminary work on categorizing the actors according to the NAICS (North American Industry Classification System) classification was done by Anderson and Manseau (1999). Figure 1 is a graphic representation of the set of industries and organizations that constitute the Canadian construction cluster. This set of actors is similar but not identical to that found in the work on the construction cluster done in other countries.

At the heart of the construction cluster are the actors involved in the construction production system — those involved in the value-added chain that leads to the final built structure. The value-added chain extends from the suppliers of raw materials to the building products suppliers to the subsystem assemblers and, finally, to the facility assemblers. Also part of the cluster are private-sector actors, which provide expertise, process equipment and financial support to the production-system actors (suppliers of machinery and equipment, professional services, and building owners and managers), and public-sector actors, which provide manpower, regulatory context and infrastructure support (educational/research institutions, government regulatory agencies, professional associations/unions and infrastructure providers).

Figure 1: Model of the Canadian Construction Cluster

Private Sector Actors	Suppliers of machinery and equipment		Professional services				Building owners and managers
Production-System Actors	Suppliers of raw materials	→	Building products suppliers	→	Subsystem assemblers	→	Facility assemblers
Public-Sector Actors	Educational/ research institutions		Government regulatory agencies		Professional associations/ unions		Infrastructure providers

A simple model of the flow of embodied knowledge, which was the object of the study presented here, is found in Figure 2.

Figure 2: Flow of Embodied Knowledge from Manufacturing Firms to Construction Firms

Generation	*Transmission*	*Use*
Manufacturing Performers	⟶	*Construction Performers*
• Building products suppliers • Suppliers of machinery and equipment	• Direct supplier/user relations • Facilitating actors	• Subsystem assemblers • Facility assemblers

Knowledge embodied in construction products is generated by two sets of manufacturing actors: those that manufacture building products used in construction (building products suppliers); and those that manufacture products used in the process of constructing built structures (suppliers of machinery and equipment). Construction products embodying knowledge are used by two sets of construction actors: those involved in assembling subsystems (subsystem assemblers); and those involved in the coordination of all the subsystems to produce the final built structures (facility assemblers). Transmission of the knowledge-embodied products can occur directly between the manufacturing actors and the construction actors or it can be facilitated by other actors. Characterizing the nature of the direct supplier/user relations and identifying the facilitating actors and characterizing their roles were the objectives of the study reported here.

3. METHODOLOGY

The data that were analyzed were based on two Statistics Canada surveys, both carried out in 1999. The Survey of Innovation, Advanced Technologies and Practices in the Construction and Related Industries surveyed construction firms classified into 13 industrial groups based on NAICS code 23. The Survey of Innovation surveyed manufacturing firms classified into 31 industrial groups based on NAICS codes 31-33.[4] The firm groups that were analyzed are presented in Table 1 (see Statistics Canada 1998a).

4. GENERATORS OF EMBODIED KNOWLEDGE

Question 21 of the Survey of Innovation 1999 asked manufacturing firms if they supplied building products to construction industries:

> During the last three years, 1997 to 1999, did your firm offer products which were **incorporated into** buildings and other engineering works such as roads, dams, bridges, sewers, transmission lines and pipelines? Some examples of building products are windows, plaster board, bricks, concrete, heating and plumbing systems, roofing, security systems, electrical systems and others.

Table 1: Definition of Firm Groups

Building Products Suppliers	• Self-identified firms in NAICS 31-33 (Manufacturing) that responded "yes" to question 21 of the Survey of Innovation 1999: Products were incorporated into built structures
Suppliers of Machinery and Equipment	• Self-identified firms in NAICS 31-33 (Manufacturing) that responded "yes" to question 22 of the Survey of Innovation 1999: Products (machinery, equipment or tools) were used during the process of constructing built structures
Dedicated Construction Manufacturers	• Building products suppliers and suppliers of machinery and equipment whose sales of construction products represented more than 50% of total revenues
Facility Assemblers	• NAICS 23121: Residential Building Construction • NAICS 23122: Non-Residential Building Construction • NAICS 2313: Engineering Construction
Subsystem Assemblers (Exterior)	• NAICS 2321: Site Preparation Work • NAICS 2322: Building Structure Work • NAICS 2323 (excluding 23233): Building Exterior Finishing Work • NAICS 23233: Roofing and Related Work • NAICS 2329: Other Special Trade Contracting
Subsystem Assemblers (Interior)	• NAICS 2324: Building Interior Finishing Work • NAICS 2325 (excluding 23251 & 23252): Building Equipment Installation • NAICS 23251: Electrical Work • NAICS 23252: Plumbing, Heating, and Air-Conditioning Installation
Facilitating Actors	• Actors involved in transmission of embodied knowledge. Facilitating actors were identified in questions on source of information for innovation

Almost one-quarter (23%) of all manufacturing firms indicated that they supplied construction firms with building products, and in eight of the 31 industries surveyed more than 40% of firms responded that they supplied building products. The highest percentage of firms supplying building products was found in Non-Metallic Mineral Product

(73%), Veneer, Plywood and Engineered Wood Product (58%), and Petroleum and Coal Products (51%) Manufacturing[5] (Figure 3).

Manufacturing firms were asked what percentage of their revenues came from the sale of building products. As shown in Figure 4, 50% of all firms supplying building products derived more than 50% of their revenues from the sale of building products and could thus be considered dedicated suppliers of building products.

Question 22 of the Survey of Innovation 1999 also asked manufacturing firms whether they supplied machinery, equipment or tools to construction industries:

> During the last three years, 1997 to 1999, did your firm offer machinery, equipment or tools which were **used during the process** of constructing buildings and other engineering works such as roads, dams, bridges, sewers, transmission lines, and pipelines? Some examples of products used during construction are bulldozers, cranes, power tools, scaffolding, survey equipment and others.

Figure 3: Percentage of Manufacturing Firms Supplying Building Products, by Selected Manufacturing Industries, 1997-99

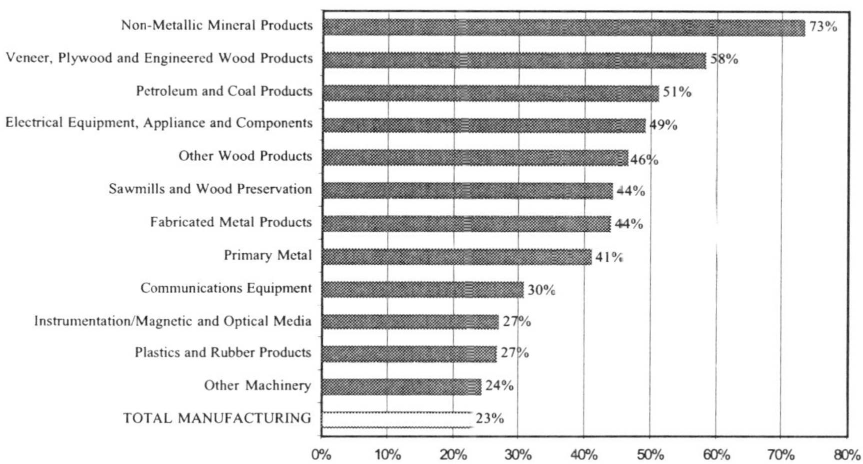

Source: Statistics Canada, Survey of Innovation 1999.

Figure 4: Percentage of Manufacturing Firms Supplying Building Products, by Percentage of Sales, 1997-99

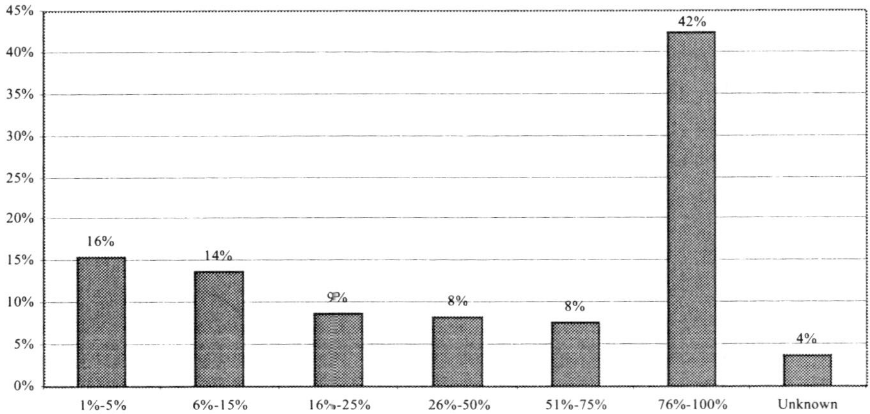

Source: Statistics Canada, Survey of Innovation 1999.

Six per cent of all manufacturing firms responded that they did offer products that were used in the construction process. The highest percentage of firms offering products used in the construction process was found in Agricultural, Construction and Mining and Industrial Machinery (17%), Electrical Equipment, Appliance and Component (16%) and Non-Metallic Mineral Product (16%) Manufacturing (Figure 5). Among these firms, only 19% were dedicated suppliers (deriving more than 50% of their revenues from the sale of construction machinery and equipment) (Figure 6). Thus, firms that offered construction machinery, equipment or tools were supplying products to other industries as well.[6]

The industries that supplied building products and construction machinery, equipment or tools varied in the extent to which they contained firms that undertook product innovation. Figure 7 presents the percentage of product innovators for the top four industries supplying building products, on the one hand, and the top four industries supplying construction machinery, equipment or tools, on the other (two of which were common). Two of the industries were above the average for the whole manufacturing sector and four were below.

Figure 5: Percentage of Manufacturing Firms Offering Construction Machinery, Equipment or Tools, by Selected Manufacturing Industries, 1997-99

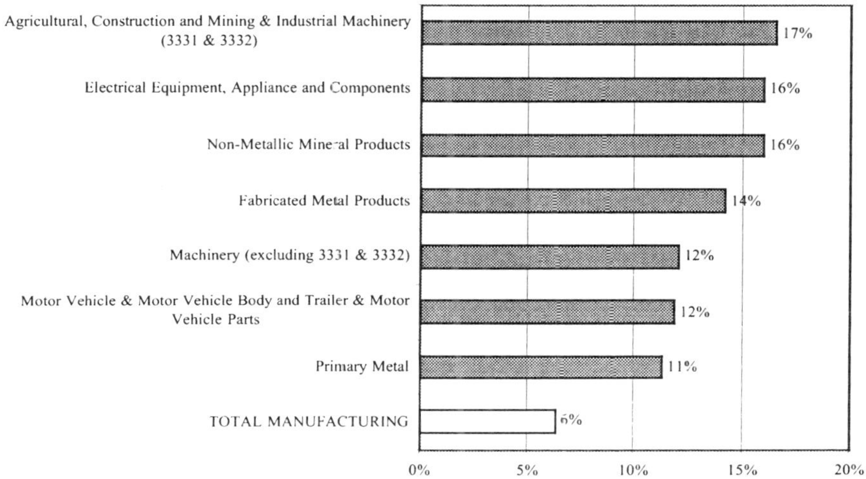

Source: Statistics Canada, Survey of Innovation 1999.

Figure 6: Percentage of Manufacturing Firms Offering Construction Machinery, Equipment or Tools, by Percentage of Sales, 1997-99

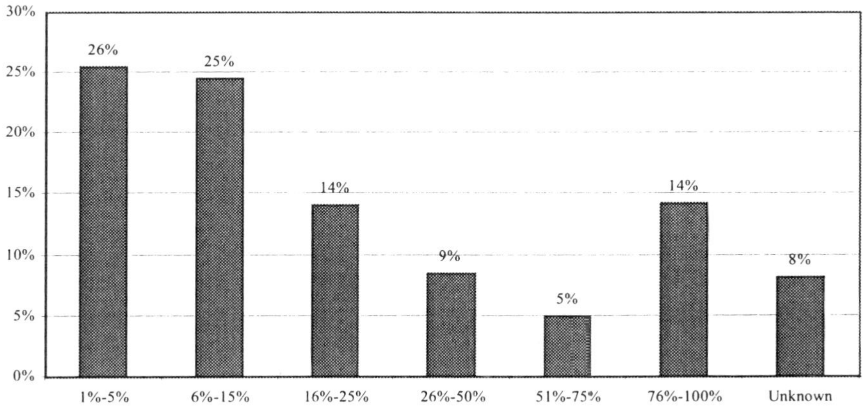

Source: Statistics Canada, Survey of Innovation 1999.

Figure 7: Percentage of Product Innovators in Industries Supplying Building Products and Construction Machinery, Equipment or Tools, by Selected Manufacturing Industries, 1997-99

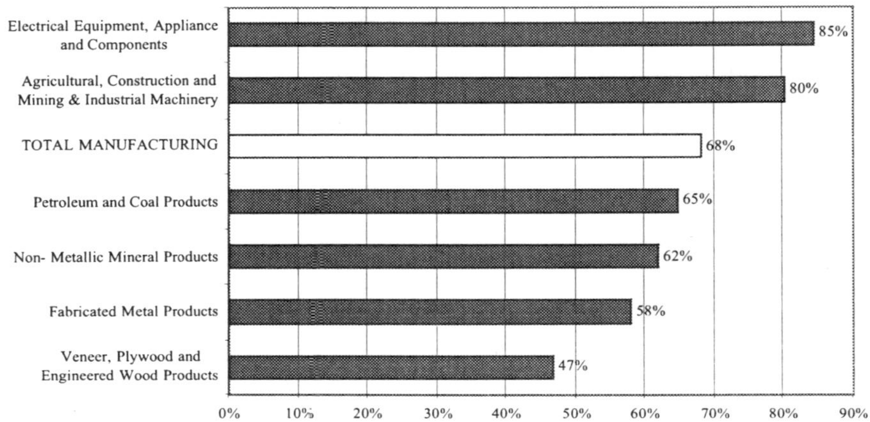

Source: Statistics Canada, Survey of Innovation 1999.

5. USERS OF EMBODIED KNOWLEDGE

The Survey of Innovation, Advanced Technologies and Practices in the Construction and Related Industries 1999 asked firms to indicate whether or not they used a particular set of 18 advanced technologies. Each of these technologies was produced by manufacturers, with the exception of the design technologies (software programs), which were produced by the service sector. The data from the survey thus provide an indicator of the extent to which construction firms were using some of the most advanced manufacturing and service-sector products.

Figure 8 shows that the use of the particular set of advanced technologies varied by type of construction industry. Two general contracting industries (Engineering Construction and Non-Residential Building Construction) and one specialized trade (Electrical Work) used the largest average number of advanced technologies.[7] With regard to the use of specific types of advanced technologies, this varied considerably from industry to industry, as can be seen in Table 2.

Figure 8: Average Number of Advanced Technologies Used by Construction Industries, 1999

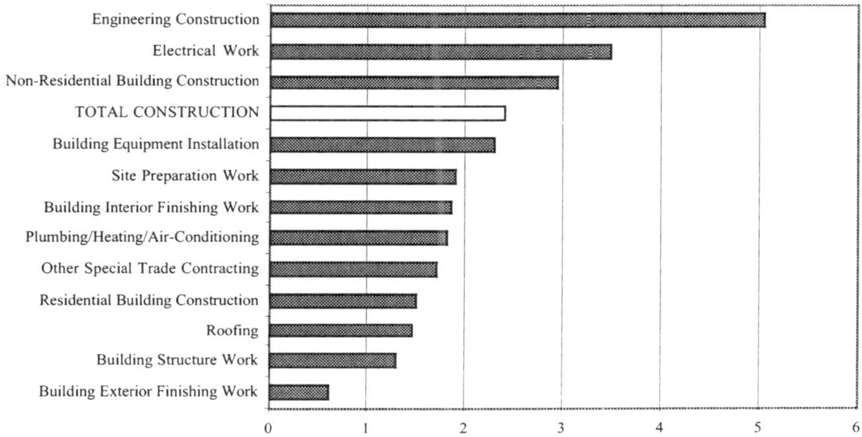

Source: Statistics Canada, Survey of Innovation, Advanced Technologies and Practices in the Construction and Related Industries 1999.

Table 2: Percentage of Construction Firms with At Least One Advanced Technology in a Category, by Top Three Industries, 1999

Category of Advanced Technology	*Rank*		
	1	*2*	*3*
Communications	Engineering Construction (72%)	Electrical Work (66%)	Building Equipment Installation (64%)
On-Site Plant and Equipment	Engineering Construction (58%)	Electrical Work (41%)	Site Preparation Work (39%)
Materials	Non-Residential Building Construction (35%)	Site Preparation Work (32%)	Engineering Construction (31%)
Systems	Electrical Work (62%)	Plumbing/Heating/Air-Conditioning (38%)	Engineering Construction (35%)
Design	Engineering Construction (52%)	Building Interior Finishing Work (40%)	Non-Residential Building Construction (33%)

Source: Statistics Canada, Survey of Innovation, Advanced Technologies and Practices in the Construction and Related Industries 1999.

The importance of the use of information and communication technologies (ICTs) applications is shown in Figure 9: the planned rate of use anticipated for the two years following the survey period ranged from 10% to 25%. The responses to a survey question asking firms to identify the most important technology or practice during the three years prior to the survey period also indicated the importance of ICTs. One-half of the firms that responded to the question indicated that ICTs had had a great impact on their business.[8]

Figure 9: Percentage of Construction Firms Planning to Use Advanced Technology Within the Following Two Years, 1999

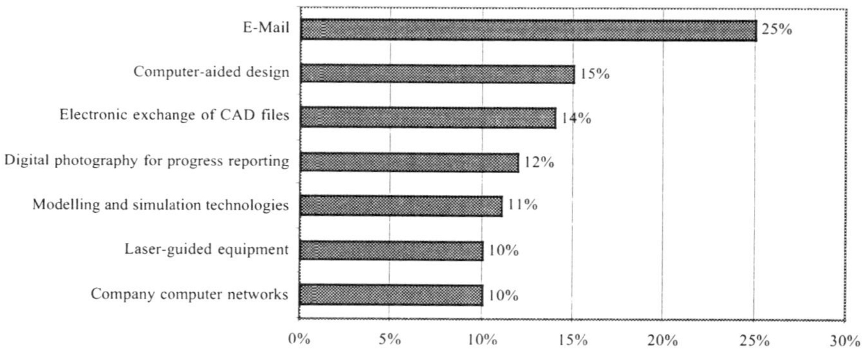

Source: Statistics Canada, Survey of Innovation, Advanced Technologies and Practices in the Construction and Related Industries 1999.

6. TRANSMISSION OF EMBODIED KNOWLEDGE

In the 1999 survey of the construction and related industries, firms were asked to indicate, from a list of 12 obstacles, which were the major obstacles to using new or improved building products, building systems and construction equipment. The results are presented in Figure 10.

The most commonly cited obstacle to using new or improved products, systems and equipment was their high cost. Interestingly, the second most common obstacle cited was a shortage of skilled workers,

Figure 10: Percentage of Construction Firms Indicating Obstacles to Using New or Improved Building Products, Building Systems and Construction Equipment, 1999

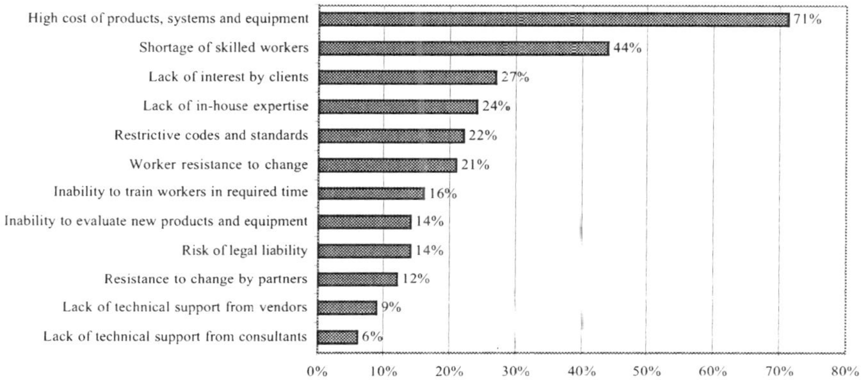

Source: Statistics Canada, Survey of Innovation, Advanced Technologies and Practices in the Construction and Related Industries 1999.

pointing to the importance of training as a key element in the transmission of new products and practices to construction industries. Restrictive codes and standards, which are often considered critical obstacles, were cited by only about one-fifth (22%) of the respondents.

In terms of obstacles related to the transmission of knowledge, Building Interior Finishing Work was among the top three industries with a perceived lack of receptor capability (lack of in-house expertise and inability to evaluate products) combined with a perceived lack of technical support from vendors (Table 3). Non-Residential Building Construction was also among those industries facing a lack of receptor capability. Plumbing, Heating and Air-Conditioning Installation and Building Equipment Installation both had a high percentage of firms indicating the lack of technical support from vendors as an obstacle.

As shown in Figure 11, construction firms were heavily reliant on suppliers as sources of information on advanced technologies and practices (72%). Approximately one-half of businesses indicated that trade journals and newsletters were a source of information. Clients,

Table 3: Percentage of Construction Firms Confronting Obstacles Related to Lack of Receptor Capability or Support from Vendors, by Top Three Industries, 1999

	Rank		
Obstacle	*1*	*2*	*3*
Lack of In-House Expertise	Building Interior Finishing Work (51%)	Non-Residential Building Construction (36%)	Building Equipment Installation (23%)
Inability to Evaluate New Products and Equipment	Non-Residential Building Construction (29%)	Building Exterior Finishing Work (22%)	Building Interior Finishing Work (18%)
Lack of Technical Support from Vendors	Plumbing/Heating/ Air-Conditioning (24%)	Building Equipment Installation (20%)	Building Interior Finishing Work (18%)

Source: Statistics Canada, Survey of Innovation, Advanced Technologies and Practices in the Construction and Related Industries 1999.

Figure 11: Percentage of Construction Firms That Indicated Source of Information on Advanced Technologies and Practices, 1999

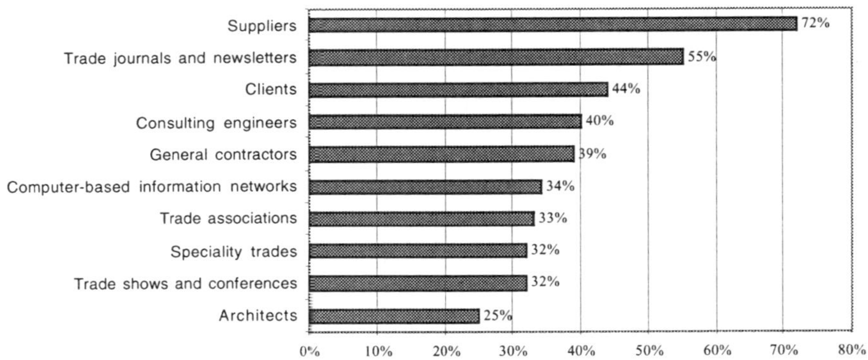

Source: Statistics Canada, Survey of Innovation, Advanced Technologies and Practices in the Construction and Related Industries 1999.

consulting engineers and general contractors were all cited as sources of information by approximately 40% of firms.

Although public organizations (Table 4) (federal government laboratories, universities and colleges) were used as sources of information by a relatively low percentage of construction firms (9% and 12%, respectively), for certain construction industries they were important sources. A relatively high percentage of firms in Building Interior Finishing Work cited federal government laboratories (34%) and universities and colleges (35%) as sources of information for innovation. A relatively high percentage of firms in Electrical Work also used federal government laboratories (12%) and universities and colleges (14%) as sources of information on advanced technologies and practices.

As shown in Figure 12, clients were considered an important source of information for innovation by industries that supplied building products.

Table 4: Percentage of Construction Firms Citing Public Organizations as Source of Information on Advanced Technologies and Practices, by Top Three Industries, 1999

Type of Public Organization	*Rank*		
	1	*2*	*3*
Federal Government Laboratories	Building Interior Finishing Work (34%)	Electrical Work (12%)	Non-Residential Building Construction (7%)
Universities and Colleges	Building Interior Finishing Work (35%)	Building Equipment Installation (22%)	Electrical Work (14%)

Source: Statistics Canada, Survey of Innovation, Advanced Technologies and Practices in the Construction and Related Industries 1999.

7. CONCLUSION

The study presented in this chapter investigated the flow of innovative products from manufacturing industries to construction industries, in an attempt to gain a better understanding of the nature of innovation in

Figure 12: Percentage of Innovative Building Products Firms Indicating Client as Source of Information for Innovation, by Selected Manufacturing Industries, 1997-99

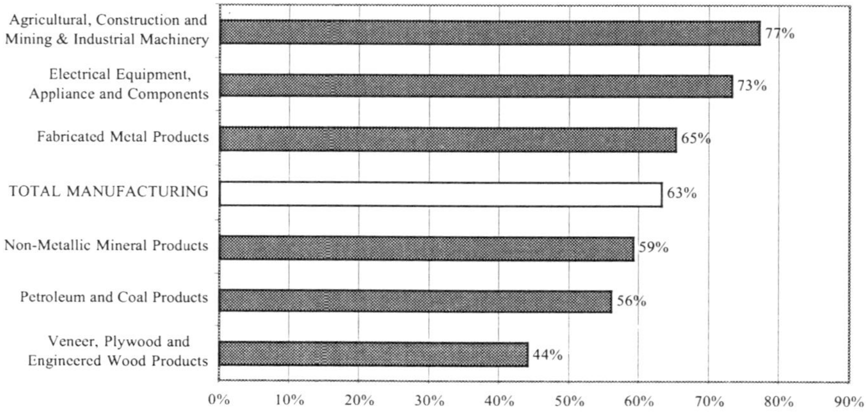

Source: Statistics Canada, Survey of Innovation 1999.

construction industries — not only the generation of knowledge, but also the transmission and use of knowledge embodied in innovative products.

The study revealed that there were a significant number of firms in the manufacturing sector involved in manufacturing products for construction industries. Twenty-three per cent of all manufacturing firms indicated that they supplied building products to construction firms, and 6% indicated that they supplied machinery, equipment or tools. One-half of the suppliers of building products were found to be dedicated suppliers — i.e., more than 50% of their revenues were derived from sales to construction industries — whereas, in the case of those firms supplying machinery, equipment or tools, only one-fifth were dedicated suppliers.

There was considerable variation in the percentage of firms in individual industries supplying building products. It is, however, evident that, for certain manufacturing industries, the construction industries were major clients. For example, 73% of firms in Non-Metallic Mineral Product Manufacturing and 58% in Veneer, Plywood and Engineered Wood Product Manufacturing supplied building products to construction firms. In contrast, manufacturing industries that supplied machinery, equipment or tools tended to have a low percentage of firms supplying products to construction industries. Thus, for example, Agricultural, Construction and Mining and Industrial Machinery Manufacturing, which had the highest percentage of firms supplying products used in the construction process, had only 17% of supplier firms. The industries that were suppliers to construction industries ranged from those with lower rates of innovation to those that were among the most innovative.

With regard to the use of advanced technologies, considerable variation was found among different construction industries. The average number of advanced technologies used by different industries varied, as did the extent to which the industries used different types of technologies (communications, on-site plant and equipment, materials, systems and design). Of the technologies listed in the survey, ICTs were consistently ranked as important. The planned rate of use of ICTs was among the highest, and, when firms were asked to identify technologies that had had the most impact during the previous three years, one-half cited ICTs.

Cost was seen as the major obstacle to the adoption of new technologies and practices (71% of firms in construction industries), followed by a shortage of skilled workers (44%). The shortage of skilled workers points to the importance of training in ensuring the transmission of new products and practices to construction industries. Lack of receptor capability and lack of technical support from vendors were perceived to be severe problems for a number of industries, including Building Interior Finishing Work and Non-Residential Building Construction.

Construction firms were heavily reliant on suppliers, trade journals and newsletters, and clients to provide information on advanced technologies and practices (72%, 55% and 44%, respectively). Public organizations (federal government laboratories, universities and colleges) were found to be particularly important sources of information on advanced technologies and practices for certain industries, particularly Building Interior Finishing Work and Electrical Work. From the supplier perspective, clients were seen as an important source of information for innovation.

In conclusion, the cluster approach to analyzing the production system of construction industries, combined with a perspective on innovation that includes the generation, transmission and use of knowledge embodied in products, provides a better understanding of the nature of innovation in construction industries.

Notes

[1]The discussion in this chapter uses the definition of innovation found in the *Oslo Manual* (OECD/Eurostat 1997).

[2]Built structures include residential and non-residential buildings, as well as engineering works such as dams, refineries, highways, bridges, sewers, power and commission lines, and similar structures and works.

[3]For a more detailed discussion of the generation, transmission and use of knowledge, see Statistics Canada (1998b).

[4]For details of the methodology of each survey, see Anderson and Schaan (2001), and Schaan and Nemes (2002).

[5]For the study presented in this chapter, the manufacturing sector was organized into 31 industrial groupings. The industrial groupings discussed here and their corresponding NAICS codes are as follows: Non-Metallic Mineral Product (NAICS 327); Veneer, Plywood and Engineered Wood Product (NAICS 3212); Petroleum and Coal Products (NAICS 324); Electrical Equipment, Appliance and Component (NAICS 335); Other Wood Product (NAICS 3219); Sawmills and Wood Preservation (NAICS 3211); Fabricated Metal Product (NAICS 332); Primary Metal (NAICS 331); Communications Equipment (NAICS 3342); Navigational, Measuring, Medical and Control Instruments and Manufacturing and Reproducing Magnetic and Optical Media (NAICS 3345 & 3346); Plastics and Rubber Products (NAICS 326); Other Machinery (NAICS 3333, 3334, 3335, 3336 & 3339); Agricultural, Construction and Mining and Industrial Machinery (NAICS 3331 & 3332); Motor Vehicle, Motor Vehicle Body and Trailer and Motor Vehicle Parts (NAICS 3361, 3362 & 3363).

[6]Within each of the industry groupings that were devised there were likely to be industries at the five- or six-digit level that had a higher percentage of firms supplying products used in the construction process. For example, NAICS 33312 covers Construction Machinery Manufacturing, which would probably have a very high percentage of supplier firms. The design of the survey did not allow a breakdown to this level.

[7]For more details on the use of advanced technologies by different industries, see Anderson and Schaan (2001).

[8]See Anderson, Bakker and Manseau (2002).

Next Steps

14

Lessons Learned and Future Directions for Innovation Research and Policy

Andrew Sharpe

1. INTRODUCTION

The chapters in this volume present new findings on the nature of innovation in Canada, largely based on the results of the Survey of Innovation 1999 conducted by Statistics Canada. The chapters greatly contribute to our understanding of the innovation process in Canada, both in terms of the determinants of innovation and the impact of innovation on economic performance. This material has been and will continue to be extremely useful for the development of effective innovation policy in this country.

This final chapter provides an overview of the lessons learned from the survey and outlines future directions for research and policy development in the innovation area. The chapter first highlights a number of the key findings of the volume, then identifies research gaps in our knowledge of innovation and the implications for future research, and finally discusses the policy implications of the research findings.

Innovation is at the heart of the federal government's economic policy agenda. In February 2002 the Government of Canada as part of its innovation strategy released two discussion papers — *Achieving Excellence: Investing in People. Knowledge and Opportunity*, drafted by Industry Canada (Government of Canada 2002a) and *Knowledge Matters: Skills and Learning for Canadians*, drafted by Human Resources Development Canada (Government of Canada 2002b). The first paper provides a comprehensive review of innovation in Canada. In particular, it identifies a number of goals and targets for innovation in this country, including by 2010 at least doubling the federal government's current investments in R&D and making Canada among the top five countries in the world in terms of R&D performance.

2. MEASUREMENT AND ANALYSIS OF INNOVATION IN CANADA: FINDINGS AND IMPLICATIONS

Measuring innovation is now a government-wide interest and priority in Canada. In addition to Statistics Canada, Industry Canada, the National Research Council, and Natural Resources Canada, other departments and agencies, both federal and provincial, are interested in tracking innovative activity in their respective sectors of interest through innovation surveys. This growing interest in the analysis of the innovation process is a very positive development. It reflects the pioneering work of Statistics Canada in the development and implementation of innovation surveys and in making the survey data available to analysts, while protecting the confidentiality of the survey responses. It should be noted that a great advantage of Statistics Canada conducting innovation surveys is the very high response rate of around 95 per cent, as firms are required to respond under the Statistics Act. This compares with much lower response rates in comparable surveys in other countries, such as the EU's Community Innovation Survey.

The measurement of innovation through innovation surveys and the analysis of survey results has produced important findings in terms of the characteristics of innovative firms, the determinants of innovation, as well as the impact of innovation on firm performance. Some of these findings are highlighted below.

2.1 Novelty of Innovation

The *Oslo Manual* (OECD/Eurostat 1997) provides guidelines for the measurement of the activity of innovation, and it advises on the boundaries between new or significantly improved products and processes and those that are not sufficiently new or improved to qualify as an innovation. The *Oslo Manual* also provides three categories of novelty of innovation, the highest of which is world-first, the lowest is firm-first and the intermediate category is left open. In the Canadian case, the choice was Canada-first. The Survey of Innovation 1999 used the lowest category of innovation for its identification of innovation and only used the three categories to classify the most important innovation of the firm. As a consequence, about four fifths of manufacturing firms in Canada were found to be innovators over the 1997-99 period, leaving only one-fifth in the non-innovator category. One implication of this high incidence of innovation is that the sample size of the non-innovators in innovation surveys may not be large enough to support detailed analysis of the difference in characteristics between innovators and non-innovators

To shed light on the nature of a firm's innovative activity, the Survey of Innovation 1999 developed three categories of innovation based on the concept of novelty of innovation. Firms were asked whether their innovative activity was a world first, a first in Canada, or a firm first. The first two categories can be combined to form a "first to the market" category while the third category is referred to as "first to the firm." In the 1999 survey, about 15 per cent of innovative manufacturing firms responded that their innovative activity was in the first category, 25 per cent in the second category and 60 per cent in the third category. It was thus the broadest or least restrictive category of innovative activity that accounts for Canada's high rate of innovation.

This suggests that the degree of novelty be restricted if useful insights are to be gained about the process of innovation. This was in fact done in a number of the chapters. For example, Mohnen and Therrien combined world first and Canada first firms to compare innovation in Canada with that in Europe, where the EU Community Innovation Survey uses "first to the market" as its measure of novelty. Corrazin also identified the characteristics of the world-first firms relative to other innovative firms. He found that the former firms hire more recent university graduates, make more use of teams, and conduct more R&D. These firms are creators of innovations, while other innovative firms are adopters of technologies and practices which lead to innovation in the firm.

2.2 Diffusion of Technologies and Practices

If the definition of innovation is restricted to "first to the market," it raises a question about how to understand the innovative activity of those firms that are excluded from this category, but are included in the Canadian category of "first to the firm" for product and process innovation. The innovative behaviour of these firms can be captured by surveys on the use, or planned use, of technologies or practices, as outlined in the *Oslo Manual* (OECD/Eurostat 1997: 76).

Surveys of the use of advanced manufacturing technologies have been conducted for the manufacturing sector in Canada (Sabourin and Beckstead 1999). There are also more targeted surveys on the use of information and communication technologies (Statistics Canada 2003), biotechnologies (McNiven, Raoub and Traoré 2003), and knowledge management practices (Earl 2003). All these surveys have contributed to the understanding of the innovative activity in firms that do not necessarily bring products or processes first to the market.

2.3 Innovation Surveys Must Reflect Industries Differences

The same framework for measuring innovative activities does not apply equally well to all industries. For example, Anderson points out that construction firms are closely linked to networks of materials and equipment suppliers and that product or process innovation may be missed if only these construction firms are surveyed and suppliers ignored. Schaan notes that in the forestry sector, machinery and equipment, which may make the firm a leader in the sector, last much longer than the three-year reference period for innovative activity standard in innovation surveys. As a result, the most advanced firm could be classified as a non-innovator if it had invested in the machinery more than three years earlier.

These examples suggest that in the development of innovation surveys it is necessary to factor in linkages between actors in the innovation system in determining the scope of innovative activity and to take account of both product and process life-cycles before deciding upon the period of observation. Consequently, an innovation survey designed for firms in manufacturing industries may be inappropriate for other sectors unless the characteristics of those sectors are taken into account in the development of the survey.

2.4 The Importance of Networks, Alliances and Partnerships

A key finding of the volume is that networks, alliances and collaboration are particularly important for innovation. For example, Landry and Amara find that Canadian manufacturing firms that are world-first in their innovative activity have the most extensive networks. While the causation may run both ways (innovative firms may find it easier to develop networks), firms that are well connected to other firms through networks have more opportunities to learn about and access new technologies. Therrien and Chang find that local collaboration also has a very positive effect on innovation performance.

These findings highlight the need for further research to measure the relationships between the linkages of the firm and their innovative activity. Some of this has been initiated in the work described in de la Mothe and Link (2002).

2.5 Skills and Innovation

A finding, emphasized in a number of chapters, is the importance of a strong skills base for innovative activity. As noted by Hamdani, the greater the knowledge intensity of the firms, the greater the likelihood of innovation. This suggests the need to collect more information on skills in relation to innovation in future surveys. There is also a need to gain a better understanding of the impact of innovative activity on skill requirements within the firm.

2.6 Intellectual Property Protection is Expensive for Small Firms

The importance of intellectual property rights such as patents and trademarks for successful innovation has been well established. If firms cannot protect their results of their innovative activity, there will be less payoff from such activities and less incentive to undertake them. Hanel in his chapter finds that small manufacturing firms, in particular, often do not have the resources to protect their intellectual property and consequently under-invest in this area, with negative implications for innovation. To counterbalance this lack of resources, Hanel suggests that government subsidize the cost of obtaining, maintaining and enforcing intellectual property rights for small firms.

2.7 Firm Size Matters for Innovation

A standard finding in the innovation literature is that the incidence of innovation is greater for large firms than small firms. This result is supported by the chapters in this volume. For example, Mohnen and Therrien find evidence that large firms innovate more. These firms are more likely to have a dedicated R&D unit to generate new ideas and to provide the capacity to absorb ideas from outside the firm. This absorptive capacity allows the large firms to make more use of government laboratories as a source of knowledge for innovation.

Le and Tang find that small firms are as likely as large firms to convert innovation inputs into innovation, but they are less likely to engage in the set of innovation inputs examined in the Survey of Innovation 1999. As a consequence, their propensity to innovate is lower than that of large firms, a result that is consistent with the findings of Mohnen and Therrien.

As different policies may be needed for firms of different size, descriptive statistics on the characteristics of innovating and non-innovating firms, by size class, are essential outputs of innovation surveys.

2.8 Competition Fosters Innovation

It is well known that firm behaviour, including innovative behaviour, is affected by the competitive environment in which the firm operates. The chapter by Tang provides a further confirmation of this view that competitive pressures motivate firms to innovate. Without such pressures, firms have little incentive to adopt new technologies or engage in other innovative activities. Less competitive industries show a lower propensity to innovate than competitive industries. This finding has important implications for competition policy.

2.9 Innovative Firms Outperform Non-innovative Firms

Given the competitive advantages that innovation can confer, one would expect that innovative firms would outperform non-innovative firms on various economic criteria. The chapter by Cozzarin confirms this, as he finds a positive link between innovation and firm performance, as measured by productivity levels. He also finds that firms which are world-first innovators enjoy superior economic performance compared to other categories of innovators.

2.10 International Comparisons of Innovation

International comparisons of innovative activity help policy development. In this regard, the finding that Canada has a high level of innovative activity relative to other countries is particularly striking. Mohnen and Therrien find that the propensity to innovate was in fact much higher in Canada than in the four European countries (France, Germany, Spain, and Ireland) covered by the EU's Community Innovation Survey — about four-fifths of firms in 1997-1999 as opposed to about one-half in 1994-6. This was largely explained by the much higher rate of innovation for small firms in Canada compared to the EU — 75 per cent versus 41 per cent.

The overall innovation intensity (innovation as a proportion of sales) was also higher in Canada than in Europe, but when sectoral composition, size and other factors were controlled for, Canadian firms had a lower share of sales from innovative products when compared with European firms. These findings highlight the fact that both descriptive statistics and econometric analysis must be combined in order to identify international differences that can be used to inform the policy development process.

3. RESEARCH GAPS AND DIRECTIONS FOR FUTURE WORK

While many insights have been gained from the empirical work based on the Survey of Innovation 1999, and earlier surveys, there are still significant gaps in our understanding of innovation. Some of these gaps will be filled in the future as new innovation surveys lead to the development of a time series on innovative activity and economic performance of Canadian firms. Other gaps will require different approaches. A number of research gaps in our understanding of innovation in Canada are highlighted below and suggest directions for future work.

3.1 Productivity

The first gap is the lack of insight into both time series and cross-section productivity trends that innovation surveys appear to shed. Innovation or technical change is a, if not the, major driver of productivity trends and much of the discussion of productivity trends focuses on changes in the pace of technological change. In the analysis of productivity trends, it is useful to think in terms of stylized periods of productivity growth. In the immediate postwar period productivity growth was strong in developed countries. A period of much slower productivity growth began in 1973 in all developed countries and has continued to date, except in the United States where productivity growth accelerated after 1995, and Canada, where productivity growth picked up after 1996.

These productivity developments could be accounted for by shifts in the pace of innovative activity. The postwar productivity boom reflected the rapid diffusion of technological discoveries and innovations developed but not exploited during the Great Depression of the 1930s and the Second World War. The post-1973 productivity slowdown reflected the erosion of the productivity-augmenting impact of these innovations. The recent productivity acceleration in the United States and Canada is linked to rapid innovation in the information and communication technology (ICT) sector and the diffusion of ICT products throughout the economy.

Equally, both productivity levels and growth rates vary significantly across sectors. For example, at the sector level real output per hour in Canada in 2000 ranged from a high of $107.56 (1997 dollars) in finance, real estate and insurance to a low of $20.98 in agriculture. Productivity growth rates also exhibit large inter-industry differences. The construction sector in fact experienced negative labour productivity growth in Canada in the 1990s, while communications enjoyed growth of over 3 per cent per year.

Unfortunately, innovation surveys offer little in the way of evidence to confirm or reject the above hypothesis to explain long-run aggregate productivity trends or to shed light on differences in productivity levels and growth rates by sector. In many ways, this is understandable. There is no long time series of consistent innovation survey results for one country, let alone a set of countries and such data are needed to track trends in the pace of innovation over time. Equally, there are few innovation surveys that provide comparable data across sectors at the national or international level so inter-sectoral comparisons of the pace of innovation are difficult. In Europe, this issue is being addressed through the Community Innovation Surveys.

Innovation surveys do not put priority on the collection of data that would allow the development of reliable estimates of productivity at the firm, enterprise, or industry level. One reason for this is to keep the response burden to a minimum, given the priority of collecting information on innovation. Statistics Canada innovation surveys can be linked to other business surveys to provide data on output and labour input for productivity analysis. However, this is not done frequently. For example, the Statistics Canada survey of innovation in the construction sector (Anderson and Schaan 2001) shed no light on why the level of output per hour in the construction sector in Canada in 2000 was lower than it have been 20 years earlier.[1]

It would be useful if in the future innovation surveys were designed with the objective of elucidating productivity issues through direct observation or through data linkage. Equally, it is important to marry both the development of innovation surveys and the interpretation of the findings with expert knowledge on historical trends in technology use and innovation in the sector. For example, the greatest insight gained by the author into the evolution of productivity trends in the construction sector came from a Canada Mortgage and Housing Corporation (CMHC) study that documented the changing production methods in the sector between the 1940s and 1980s in Canada. It found that from the 1940s to the 1960s, new production methods greatly reduced the labour requirements needed to build new houses. However, the new methods of production introduced from the 1960s to the 1980s had less impact on improving labour productivity.

3.2 Skills

A second knowledge gap concerns the role that that skill sets play in the innovation process. Does a lack of certain skills constitute an effective barrier to successful innovation or can firms generally find suitably qualified workers when needed. The private sector certainly argues that the former is the case. Much more research is needed on the impact of changing innovation patterns on human resource requirements and on the importance of knowledge workers for innovative activity.

3.3 Organizational Structures and Practices

A third knowledge gap, somewhat related to the skills issue, is what are the most appropriate organizational structures and management practices to foster innovation. In the past, innovation surveys have not included innovations in organizational structures and practices. But with the growing recognition of the importance of these types of innovation for firm performance, innovation surveys are now attempting to capture these types of innovative activity. An example is the inclusion of questions on knowledge management practices in the French version of the Community Innovation Survey (Kremp and Mairesse 2003).

3.4 Concentration of R&D

A fourth knowledge gap is the implications of the concentration of business sector R&D among a small number of sectors and firms. The advantages and disadvantages of such R&D concentration, in contrast to a situation of a more balanced distribution, merit study, although it is unclear whether public policy can and should influence this sectoral and firm distribution of R&D expenditure.

3.5 Regional Differences

A fifth knowledge gap concerns the provincial variation in innovation activity and outcomes. Do such variations in innovative activity reflect structural factors or policy regimes. For example, certain provinces such as Quebec have more generous tax regimes for R&D than other provinces. It would be useful to ascertain if such favourable treatment of R&D has positive effects on R&D expenditures and eventually on innovation.

3.6 Foreign Investment

A sixth knowledge gap that has received much attention in the past concerns the role of foreign investment on the innovation process. The effect on domestic innovation can be both negative and positive. For example, it has been shown that high levels of foreign ownership can lead to the concentration of R&D activity in the home country. The classic case is the Canadian auto industry, which is completely foreign-owned and does minimal R&D in Canada. The situation is less clear cut in other sectors. On the other hand, there is some evidence that foreign direct investment can be a stimulus to innovation through intra-firm technology transfer from the home country to the host country. Such transfers may also have spillover effects on Canadian firms. Foreign direct investment can also increase the degree of competition in the sector, with positive effects for innovation. Work is needed on the global context for innovation.

3.7 Impact of Innovation on Aggregate Economic Performance

A final knowledge gap is the link between innovative activity and various measures of economic performance at the aggregate or national level. The general presumption has been that innovation is important for economic performance and that highly innovative countries will have the best economic performance. But reality is different. Sweden and Finland rank at or near the top in almost all measures of innovative activity, but neither country ranks at or near the top in productivity levels or growth rates or in living standards, defined in terms of GDP per capita. This of course does not mean that innovation is not important for successful economic performance, only that it is only part of a very complex story. There may be ways for countries to achieve a high standards of living without being highly innovative. For example,

Luxembourg and Italy have moved toward high-end banking and other services. Both countries have done very well in these areas despite a low level of innovation as traditionally measured. These low-innovation paths to prosperity merit closer examination, especially those involving new practices as well as new technologies.

3.8 Importance of Case Studies

Innovation surveys provide information about the innovative activity of particular sectors or industries based on firm data, but they do not provide insight into the innovation experience of particular firms. Case studies are needed for this purpose. Consequently, case studies providing firm-specific insights into the innovation process should be used to complement the quantitative and econometric evidence from innovation surveys to obtain a complete understanding of the innovation process. In this respect, it is encouraging to note the work of the Canadian Innovation Systems Research Network (ISRN) which grew out of a Statistics Canada research workshop on local and regional and innovation (de la Mothe and Paquet 1998). The ISRN is publishing more and more analyses of case studies (Holbrook and Wolfe 2000, 2002 and Wolfe 2003) which complement work based on Statistics Canada surveys

4. POLICY ISSUES AND IMPLICATIONS

The federal government considers innovation a top priority. It sees it as a means to improve the standard of living and quality of life of Canadians. The government has a number of goals and targets in the innovation area, including by 2010 at least doubling the federal government's current investments in R&D, making Canada among the top five countries in the world in terms of R&D performance, improving the innovation performance of communities, leading the world in the share of private sector sales attributable to innovation and developing the most skilled and talented labour force in the world. As a follow-up to the November 2002 Innovation Summit, the federal government is developing ways of benchmarking Canada's innovation performance to international best practice. This will have implications for policy development. This section identifies a number of policy issues related to innovation.

4.1 Policy Relevance of Innovation Gaps

Policy analysts have been very good at identifying gaps in our innovation performance. They have documented that Canada has an R&D gap, a machinery and equipment investment gap, and a patents gap, relative to a number of other advanced industrial countries, particularly the United States. While it is relatively easy to identify gaps in innovation performance, the gaps raise two issues. First, is the gap important? If the gap has little effect on key economic variables such as productivity performance and product quality, then it may not be a concern. Second, if the gap is indeed of import, to what degree is it amenable to public policy? If the gap is related to structural factors over which public policy has little or no influence, then the gap should not be a policy concern.

This means that understanding the causes of the gap is crucial. But this is difficult and experts disagree. For example, despite decades of research, there is still no consensus on why Canada's business sector performs proportionately much less R&D than the business sector in many other OECD countries. While public policy initiatives should not await definitive results on this issue, which may never arrive, they should at a minimum be informed by and make use of the most comprehensive and up-to-date information on the causes of the gaps.

4.2 Types of Innovation Meriting Public Support

A second public policy issue is whether government has an interest in fostering all types of innovation. The case for public policy action to support private sector innovation efforts may lie more in the area of process innovation that increases productivity than product innovation designed to increase market share. However, if product innovation is defined in a broader context, there may be a case for promoting the development of new products that improve health care, foster distance learning and enhance quality of life. Of course, this raises a question about the market aspect of product innovation.

4.3 Focus of Innovation Strategy

A third public policy issue related to innovation strategy involves its focus. Should such a strategy foster process innovation in all industries, including traditional sectors that are considered not particularly innovative, or should it focus on providing support for the high-tech sector. In other words, is there more "bang for the buck" in supporting the widespread diffusion of technology into all sectors of the economy or in concentrating resources on support of research and development and other inputs to innovation in the high-tech sectors? This is an issue of priority setting and resource allocation.

A second element of focus concerns the relative importance given basic and applied research and development. In the United States, public support for innovation focuses on basic research, with the private sector responsible for the development and application of the results of this basic research. This policy has appeared to work very well, as the United States is by most measures the most innovative country in the world. However, it is unclear whether such a strategy is the most appropriate for Canada, a country that produces only a very small proportion of the world's supply of basic research. It may make more sense for governments in Canada to focus their limited financial resources on assisting the private sector in the commercialization of basic research, and free-riding on the results of the basic research done in other countries, except in areas where Canada is a leader.

4.4 Policy Instruments for an Innovation Strategy

The specific nature of the policy instruments to promote innovation is obviously a key public policy issue. The federal government already devotes significant resources to innovation, including: direct federal research and development activities through such institutions as the National Research Council; the support of university research through the granting councils, the Canada Research Chairs program, and the Canadian Foundation for Innovation; and financial support to the private sector to finance its R&D and other innovation-related activities through direct subsidies and tax incentives. Much work has already been done on what policies work best, but additional insight continues to develop.

4.5 The Need for a Broad Perspective on Innovation

Innovation is not just an economic phenomenon. It concerns people, culture and context. In the past, the federal government has adopted a board perspective on innovation and innovation policy. For example in 1997, it emphasized that "Innovation is critical not only to economic success, but also to preserving the environment and renewing social fabric" (Industry Canada 1997b: 36). It is important that this multi-dimensional approach to the definition, determinants, and impact of innovation continue and that innovation be more than a preoccupation of economic policy.

5. CONCLUDING REMARKS

Innovation is high on the federal government's policy agenda. The development of innovation policy requires a detailed understanding of the innovation process. This in turn requires that the extent of innovative activity be accurately measured through innovation surveys and that the results of the surveys be rigorously analyzed. The volume accomplishes this task well. The chapters document the nature of the measurement of innovation in Canada, as captured by the Survey of Innovation 1999 conducted by Statistics Canada, and shed light on the characteristics of innovative firms, the determinants of innovation, and the impact of innovation on firm performance. Consequently, the volume makes a very important contribution to public policy, as the new knowledge it presents is essential for the development of effective innovation policy in Canada.

Notes

[1]According to official Statistics Canada estimates, output per hour in the total construction sector in 2000 was $22.61 in 1992, compared to $23.34 in 1980. See Sharpe (2001) for a detailed analysis of productivity trends in the construction sector.

Bibliography

Bibliography

Abernathy, William and Kim B. Clark. 1985. "Innovation: Mapping the Winds of Creative Destruction." *Research Policy*. 14, 1 (February): 3-22.

Acs, Zoltan J., ed. 2000. *Regional Innovation, Knowledge and Global Change*. New York. Pinter.

Acs, Zoltan J. and David B. Audretsch. 1988. "Innovation in Large and Small Firms: An Empirical Analysis." *American Economic Review*. 78, 4: 678-90.

Acs, Zoltan J. and David B. Audretsch. 1990. *Innovation and Small Firms*. Cambridge, USA. MIT Press.

Acs, Zoltan J. and Sharon Gifford. 1996. "Innovation of Entrepreneurial Firms." *Small Business Economics*. 8, 3: 203-18.

Acs, Zoltan J., David B. Audretsch and Maryann P. Feldman. 1994. "R&D Spillovers and Innovative Activity." *Managerial and Decision Economics*. 15, 2 (May): 678-90.

Amable, B., R. Barré and R. Boyer. 1997. *Les systèmes d'innovation à l'ère de la globalisation*. Paris. Economica.

Anderson, Frances and André Manseau. 1999. "A Statistical Approach to the Measurement of Innovation in Construction." Presented at the Third International Conference on Technology Policy and Innovation. Austin, Texas. August 30 - September 2.

Anderson, Frances and Susan Schaan. 2001. *Innovation, Advanced Technologies and Practices in the Construction and Related Industries: National Estimates*. Catalogue No. 88F0006XIB2001004. Ottawa. Statistics Canada.

Anderson, Frances, Cathy Bakker and André Manseau. 2002. "Does Organizational Innovation Provide Competitive Advantage in the Construction Industry?" Paper presented at Statistics Canada's Economic Conference 2002. Ottawa. May 6-7.

Archibugi, Daniele and Mario Pianta. 1996. "Measuring Technological Change Through Patents and Innovation Surveys." *Technovation*.16, 9: 451-68.

Armour, Henry O. and David J. Teece. 1980. "Vertical Integration and Technological Innovation." *Review of Economics and Statistics*. 62, 3 (Aug): 470-74.

Arthur, W. Brian. 1990. "Silicon Valley Locational Clusters: When Do Increasing Returns Imply Monopoly." *Mathematical Social Sciences*. 19, 3: 235-51.

Association of Consulting Engineers of Canada (ACEC). 1998. *Canadian Consulting Engineering Industry Profile*. Ottawa.

Association of Consulting Engineers of Canada (ACEC). 1999. *Canadian Consulting Engineering Industry Profile*. Ottawa.

Athey, Susan and Scott Stern. 1998. *An Empirical Framework for Testing Theories About Complementarity in Organizational Design*. Cambridge, USA. National Bureau of Economic Research (NBER) W6600.

Audretsch, David B. 1995. "Firm Profitability, Growth and Innovation." *Review of Industrial Organization*. 10: 579-88.

Australian Bureau of Statistics. 1998. *Innovation in Mining*. (Australian Bureau of Statistics Catalogue No. 8121.0, pp 16.) Canberra.

Baldwin, John R. 1997. *Innovation and Intellectual Property*. Catalogue No. 88-515-XPE. Ottawa. Statistics Canada.

Baldwin, John R. 1997. *The Importance of Research and Development for Innovation in Small and Large Canadian Manufacturing Firms*. Catalogue No. 11F0019MIE1997107. Ottawa. Statistics Canada.

Baldwin, John R. and Petr Hanel. 2003. *Innovation and Knowledge Creation in an Open Economy*. Cambridge UK. Cambridge University Press.

Baldwin, John R. and David Sabourin. 1996. "Technology and Competitiveness in Canadian Manufacturing Establishments." *Canadian Economic Observer*. 9, 5: 3.1-3.15. Catalogue No. 11-010-XPB. Ottawa. Statistics Canada.

Baldwin, John R., Petr Hanel and David Sabourin. 2000. *Determinants of Innovative Activity in Canadian Manufacturing Firms: The Role of Intellectual Property Rights*. Cataloque No. 11F0019MPE2000122. Ottawa. Statistics Canada.

Baldwin, John R., Ed Rama and David Sabourin. 1999. *Growth of Advanced Technology Use in Canadian Manufacturing During the 1990's*. Catalogue No. 11F0019MIE1999105. Ottawa. Statistics Canada.

Baldwin, John R., Guy Gellatly, Joanne Johnson and Valerie Peters. 1998. *Innovation in Dynamic Service Industries*. Catalogue No. 88-516-XPB. Ottawa. Statistics Canada.

Baptista, Rui and Peter Swann. 1998. "Do Firms in Clusters Innovate More?" *Research Policy*. 27, 5 (September): 525-40.

Bartel, Ann P. and Frank Lichtenberg. 1987. "The Comparative Advantage of Educated Workers in Implementing New Technology." *Review of Economics and Statistics*. LXIX(69), 1: 1-11.

Berman, Eli, John Bound and Zvi Griliches. 1994. "Changes in the Demand for Skilled Labor Within U.S. Manufacturing: Evidence from the Annual Survey of Manufacturers." *Quarterly Journal of Economics*. CIX, 2: 367-397.

Booz, Allen and Hamilton. 1982. *New Product Management for the 1980s*. New York. Booz, Allen and Hamilton Inc.

Bound, John, Clint Cummins, Zvi Griliches, Bronwyn H. Hall and Adam Jaffe. 1984. "Who Does R&D and Who Patents?" In *R&D, Patents and Productivity*. Z. Griliches (ed). Chicago. University of Chicago Press for the National Bureau of Economic Research. 21-54.

Braczyk, Hans-Joachim., Philip Cooke and Martin Heidenreich, eds. 1999. *Regional Innovation Systems: The Role of Governance in a Globalized World*. London. UCL Press.

Brouwer, Erik and Alfred Kleinknecht. 1996. "Firm Size, Small Business Presence and Sales of Innovative Products: A Micro-Econometric Analysis." *Small Business Economics*. 8, 3 (June): 189-201.

Burt, Ronald S. 1992. *Structural Holes: The Social Structure of Competition*. Boston. Harvard University Press.

Burt, Ronald S. 2000. "Contingent Value of Social Capital." In *Knowledge and Social Capital: Foundations and Applications*. E. Lesser (ed). Boston. Butterworth Heinemann. 255-86

Cohen, Wesley M. and Steven Klepper. 1996. "A Reprise of Size and R&D." *Economic Journal* 106, (July): 925-51.

Cohen, Wesley M., and Richard C. Levin. 1989. "Empirical Studies of Innovation and Market Structure." In *Handbook of Industrial Organization. Vol. II*. Richard Schmalensee and Robert D. Willig (eds.). Amsterdam and New York: North-Holland. 1059-1107.

Cohen, Wesley M. and David A. Levinthal. 1989. "Innovation and Learning: The Two Faces of R&D." *The Economic Journal*. 99, (September): 569-96.

Cohen, Wesley M. and David A. Levinthal. 1990. "Absorptive Capacity: A New Perspective on Learning and Innovation." *Administrative Science Quarterly*. 35, 1 (March): 128-52.

Cohen, Wesley M., Richard C. Levin and David C. Mowery. 1987. "Firm Size and R&D Intensity: A Re-examination." *Journal of Industrial Economics*. 35, 4 (June): 543-63.

Cohen, Wesley M., Richard R. Nelson and John P. Walsh. 2000. *Protecting Their Intellectual Assets: Appropriability Conditions and Why U.S. Manufacturing Firms Patent (or Not)*. Cambridge USA. NBER Working Paper No. 7552. February.

Collecchia, Alessandra and George Papaconstantinou. 1996. *The Evolution of Skills in OECD Countries and the Role of Technology*. STI Working Papers 1996-97. (OECD/GD(96)183). Paris. OECD.

Conference Board of Canada, The. 2000. *2nd Annual Innovation Report: Collaborating for Innovation*. Ottawa.

Conference Board of Canada The. 2001a. "Investing in Innovation in the Resource Sector: Industry Needs." In *Barriers and Opportunities for Action*. Ottawa.

Conference Board of Canada, The. 2001b. *Investing in Canada: 3rd Annual Innovation Report* 336-01 Detailed Findings. Ottawa.

Cooke, Philip, Patries Boekholt and Franz Tödtling, eds. 2000. *The Governance of Innovation in Europe: Regional Perspectives on Global Competitiveness*. New York. Pinter.

Coombs, R., P. Saviotti and V. Walsh, eds. 1996. *Technological Collaboration*. Cheltenham. Elgar Publishers.

Cozzarin, Brian. 2001. "Innovation Quality and Firm Performance." Unpublished report prepared for Industry Canada. Ottawa.

Crépon Bruce, Emmanuel Duget and Jacques Mairesse. 1998. "Research, Innovation, and Productivity: An Econometric Analysis at the Firm Level." *Economics of Innovation and New Technology*. 7, 2: 115-58.

Dahl, Michael and Bent Dalum. 2001. "The Construction Cluster in Denmark." In *Innovative Clusters: Drivers of National Innovation Systems*. Paris. OECD. 179-202.

De Bresson, Christian. 1980. "The Direct Measurement of Innovation." The Science and Technology Indicators Conference. Paris. OECD. September 15-19.

de la Mothe, John. 2001. "Knowledge, Learning and Innovation Policy." In *Knowledge Management in the Innovation Process*. John de la Mothe and Dominique Foray (eds.) Boston. Kluwer Academic Publishers. 205-13.

de la Mothe, John and Albert N. Link, eds. 2002. *Networks, Alliances and Partnerships in the Innovation Process*. Boston. Kluwer Academic Publishers.

de la Mothe, John and Gilles Paquet, eds. 1998. *Local and Regional Systems of Innovation*. Boston. Kluwer Academic Publishers.

De Melto, Denis P., Kathryn E. McMullen and R.M. Wills. 1980. *Preliminary Report: Innovation and Technological Change in Five Canadian Industries*. Discussion Paper No. 176. Ottawa. Economic Council of Canada.

den Hertog, Pim and Erik Brouwer. 2001. "Innovation in the Dutch Construction Cluster." In *Innovative Clusters: Drivers of National Innovation Systems*. Paris. OECD. 203-28.

Dixon, A.J. and H.R. Seddighi. 1996. "An Analysis of R&D Activities in North East England Manufacturing Firms: The Results of a Sample Survey." *Regional Studies*. 30, 3 (June): 287-94.

Dodgson, Mark. 1991. *The Management of Technological Learning: Lessons from a Biotechnology Company*. Berlin. Walter & Gruyter.

Dosi, Giovanni. 1982. "Technological Paradigms and Technological Trajectories: a Suggested Interpretation of the Determinants and Directions of Technical Change." *Research Policy*. 11, 3: 147-62.

Dosi, Giovanni. 1996. "Employment and Growth in the Knowledge-Based Economy." In *The Contribution of Economic Theory to the Understanding of a Knowledge-Based Economy*. Paris. OECD. 81-92.

Earl, Louise. 2002. *Knowledge Management in Practice in Canada, 2001*. Catalogue No. 88F00036XIE2003007. Ottawa. Statistics Canada

Edquist, Charles. ed. 1997. *Systems of Innovation: Technologies, Institutions and Organizations*. London. Pinter.

Edquist, Charles and Leif Hommen. 1999. "Systems of Innovation: Theory and Policy for the Demand Side." *Technology in Society*. 21, 1: 63-79.

European Commission. 2001. *Building an Innovative Economy in Europe: A Review of 12 Studies of Innovation Policy and Practice in Today's Europe*. Luxembourg.

Evangelista, Rinaldo, Guilio Perani, Fabio Rapiti and Daniele Archibugi. 1997. "Nature and Impact of Innovation in the Manufacturing Industry: Some Evidence from the Italian Innovation Survey." *Research Policy*. 26, 4-5 (December): 521-36.

Feldman, Maryann P. 1994. "Why Location Matters for Innovative Activities." Chapter 2 in *The Geography of Innovation*. Dordrecht. The Netherlands. Kluwer Academic Publisher XX. 13-29.

Foray, Dominique. 1998. *Comment mesurer l'économie de la connaissance? Un cadre analytique*. Working Paper IMRI–WP 9810. Paris. Université Paris Dauphine.

Foray, Dominique. 2000. "Characterizing the Knowledge Base: Available and Missing Indicators." In *Knowledge Management in the Learning Society*. Paris. OECD. 239-57.

Foray, Dominique and Paul David. 1995. "Accessing and Expanding the Science and Technology Knowledge Base." *STI Science, Technology, Industry Review*. 16: 13-68.

Foyn, Frank. 1999. "Community Innovation Survey 1997-1998." *Statistics in Focus*. Eurostat. Theme 9. 2/1999.

Foyn, Frank. 2000. "Community Innovation Survey 1997-1998." *Statistics in Focus*. Eurostat. Theme 9. 2/2000.

Freeman, Christopher. 1987. *Technology Policy and Economic Performance: Lessons from Japan*. London. Pinter Publishers.

Friedman, David D., William M. Landes and Richard A. Posner. 1991. "Some Economics of Trade Secret Law." *Journal of Economic Perspectives*. 5, 1: 61-72.

Gault, Fred and Daood Hamdani. 2000. "Measuring Novelty of Innovation: Additional Innovation Indicators." Presented at the Conference on the Economics and Socio-economics of Services: International Perspective. Lille-Roubaix, France. June 22-23.

Gault, Fred and Wendy Hansen. 1999. "Knowledge Flows in the S&T System." In *Innovation and Structural Change in Post-Socialist Countries: A Quantitative Approach*. D.A. Dyker and S. Radosevic (eds.). Boston. Kluwer Academic Publishers. 411-17.

Gibbons, Michael and Ron Johnston. 1974. "The Roles of Science in Technological Innovation." *Research Policy*. 3, 3 (November): 220-42.

Government of Canada. 1996. *Science and Technology for a New Century: A Federal Strategy*. Ottawa.

Government of Canada. 2002a. *Achieving Excellence: Investing in People, Knowledge and Opportunity*. Ottawa. Industry Canada.

Government of Canada. 2002b. *Knowledge Matters: Skills and Learning for Canadians*. Ottawa. Human Resources Development Canada.

Government of Canada. 2002c. *Canadians Speak on Innovation and Learning*. Ottawa.

Government of Canada. 2002d. *National Summit on Innovation and Learning: Summary*. Ottawa

Granovetter, Mark S. 1973. "The Strength of Weak Ties." *American Journal of Sociology*. 78, 6 (May): 1360-80.

Greene, William H. 1997. *Econometric Analysis*. Third Edition. New York. Prentice Hall.

Greene, William H. 2000. *Econometric Analysis*. Fourth Edition. New Jersey. Prentice Hall.

Griliches, Zvi 1986. "Productivity, R&D and Basic Research at the Firm Level in the 1970s." *American Economic Review*. 76 (March): 141-54.

Gu, Wulong and Jianmin Tang. 2001. "A New Perspective on the Canadian Manufacturing Innovation-Productivity Linkage." Unpublished manuscript. Ottawa. Statistics Canada and Industry Canada.

Guellec, Dominique and William Pattinson. 2001. "Innovation Surveys: A Few Lessons from OECD Countries' Experience." *STI Review: Special Issue on New Science and Technology Indicators*. Paris. OECD. 27: 77-102.

Hair, Joseph F. Jr., Rolf E. Anderson, Ronald L. Tatham (contributor), William C. Black and Ralph E. Anderson (contributor). 1995. *Multivariate Data Analysis with Readings*. Fourth Edition. Englewood Cliffs. Prentice Hall.

Hamdani, Daood. 1996. "The Temporary Help Service Industry: Its Role, Structure and Growth." *Services Indicators*. 3, 1: 73-91. Catalogue No. 63-016-XPB. Ottawa. Statistics Canada.

Hamdani, Daood. 1998. "Innovation and Labour Skills: A Study of the Consulting Engineering Industry." Presented at the Workshop on Conceptualizing and Measuring Innovation in Services, at the Centre for Research on Innovation and Competitiveness. University of Manchester. May.

Hamdani, Daood. 1999. "Measuring Innovation: The Canadian Experience." Unpublished manuscript. Ottawa. Statistics Canada.

Hamdani, Daood. 2000. "Why Do the Surveys on Innovation and R&D Diverge?" *Innovation Analysis Bulletin*. Vol. 2, No. 3 (September): 8-9. Catalogue No. 88-003-XIE. Ottawa. Statistics Canada.

Hamdani, Daood. 2001. *Capacity to Innovate, Innovation and Impact: The Canadian Engineering Services Industry*. Research Paper No. 11, Catalogue 88F0017MIE. Ottawa. Statistics Canada.

Hanel, Petr. 2001. "Current Intellectual Protection Practices by Manufacturing Firms in Canada." Presented at the International Conference on Intellectual Property and Innovation in the Knowledge-Based Economy, Toronto. May 23-4.

Hansen, Morton H., Nitin Nohria and Thomas Tierney. 1999. "What's Your Strategy for Managing Knowledge?" *Harvard Business Review*. 77, 2 (March-April): 106-16.

Hansen, Wendy. 1997. "Developing Indicators for a Knowledge-Based Society: Human Capital." Paper presented at the NATO Advanced Workshop on Quantitative Studies for S&T Policy for Economies in Transition. October 23-25. Moscow.

Hatzichronoglou, T. 1997. *Revision of the High-Technology Sector and Product Classification*. STI Working Papers. Paris. OECD.

Holbrook, J.Adam D. and L.P. Hughes. 2000. "Innovation in Enterprises in British Columbia." In *Local and Regional Systems of Innovation*. John de la Mothe and Gilles Paquet (eds.). Boston. Kluwer Academic Publishers. 173-90.

Holbrook, J.Adam D. and David A. Wolfe, eds. 2000. *Innovation, Institution and Territory: Regional Innovation Systems in Canada*. Montreal and Kingston. McGill-Queen's University Press.

Holbrook, J.Adam D. and David A. Wolfe, eds. 2002. *Knowledge, Clusters and Regional Innovation*. Montreal and Kingston. McGill-Queen's University Press.

Holmes, J., P. Hutton and E. Weber. 1991. "A Functional-Form-Free Test of the Research and Development Firm Size Relationship." *Journal of Business and Economics Statistics*. 9, 1 (January): 85-90.

Horibe, F. 2001. "The Efficiency-Innovation Dichotomy." In *Creating the Innovation Culture: Leveraging Visionaries, Dissenters and Other Useful Troublemakers in Your Organization*. Toronto. John Wiley and Sons. 9-21.

Industry Canada. 1997a. *Sector Competitiveness Framework Series, Consulting Engineering Industry.* http:strategis.ic.gc.ca

Industry Canada. 1997b. *Minding our Future: A Report on Federal Science and Technology – 1997.* Ottawa.

Industry, Science and Technology Canada. 1989. *Appendices for the Survey of Intellectual Property Rights in Canada, Final Report*. Ottawa.

Innis, Harold A. 1988. "The Fur Trade." (First published in 1956). In *Approaches to Canadian Economic History: A Selection of Essays*. W.T. Easterbrook and M.H. Watkins E (eds.). Ottawa. Carleton University Press. 20-27.

Johnson, Bjorn. 1995. "Institutional Learning." In *National Systems of Innovation*. Bengt-Åke. Lundvall, (ed.). London. Pinter. 23-44.

Jöreskog, Karl G. and Dag Sörbom. 1984. *LISREL VI: Analysis of linear structural relationships by maximum likelihood, instrumental variables, and least squares methods*. Mooresville, Ind. Scientific Software, Inc.

Jöreskog, Karl G.and Dag Sörbom. 1996. *LISREL 8: User's Reference Guide*. Mooresville. Ind. Scientific Software International, Inc.

Jöreskog, Karl G. and Dag Sörbom. 1999. *LISREL 8: Structural Equation Modeling with the SIMPLIS Command Language*. Mooresville. Ind. Scientific Software International, Inc.

Kenney, M. 2000. "A Note on the Comparison Between Cambridge, England and Silicon Valley, Berkeley." Roundtable on the International Economy (BRIE). Research Notes #6. Berkeley. U.S.A. University of California.

Kleinknecht, Alfred. ed. 1996. *Determinants of Innovation: The Message from New Indicators*. London. Macmillan and New York. St. Martin's Press.

Kline, Stephen J. and Nathan Rosenberg. 1986. "An Overview of Innovation." In *The Positive Sum Strategy: Harnessing Technology for Economic Growth*. R. Landau and N. Rosenberg (eds.). Washington, DC. National Academy Press. 275-306.

Klomp, Luuk and George van Leeuwen. 2001. "Linking Innovation and Firm Performance: A New Approach." *International Journal of the Economics of Business*. 8: 343-64.

Kremp, Elisabeth and Jacques Mairesse. 2003. "Knowledge Management, Innovation and Productivity: A Firm Level Exploration Based on the

French CIS3 Data." In *Measuring Knowledge Management in the Business Sector: First Steps*. Dominique Foray and Fred Gault (eds.). Paris. OECD. 135-158.

Laestadius, Staffan. 1998. "The Relevance of Science and Technology Indicators: The Case of Pulp and Paper." *Research Policy*. 27, 4 (August): 385-95.

Landry, Réjean. 1996. "Enquête sur les capacités technologiques et les besoins des entreprises innovantes de la région Chaudière-Appalaches." Unpublished report prepared for l'Antenne technologique Chaudière-Appalaches.

Landry, Réjean. 1998. "L"innovation de produits et de procédés de fabrication dans les entreprises manufacturières de l'Estrie: état de la situation et pistes d'action." Unpublished report prepared for le Groupe d'action technologique de l'Estrie [GATE].

Landry, Réjean and Nabil Amara. 1998. "The Chaudière-Appalaches System of Industrial Innovation." In *Local and Regional Systems of Innovation*. John de la Mothe and Gilles Paquet. (eds.). Boston. Kluwer Academic Publishers. 257-76.

Landry, Réjean and Nabil Amara. 2001. "Creativity, Innovation and Business Practices in the Matter of Knowledge Management." In *Knowledge Management in the Innovation Process*. John de la Mothe and Dominique Foray (eds.). Boston. Kluwer Academic Publishers. 55-79.

Landry, Réjean and Pierre Fréchette. 1998. "L"innovation dans les entreprises manufacturières de la région Québec-Chaudière-Appalaches." Unpublished report prepared for le Groupe d'action technologique et industriel de la région de Québec-Chaudière-Appalaches [GATICA].

Landry, Réjean, Moktar Lamari and Nabil Amara. 2000a. "Bilan technologique de la Mauricie: analyse des résultats de l'enquête quantitative auprès des entreprises manufacturières." Unpublished study carried out for la Technopole de la Vallée du St-Maurice. 102 p.

Landry, Réjean, Moktar Lamari and Nabil Amara. 2000b. "Innovation et technologies avancées dans les entreprises manufacturières de la Montérégie: bilan et pistes d'action." Unpublished study carried out for Valo Tech Inc. 221 p.

Landry, Réjean, Moktar Lamari and Nabil Amara. 2002. "Does Social Capital Determine Innovation? To What Extent?" *Technological Forecasting and Social Change*. 69, 7: 681-701.

Lavoie, Marie and Richard Roy. 1998. *Employment in the Knowledge-Based Economy: A Growth Accounting Exercise for Canada*. Research Paper # R-98-8E. Ottawa. Human Resources Development Canada.

Lavoie, Marie and Pierre Therrien. 1999. *Employment Effects of Computerization, 1971-1991*. Research Paper # R-99-2E. Ottawa. Human Resources Development Canada.

Le, Can D. 2001. *A Profile of the Canadian Manufacturing High-Technology Industries*. Research Paper. Ottawa. Industry Canada.

Le, Can D. and Jianmin Tang. 2001. "Why Does Canada Spend Less on R&D Than Its Key Trade Competitors?" Unpublished manuscript. Ottawa. Industry Canada.

Leiponen, Aija. 2000a. "Collaboration, and Firm Performance, Increasing Returns from Knowledge Complementarities?" Mimeo. Cirano conference, June 2000.

Leiponen, Aija. 2000b. "Competencies, Innovation and Profitability of Firms." *Economics of Innovation and New Technology*. 9: 1-24.

Lengrand, Louis and Isabelle Chatrie. 1999. *Business Networks and the Knowledge-Driven Economy*. Brussels. European Commission.

Levin, Richard C., Wesley M. Cohen and David C. Mowery. 1985. "R&D Appropriability, Opportunity, and Market Structure: New Evidence on Some Schumpeterian Hypotheses." *American Economic Review*. 75, 2: 20-24.

Levin, Richard C., Alvin K. Klevorick, Richard R. Nelson and Sidney G. Winter. 1987. "Appropriating the Returns from Industrial Research and Development." *Brookings Papers on Economic Activity*. 1987, 3: 783-831.

Link, Albert N. 1980. "Firm Size and Efficient Entrepreneurial Activity: A Reformulation of the Schumpeter Hypothesis." *Journal of Political Economy*. 88, 4 (August): 771-82.

Link, Albert N. 1981. *Research and Development in U.S. Manufacturing*. New York. Praeger.

Loeb, Peter D. 1983. "Further Evidence of the Determinants of Industrial Research and Development Using Single and Simultaneous Equation Models." *Empirical Economics*. 8: 203-14.

Lööf, H. and A. Heshmati. 2003. "On the Relationship Between Innovation and Performance: A Sensitivity Analysis." *Economics of Innovation and New Technology*. (forthcoming).

Lucas, Robert E. 1988. "On the Mechanics of Economic Development." *Journal of Monetary Economics*. 22, 1 (July): 3-42.

Lundvall, Bengt-Åke. 1988. "Innovation as an Interactive Process: From User-Producer Interaction to the National System of Innovation." In *Technical Change and Economic Theory*. Giovanni Dosi, Christopher Freeman, Richard Nelson and Luc Soete (eds.). London. Pinter. 349-69.

Lundvall, Bengt-Åke. 1992a. "Explaining Interfirm Cooperation: The Limits of Transaction Cost Approach." In *The Embedded Firm: On the Socioeconomics of Industrial Networks*. Gernot Grabher (ed.). London. Routledge. 52-64.

Lundvall, Bengt-Åke. 1992b. "Introduction, National Systems of Innovation: Towards a Theory of Innovation and Interactive Learning." In *The Political Economy of Science, Technology and Innovation*. Ben R. Martin and Paul Nightingale (eds.). Cambridge. UK. Cambridge University Press. 524-43.

Lundvall, Bengt-Åke. ed. 1995. *National Systems of Innovation*. London. Pinter.

Lundvall, Bengt-Åke. 1998. "Why Study National Systems and National Styles of Innovation?" *Technology Analysis and Strategic Management*. 10: 407-21.

MacDonald, Paul and Michael Clow. 1999. "Just One Damn Machine After Another: Situating Technological Innovation in the Industrialization of Tree Harvesting Systems." *Technology in Society*. 21, 3 (August): 323-44.

Maddala, G.S. 1983. *Limited-dependent and Qualitative Variables in Econometrics*. Cambridge University Press. Cambridge.

Maillat, D. 1995. *Systèmes territoriaux de production, milieux innovateurs et politiques régionales*. Switzerland. Université de Neuchâtel. Unpublished manuscript.

Mairesse, Jacques and Pierre Mohnen. 2001. "To Be or Not to Be Innovative: An Exercise in Measurement." *STI Review: Special Issue on New Science and Technology Indicators*. Paris. OECD. 27: 103-29.

Malerba, Franco. 1992. "Learning by Firms and Incremental Technical Change." *Economic Journal*. 102, 413: 845-59.

Mardia, Kanti V. 1980. "Tests of Univariate and Multivariate Normality." In *Handbook of Statistics*. Vol. 1, Paruchuri R. Krishnaiah (ed.). Amsterdam and New York. North-Holland. 279-320.

Marshall, Alfred. 1890. *Principles of Economics*. London. Macmillan.

Maxwell, Judith. 2003. *Innovation is a Social Process*. Catalogue No. 88F0006XIE2003006, Ottawa: Statistics Canada.

McFetridge, Donald G. 1993. "The Canadian System of Industrial Innovation." In *National Innovation Systems: A Comparative Analysis*. Richard Nelson (ed). New York. Oxford University Press. 299-323.

McNiven, Chuck, Lara Raoub and Namatié Tratoré. 2003. *Features of Canadian Biotechnology Innovative Firms: Results from the Biotechnology Use and Development Survey, 2001*. Catalogue No. 88F0006XIE2003005. Ottawa. Statistics Canada.

Meisel, J.B. and S.A.Y. Lin. 1983. "The Impact of Market Structure on the Firm's Allocation of Resources to Research and Development." *Quarterly Review of Economics and Business*. 23: 28-43.

Melin, Göran. 2000. Pragmatism and Self-Organization: Research Collaboration on the Individual Level. *Research Policy*. 29, 1 (January): 31-40.

Mohnen, Pierre and Marcel Dagenais. 2002. "Towards an Innovation Intensity Index: The Case of CIS-I in Denmark and Ireland." In *Innovation and Firm Performance: Econometric Explorations of Survey Data*. Alfred Kleinknecht and Pierre Mohnen (eds.). Hampshire and New York. Palgrave. 3-30.

Mohnen, Pierre and Pierre Therrien. 2001. "Comparison of Innovative Performance by Canadian and European Firms: Explaining the Differences." Mimeo. CIRANO and Industry Canada.

Mohnen, Pierre, Jacques Romain and Jean-Sébastian Gallant. 1996. "Productivity and Research and Development in Two Canadian Forest Product Industries." *Forest Science*. 42, 4 (November): 487-97.

Nahapiet, Janine and Sumantra Ghoshal. 1998. "Social Capital, Intellectual Capital, and the Organizational Advantage." *Academy of Management Review*. 23, 2: 242-66.

Nauwelaers, Claire and René Wintjes. 2000. *SME Policy and the Regional Dimension of Innovation: Towards a New Paradigm for Innovation Policy?* Working Paper mer00023. The Netherlands. Maastricht Economic Research Institute on Innovation and Technology (MERIT).

Nelson, Richard. ed. 1993. *National Innovation Systems: A Comparative Analysis*. New York. Oxford University Press.

Nelson, Richard and Edmund Phelps. 1966. "Investment in Humans, Technological Diffusion and Economic Growth." *American Economic Review*. 61, 2: 69-75.

Nelson, Richard R. and Sidney G. Winter. 1982. *An Evolutionary Theory of Economic Change*. Cambridge USA. Belknap Press.

Niosi, Jorge. 2000. *Canada's National System of Innovation*. Montreal and Kingston. McGill-Queen's Press.

Niosi, Jorge, Paolo Saviotti, Bertrand Bellon and Michael Crow. 1993. "National Systems of Innovation: in Search of a Workable Concept." *Technology in Society*. 15, 2: 207-27.

Nonanka, Kujiro and Hirotaka Takeuchi. 1995. *The Knowledge Creating Company*. New York. Oxford University Press.

Nooteboom, Bart. 1999. "Innovation and Inter-Firm Linkages: New Implications for Policy." *Research Policy*. 28, 8 (November): 793-805.

OECD (Organisation for Economic Co-operation and Development). 1992. *Proposed Guidelines for Collecting and Interpreting Technological Innovation Data: Oslo Manual*. (OECD/GD(92)26). Paris.

OECD. 2000a. *A New Economy? The Changing Role of Innovation and Information Technology in Growth*. Paris.

OECD. 2000b. *Role of Competition and Cooperation in Innovation and Growth*. Paris.

OECD/Eurostat. 1997. *Oslo Manual: Proposed Guidelines for Collecting and Interpreting Technological Innovation Data*. Paris.

Oerlemans, Leon A.G., Marius T.H. Meeus and Frans W.M. Boekema. 1998. *Learning, Innovation and Proximity*. ECIS Working Paper 98/3. Eindhoven. Faculty of Technology Management. Eindhoven University of Technology.

Okubo, Yoshiko and Cecilia Sjoberg. 2000. "The Changing Pattern of Industrial Scientific Research Collaboration in Sweden." *Research Policy*. 29, 1 (January): 81-98.

Padmore, Tim, Hans Schuetze and Hervey Gibson. 1998. "Modelling Systems of Innovation: An Enterprise-Centred View." *Research Policy*. 26, 6 (February): 605-24.

Pianta, Mario and Giorgio Sirilli. 1998. *The Use of Innovation Surveys for Policy Evaluation*. IDEA Paper Series. Oslo. No 2. STEP Group (Studies in Technology, Innovation and Economic Policy).

Porter, Michael E. 1990. *The Competitive Advantage of Nations*. New York. Free Press.

Porter, Michael E. 1998. *On Competition*. Boston. Harvard Business School Publishing.

Porter, Michael. 1999. "Clusters and the New Economics of Competition." *Harvard Business Review*. 77, 6 (Nov.-Dec): 77-90.

Porter, Michael. 2000. "Location, Competition and Economic Development: Local Clusters in a Global Economy." *Economic Development Quarterly*. 14, 1: 15-34.

Powell, Walter W. 1990. "Neither Market Nor Hierarchy: Network Forms of Organization." *Research in Organizational Behavior*. 12: 295-336.

Prahalad, Coimbatore K. and Gary Hamel. 1990. "The Core Competence of the Corporation." *Harvard Business Review*. 68, 3 (May-June): 79-91.

Quah, Daniel T. 1995. *Regional Convergence Clusters Across Europe*. Discussion Paper No. 274. London. Centre for Economic Performance.

Quinn, James Brian, Thomas L. Dooley and Penny C. Paquette. 1991. "The Intellectual Holding Company: Structuring Around Core Activities." In *The Strategy Process*. Henry Mintzberg and James Brian Quinn (eds.). Englewood Cliffs. N.J. Prentice Hall.

Rafiquzzaman, Mohammed and Lorie Whewell. 1999. *Trends in Cross-Country Innovative Performance and the Determinants of International Patenting Activities*. Ottawa. Industry Canada.

Rao, Someshwar, Ashfaq Ahmad, William Horsman and Phaedra Kaptein-Russell. 2001. "The Importance of Innovation for Productivity." *International Productivity Monitor*. 2: 11-18.

Robson, M., J. Townsend and K. Pavitt. 1988. "Sectoral Patterns of Production and Use of Innovations in the UK: 1945-1983." *Research Policy*. 17, 1 (February): 1-14.

Romer, Paul. 1986. "Increasing Returns and Long-Run Growth." *Journal of Political Economy*. 94, 5: 1002-1037.

Romer, Paul 1990. "Endogenous Technological Change." *Journal of Political Economy*. 98, 5.2: S71-S102.

Romijn, H. and M. Albu. 2001. *Explaining Innovativeness in Small High-Technology Firms in the United Kingdom*. Working Paper 01-01. Eindhoven Centre for Innovation Studies. Eindhoven University of Technology.

Rosenberg, Nathan. 1982. *Inside the Black Box: Technology and Economics*. New York. Cambridge University Press.

Rothwell, Roderick. 1977. "The Characteristics of Successful Innovators and Technically Progressive Firms." *R&D Management*. 7, 3: 191-206.

Rothwell, Roderick. 1994. "Industrial Innovation: Success, Strategy, Trends." In *The Handbook of Industrial Innovation*. Mark Dodgson and Roderick Rothwell (eds.). Hants UK. Edward Elgar. 33-53.

Sabourin, David and Desmond Beckstead. 1999. *Technology Adoption in Canadian Manufacturing: Survey of Advanced Technology in Canadian Manufacturing, 1998*. Catalogue No. 88F0006XIB1999005. Ottawa. Statistics Canada.

Sandven, Tore. 2000. *Innovation and Economic Performance at the Enterprise Level*. Oslo. STEP Group (Studies in Technology, Innovation and Economic Policy). February.

Sandven, Tore and K. Smith. 1998. *Using Community Innovation Survey Data for Empirical Analysis: Data Reliability and Issues for Analysts*. IDEA Paper Series. No. 4. Oslo. STEP Group (Studies in Technology, Innovation and Economic Policy).

Schaan, Susan and Frances Anderson. 2001. *Innovation in Canadian Manufacturing: National Estimates-Survey of Innovation 1999*. Catalogue No. 88F0006XIB2001010. Ottawa. Statistics Canada.

Schaan Susan and Brian Nemes. 2002. *Survey of Innovation 1999, Methodological Framework: Decisions Taken and Lessons Learned*. Catalogue No. 88F0006XIE2002012. Ottawa. Statistics Canada.

Scherer, F.M. 1984. *Innovation and Growth: Schumpeterian Perspectives*. Cambridge USA. MIT Press.

Schumpeter, Joseph A. 1942. *Capitalism, Socialism and Democracy*. New York. Harper & Brothers Publishers.

Scott, J.T. 1984. "Firm Versus Industry Variability in R&D Intensity." In *R&D, Patents and Productivity*. Zvi Griliches, (ed.). Chicago. University of Chicago Press for the National Bureau of Economic Research. 233-45.

Sharpe, Andrew. 2001. "Productivity Trends in the Construction Sector in Canada: A Case of Lagging Technical Progress." *International Productivity Monitor*. Number 3, Fall: 52-68.

Soete, Luc. 1979. "Firm Size and Innovative Activity: The Evidence Reconsidered." *European Economic Review*. 12, 4 (October): 319-40.

Stalk, George, Philip Evans and Lawrence E. Shulman. 1992. "Competing on Capabilities: The New Rules of Corporate Strategy." *Harvard Business Review*. 70, 2 (March-April): 57-69.

Statistics Canada. 1979. *Concepts and Definitions of the Census of Manufacturers*. Catalogue No. 31-528. Ottawa.

Statistics Canada. 1985. *Manufacturing Industries of Canada: National and Provincial Areas*. Catalogue No. 31-203. Ottawa.

Statistics Canada. 1996a. *Architectural, Engineering and Scientific Services in Canada*. Catalogue No. 63-234. Ottawa.

Statistics Canada. 1996b. *Survey of Innovation, 1996*. Ottawa.

Statistics Canada. 1998a. *North American Industry Classification System, Canada, 1997*, Catalogue No. 12-501-XPE. Ottawa.

Statistics Canada. 1998b. *Science and Technology Activities and Impacts: A Framework for a Statistical Information System*. Catalogue No. 88-522-XPB. Ottawa.

Statistics Canada. 1998c. *Services Indicators*. Vol 5. No. 4. Catalogue No. 63-016-XPB. Ottawa.

Statistics Canada. 2001. *Innovation in Canadian Manufacturing: National Estimates*. Catalogue No. 88F0006XIE01010. Ottawa.

Statistics Canada. 2003 "Electronic Commerce and Technology, 2002." *The Daily*. April 2, 2003. Catalogue No. 11-001-XIE, Ottawa.

Stern, Scott, Michael E. Porter and Jeffrey L. Furman. 2000. *The Determinants of National Innovative Capacity*. Cambridge USA. NBER Working Paper No. 7876.

Stewart, Thomas. 1998. *Intellectual Capital: The New Wealth of Organizations*. New York. Doubleday.

Storper, Michael. 1997. *The Regional World: Territorial Development in a Global Economy*. New York. The Guilford Press.

Tether, Bruce. 2000. *Who Co-Operates for Innovation Within the Supply-Chain, and Why?: An Analysis of the United Kingdom's Innovation Survey*. CRIC Discussion Paper No. 35. Centre for Research on Innovation and Competition. University of Manchester.

Therrien, Pierre. 2002. "What Distinguishes Innovative Firms from Other Firms: Results from the 1999 Survey of Innovation." In *Knowledge, Clusters and Regional Innovation*. J.A. Holbrook and D.A. Wolfe (eds.). Montreal and Kingston. McGill-Queen's University Press. 59-84.

Therrien, Pierre and Pierre Mohnen. 2003. "How Innovative Are Canadian Firms Compared to Some European Firms?: A Comparative Look at Innovation Surveys." *Technovation*. 23, 4: 359-69.

Thomas, A. 2000. *Économétrie des variables qualitatives*. Paris. Dunod.

Topkis, Donald M. 1998. *Supermodularity and Complementarity*. Princeton, U.S.A. Princeton University Press.

Townsend, J., Flis Henwood, G. Thomas, Keith Pavitt and Sally Wyatt. 1981. *Innovations in Britain Since 1945*. Occasional Paper No. 16. SPRU. University of Sussex.

Trajtenberg, Manuel. 2000. *Is Canada Missing the "Technology Boat"?: Evidence from Patent Data*. Discussion Paper No. 9. Research Publications Program. Industry Canada.

UK (United Kingdom) Department of Trade and Industry. 1999. *UK Competitiveness Indicators*. December.

Utterback, James M. and William Abernathy. 1975. "A Dynamic Model of Process and Production Innovation." *OMEGA, the International Journal of Management Science* 3, 6: 639-56.

Venkataraman, S. 1997. "The Distinctive Domain of Entrepreneurship Research: An Editor's Perspective." In *Advances in Entrepreneurship, Firm Emergence, and Growth*. Vol. 3. J. Katz and R. Brokhaus (eds.). San Francisco. JAI Press. 119-38.

Vock, Patrick. 2001. "An Anatomy of the Swiss Construction Cluster." In *Innovative Clusters: Drivers of National Innovation Systems*. Paris. OECD. 229-47.

von Hippel, Eric. 1988. *The Sources of Innovation*. New York. Oxford University Press.

Wallmark, J.Torkel and Douglas H. McQueen. 1991. "One Hundred Major Swedish Technical Innovations, from 1945 to 1980." *Research Policy*. 20, 4 (August): 325-44.

Wintjes, René and Jan Cobbenhagen. 2000. *Knowledge Intensive Industrial Clustering Around Oce*. Working Paper No. mer00006. The Netherlands.

Maastricht Economic Research Institute on Innovation and Technology (MERIT).

Wolfe, David A. ed. 2003. *Clusters Old and New: The Transition to a Knowledge Economy in Canada's Regions.* Montreal and Kingston. McGill-Queen's University Press.

At time of publication, all articles and questionnaires from Statistics Canada were available at www.statcan.ca.

Index

Index of Authors

Index of Topics

Queen's Policy Studies Recent Publications

The Queen's Policy Studies Series is dedicated to the exploration of major policy issues that confront governments in Canada and other western nations. McGill-Queen's University Press is the exclusive world representative and distributor of books in the series.

School of Policy Studies

Delicate Dances: Public Policy and the Nonprofit Sector, Kathy L. Brock (ed.), 2003
Paper ISBN 0-88911-953-8 Cloth ISBN 0-88911-955-4

Beyond the National Divide: Regional Dimensions of Industrial Relations, Mark Thompson, Joseph B. Rose and Anthony E. Smith (eds.), 2003
Paper ISBN 0-88911-963-5 Cloth ISBN 0-88911-965-1

The Nonprofit Sector in Interesting Times: Case Studies in a Changing Sector, Kathy L. Brock and Keith G. Banting (eds.), 2003
Paper ISBN 0-88911-941-4 Cloth ISBN 0-88911-943-0

Clusters Old and New: The Transition to a Knowledge Economy in Canada's Regions, David A. Wolfe (ed.), 2003 Paper ISBN 0-88911-959-7 Cloth ISBN 0-88911-961-9

The e-Connected World: Risks and Opportunities, Stephen Coleman (ed.), 2003
Paper ISBN 0-88911-945-7 Cloth ISBN 0-88911-947-3

Knowledge, Clusters and Regional Innovation: Economic Development in Canada, J. Adam Holbrook and David A. Wolfe (eds.), 2002
Paper ISBN 0-88911-919-8 Cloth ISBN 0-88911-917-1

Lessons of Everyday Law/Le droit du quotidien, Roderick Alexander Macdonald, 2002
Paper ISBN 0-88911-915-5 Cloth ISBN 0-88911-913-9

Improving Connections Between Governments and Nonprofit and Voluntary Organizations: Public Policy and the Third Sector, Kathy L. Brock (ed.), 2002
Paper ISBN 0-88911-899-X Cloth ISBN 0-88911-907-4

Governing Food: Science, Safety and Trade, Peter W.B. Phillips and Robert Wolfe (eds.), 2001 Paper ISBN 0-88911-897-3 Cloth ISBN 0-88911-903-1

The Nonprofit Sector and Government in a New Century, Kathy L. Brock and Keith G. Banting (eds.), 2001 Paper ISBN 0-88911-901-5 Cloth ISBN 0-88911-905-8

Institute of Intergovernmental Relations

The Impact of Global and Regional Integration on Federal Systems: A Comparative Analysis, Harvey Lazar, Hamish Telford and Ronald L. Watts (eds.), 2003
Paper ISBN 1-55339-002-4 Cloth ISBN 1-55339-003-2

Canada: The State of the Federation 2001, vol. 15, *Canadian Political Culture(s) in Transition,* Hamish Telford and Harvey Lazar (eds.), 2002
Paper ISBN 0-88911-863-9 Cloth ISBN 0-88911-851-5

Federalism, Democracy and Disability Policy in Canada, Alan Puttee (ed.), 2002
Paper ISBN 0-88911-855-8 Cloth ISBN 1-55339-001-6, ISBN 0-88911-845-0 (set)

Comparaison des régimes fédéraux, 2e éd., Ronald L. Watts, 2002
ISBN 1-55339-005-9

Health Policy and Federalism: A Comparative Perspective on Multi-Level Governance, Keith G. Banting and Stan Corbett (eds.), 2001
Paper ISBN 0-88911-859-0 Cloth ISBN 1-55339-000-8, ISBN 0-88911-845-0 (set)

Disability and Federalism: Comparing Different Approaches to Full Participation, David Cameron and Fraser Valentine (eds.), 2001
Paper ISBN 0-88911-857-4 Cloth ISBN 0-88911-867-1, ISBN 0-88911-845-0 (set)

Federalism, Democracy and Health Policy in Canada, Duane Adams (ed.), 2001
Paper ISBN 0-88911-853-1 Cloth ISBN 0-88911-865-5, ISBN 0-88911-845-0 (set)

John Deutsch Institute for the Study of Economic Policy

Canadian Immigration Policy for the 21st Century, Charles M. Beach, Alan G. Green and Jeffrey G. Reitz (eds.), 2003 Paper ISBN 0-88911-954-6 Cloth ISBN 0-88911-952-X

Framing Financial Structure in an Information Environment, Thomas J. Courchene and Edwin H. Neave (eds.), Policy Forum Series no. 38, 2003
Paper ISBN 0-88911-950-3 Cloth ISBN 0-88911-948-1

Towards Evidence-Based Policy for Canadian Education/Vers des politiques canadiennes d'éducation fondées sur la recherche, Patrice de Broucker and/et Arthur Sweetman (eds./dirs.), 2002 Paper ISBN 0-88911-946-5 Cloth ISBN 0-88911-944-9

Money, Markets and Mobility: Celebrating the Ideas of Robert A. Mundell, Nobel Laureate in Economic Sciences, Thomas J. Courchene (ed.), 2002
Paper ISBN 0-88911-820-5 Cloth ISBN 0-88911-818-3

The State of Economics in Canada: Festschrift in Honour of David Slater, Patrick Grady and Andrew Sharpe (eds.), 2001
Paper ISBN 0-88911-942-2 Cloth ISBN 0-88911-940-6

The 2000 Federal Budget: Retrospect and Prospect, Paul A.R. Hobson and Thomas A. Wilson (eds.), Policy Forum Series no. 37, 2001
Paper ISBN 0-88911-816-7 Cloth ISBN 0-88911-814-0

Available from: McGill-Queen's University Press
c/o Georgetown Terminal Warehouses
34 Armstrong Avenue
Georgetown, Ontario L7G 4R9
Tel: (877) 864-8477
Fax: (877) 864-4272
E-mail: orders@gtwcanada.com